Advances in
Insect Physiology

Volume 36

Advances in Insect Physiology

Series Editors

Stephen J. Simpson

School of Biological Sciences, The University of Sydney, Sydney, Australia

Jérôme Casas

University of Tours, Institute de Recherche sur la Biologie de l'Insecte, UMR, CNRS, Tours, France

Volume 36

ELSEVIER

Amsterdam • Boston • Heidelberg • London • New York • Oxford
Paris • San Diego • San Francisco • Singapore • Sydney • Tokyo
Academic Press is an imprint of Elsevier

ACADEMIC PRESS

Academic Press is an imprint of Elsevier
32 Jamestown Road, London NW1 7BY, UK
Radarweg 29, PO Box 211, 1000 AE Amsterdam, The Netherlands
Linacre House, Jordan Hill, Oxford OX2 8DP, UK
30 Corporate Drive, Suite 400, Burlington, MA 01803, USA
525 B Street, Suite 1900, San Diego, CA 92101-4495, USA

First edition 2009

ISBN: 978-0-12-374828-7
ISSN: 0065-2806

For information on all Academic Press publications visit our website at elsevierdirect.com

Printed and bound in United Kingdom

09 10 11 12 13 10 9 8 7 6 5 4 3 2 1

Contents

Contributors

M. P. Pener

Department of Cell and Developmental Biology, The Hebrew University of Jerusalem, Jerusalem, Israel

S. J. Simpson

School of Biological Sciences, The University of Sydney, NSW, Australia

Locust Phase Polyphenism: An Update

Meir Paul Pener[a] and Stephen J. Simpson[b]

[a]*Department of Cell and Developmental Biology, The Hebrew University of Jerusalem, Jerusalem, Israel*

[b]*School of Biological Sciences, The University of Sydney, NSW, Australia*

ADVANCES IN INSECT PHYSIOLOGY VOL. 36
ISBN 978-0-12-374828-7
DOI: 10.1016/S0065-2806(08)36001-9

1 Introduction

1.1 SCOPE OF THE PRESENT REVIEW

The present review updates two former reviews by the senior author (Pener, 1991; Pener and Yerushalmi, 1998). It is devoted mostly to recent findings; within the past 9- to 10-year period, well over 200 articles were published in scientific journals on various aspects of locust phase polyphenism, markedly advancing the knowledge of the subject. However, we refer also to the older literature when background information is necessary for complementary and better treatment, or because of historical importance. We already stress here that some of the recent publications report contradictory findings, and such contradictions are especially emphasized in the present review.

1.2 LOCUST MENACE

From the practical standpoint, the locust menace is far from being over (as an example for a particular year, see Thomas *et al.*, 2000). After the major plague of the desert locust, *Schistocerca gregaria* (Forskål), in 1986–1989 (Skaf, 1990; Skaf *et al.*, 1990; Showler and Potter, 1991; Showler, 2002), several outbreaks/upsurges occurred (Showler, 1995). A recent outbreak led to an upsurge/plague of the desert locust, in 2003–2004, from the Sahel of western Africa, with swarms reaching northern Africa (Bell, 2005), southern Europe, Cyprus and the Middle East (Lecoq, 2005; FAO, 2006, p. 11 and pp. 46–50; Ceccato *et al.*, 2007). It may be mentioned that the FAO differentiates between the three terms: 'outbreak', 'upsurge' and 'plague', respectively, connoting mild, major and extremely massive concentrations of locusts, their damage and control measures (for more detailed definitions and discussion, see Van Huis *et al.*, 2007). However, not all authors adopt the hierarchical meaning of these terms. Some authors do not use the term upsurge, some others use outbreak and upsurge as synonyms. Some authors employ different kinds of classification to characterize the intensity of locust concentration (Hunter, 2004).

The general biology, with field data, and recession and invasion areas of the desert locust are detailed by Steedman (1990), Duranton and Lecoq (1990) and Latchininsky and Launois-Luong (1997). Lessons from the recent upsurge/plague of the desert locust are discussed by Lecoq (2005). Cheke and Tratalos (2007) analyse the migrations of the desert locust. Steedman (1990) also deals with other African locusts, namely, the African migratory locust, *Locusta migratoria migratorioides* (Reiche and Fairmare), the red locust, *Nomadacris septemfasciata* (Serville) and the South African brown locust, *Locustana pardalina* (Walker). The outbreaks and population dynamics of the brown locust are reported and analysed in relation to climatic factors by Todd *et al.* (2002). Price and Brown (2000) summarize a century of locust control in southern Africa, emphasizing the history of the brown locust. Lecoq (1995) deals with field conditions leading to gregarization of the Malagasy strain of *Locusta migratoria* (L.), also known as *Locusta migratoria capito* (Saussure), and presents some details on a recent outbreak of this strain. There are different opinions on the subspecies status of different geographic races of *L. migratoria*. Uvarov (1966) recognized eight geographically distinct subspecies, whereas Farrow and Colless (1980) advocated that all subtropical and tropical non-diapausing geographical strains are best regarded as spatially and temporally variable populations of a single subspecies, *L. m. migratorioides*. Instead of naming other subspecies (with egg diapause, or with adult diapause, or with both), these authors characterize the different strains by their geographic areas.

Chen's (1999) booklet provides a review on morphology, life cycle, habitats, food plants, natural enemies, breeding areas, fecundity and distribution of *L. migratoria* in China, distinguishing three subspecies: *L. migratoria manilensis* (Meyen) that is polyvoltine at favourable temperatures, *L. m. migratoria* (L.) and

L. migratoria tibetensis (Chen) (synonymized in Uvarov (1966) as *L. m. burmana* Ramme), which are usually univoltine. Although most of the references are in Chinese, or Chinese with English summary, and they cover the period 1930–1986 (except for some of the author's own papers and three other articles from the late 1980s to the early 1990s), the booklet is a useful summary in English on *L. migratoria* and some other acridids in China.

More recent publications on *L. migratoria* in China are also available. In Central and South China, outbreaks of *L. migratoria*, considered from this geographic area as *L. m. manilensis*, were reported (Ji *et al.*, 2004a; Ma *et al.*, 2005; and references therein). Zhang and Kang (2005), employing random amplified polymorphic DNA technique, analysed the genetic divergence within and among 11 geographical populations of *L. migratoria* in China. They found four regional groups: the populations from Xinjiang and Inner Mongolia (north-eastern China) may be considered as *L. m. migratoria*, that from Hainan as *L. m. manilensis*, and two populations from Tibet as *L. m. tibetensis*. The fourth group was found in the Great Plains of northern China and the authors postulated that it resulted from mutual pervasions of *L. m. migratoria* and *L. m. manilensis*, constituting a kind of hybrid zone. Stige *et al.* (2007) analysed a 1000-year-long Chinese record of the annual abundance of *L. m. manilensis* and revealed climatic forcing of decadal dynamics of this locust. They also paid attention to global warming studies relevant to such dynamics. *L. migratoria*, also considered as *L. m. migratoria* (see earlier), causes damage in the northern parts of China (Tanaka and Zhu, 2005), in south Russia and in some southern countries of the former USSR, such as Kazakhstan (Sivanpillai *et al.*, 2006), though its economic importance decreased in the Aral Sea basin as a result of man made ecological interference (Gapparov and Latchininsky, 2000). In the former Soviet countries, Kazakhstan, Turkmenistan, Uzbekistan, Kyrgyzstan, Tajikistan and Georgia, as well as in the southern Siberian plains of Russia and Afghanistan, the univoltine Moroccan locust, *Dociostaurus maroccanus* (Thunberg), and the Italian locust, *Calliptamus italicus* (L.), are serious pests (Abashidze *et al.*, 1998; Evdokimov *et al.*, 1999; Abashidze, 2000; Latchininsky, 2000; Anarbaev, 2001; Gapparov, 2001; Khasenov, 2001; Wilps *et al.*, 2002; Stride *et al.*, 2003; Sergeev and Vanjakova, 2005 and further references in these publications); for a comprehensive review on the Moroccan locust, see the book by Latchininsky and Launois-Luong (1992) and an update by Latchininsky (1998).

Plagues of the Australian plague locust, *Chortoicetes terminifera* (Walker), also occur, though the Australian Plague Locust Commission successfully prevents most plagues in an early stage of an outbreak as reported by Hunter (2004), who uses three grades of outbreaks and two grades of plagues, hierarchically superimposed upon one another, from mild to extremely massive concentrations of *C. terminifera*. Aerial detection of nymphal bands much improves the preventative control measures against this locust (Hunter *et al.*, 2008). Occasional outbreaks of the Australian spur-throated locust, *Austracris guttulosa* (Walker), known for a while in the past as *Nomadacris guttulosa* (Key and Rentz, 1994), were also observed (Hunter and Elder, 1999). Baker (1993) summarizes the biology, plague dynamics and economic impacts of these two species and of other Australian locusts, *Austroicetes cruciata* (Saussure), *Gastrimargus musicus* (Fabricius), the Australian strain of *L. migratoria*, as well as a few acridid species on the verge of the definition of locusts.

There seems to be no recent update on the South American and Central American locusts, *Schistocerca cancellata* (Serville) and *Schistocerca piceifrons* (Walker), respectively. However, an older article by Hunter and Cosenzo (1990 and references therein) may be consulted for the former and Dirección General de Sanidad Vegetal (1998) for the latter. The names of these two species are presented herewith according to Harvey's (1981) reclassification of the *Schistocerca americana* complex. Before Harvey's study, *S. piceifrons* was designated as *Schistocerca paranensis*.

Better understanding of locust phase polyphenism has an applied potential; interference with gregarization, migration, or band or swarm formation by physiological or ecophysiological manipulations may lead to non-conventional control of locusts. This is especially important because present control measures are based mostly on chemical pesticides that conflict with environmental conservation (Peveling, 2001, and references therein), though recently biological control with mycopesticides became promising (Milner and Hunter, 2001; Long and Hunter, 2005). Such mycopesticides act slowly and are best applied to an early outbreak of a locust population (Lomer *et al.*, 2001). The best known mycopesticide against locusts and grasshoppers is *Metarhizium anisopliae* var. *acridum*; its state in integrated pest management is discussed by Hunter (2005). Klass *et al.* (2007a,b) developed a model on effectiveness of the control of several locust species by *M. anisopliae* var. *acridum*.

2 Insect polyphenism

The meaning and considerations related to the term 'polyphenism' in insects, equivalent to the former terms 'facultative polymorphism', or 'environmentally regulated polymorphism', were discussed in detail by Pener (1991, pp. 1–2). Briefly, Michener (1961) suggested the term 'polyphenic', instead of 'polymorphic', for connoting environmentally induced major physiological or behavioural differences or both in the same insect species. Lüscher (1976) extended the term to include morphological differences. Hardie and Lees (1985, p. 473) defined polyphenism as ''occurrence of two or more distinct phenotypes which can be induced in individuals of the same genotype by extrinsic factors''. This definition emphasizes the causative role of the extrinsic factors in the same genotype. Most of the relevant recent literature uses the term 'polyphenism' that gradually replaces the older term 'polymorphism'. From the etymological standpoint, polyphenism is more correct. Moreover, with recent advances in molecular biology, such as genomics, proteomics and peptidomics, polymorphism has gained a somewhat different and much wider meaning. Therefore, accepting the present trend, we use the term polyphenism.

According to Nijhout (1994, p. 176), ''Polyphenisms are formally distinguished from polymorphisms by the fact that in the latter the alternative phenotypes are due to genetic differences'', whereas in polyphenism, ''the alternative phenotypes are developmental variants that are genetically identical''. However, the strict distinction is somewhat blurred by the fact that some bivoltine species, exhibiting two distinctly different phenotypes in different seasons, are capable of producing intermediate phenotypes by experimentally created intermediate environmental conditions, or by experimental manipulation of the physiological mechanism (Nijhout, 1994, 2003 – see especially p. 13 and Fig. 4 – and references therein). Therefore, different phenotypes of the same species that are presently considered as polymorphic may actually turn out to be polyphenic.

The present review does not deal with sequential polyphenism such as different forms (larva, pupa, imago) of Holometabola and refers to sexual diphenism (male versus female) only when it is relevant to locust phases. Also, sexual diphenism may be better considered as sexual dimorphism, because generally male and female insects of the same species are not identical genetically.

Insect polyphenism received considerable attention in recent years. Applebaum and Heifetz (1999) reviewed density-dependent physiological phase polyphenism in insects. Nijhout (1999, 2003), as well as Evans and Wheeler (2001), contributed theoretical considerations, with examples, to insect polyphenism. Hartfelder (2000) reviewed polyphenism in social insects, Miura (2004) added to this subject and Roisin (2000) published a review on caste polyphenism in termites. Zera and Denno (1997), then Zera (2004), devoted reviews to dispersal and wing polyphenism in insects. Emlen and Nijhout (2000) dealt with continuous and discontinuous polyphenism of exaggerated morphologies in insects. Dingle (2002) treated endocrine aspects of insect polyphenism in a chapter on hormonal mediation of insect life histories. More recently, Hartfelder and Emlen (2005) reviewed the whole subject of insect polyphenism, and their section on aphid polyphenism seems to be the sole recent summary for this group of insects. Zera (2007) devoted an article to the methods employed for investigating endocrine (mostly juvenile hormone [JH]) regulation of insect polyphenism, outlining the limitations of simple 'hormone manipulations' that usually means application of a hormone or a hormone analogue.

Although locust phase polyphenism is among the most striking examples of phenotypic plasticity in the Animal Kingdom, it has not featured prominently in the burgeoning literature in this area, perhaps because of perceived applied theme of much of the older work (Simpson and Sword, 2009). For example, there are no references to locusts in four major texts on phenotypic plasticity

(Schlichting and Pigliucci, 1998; Pigliucci, 2001; DeWitt and Scheiner, 2004; Pigliucci and Preston, 2004). An exception is the book by West-Eberhard (2003), where locusts are covered briefly but are, at least, considered to be ''among the most striking coordinated alternative phenotypes known'' (West-Eberhard, 2003, p. 132).

3 Density-dependent phase polyphenism

3.1 LOCUSTS

The theory of locust phases was formed by Uvarov (1921) in a taxonomic revision of the genus *Locusta*. He found that *L. migratoria* and *Locusta danica* are respectively the swarming and the solitary forms or 'phases' of the same species. In the same article, he also indicated that the South African brown locust, *L. pardalina*, has similar swarming and solitary phases. The advancement and generalization of the locust phase theory was reviewed by Uvarov (1966 and earlier references therein) and extensively discussed by Pener (1991, pp. 4–7). According to present concepts, locusts show density-dependent phase polyphenism in morphology, anatomy, colouration, development, reproduction, physiology, biochemistry, molecular biology, behaviour, chemical ecology (pheromones) and other aspects of ecology. The two phases are named 'gregarious' and 'solitarious'. The term 'solitarious' was introduced by Uvarov (1966, p. 332) instead of the former term 'solitary', to avoid the ambiguity of the latter (a gregarious phase insect may be alone, in which case it would be solitary but not yet solitarious), but both terms are used in the recent literature. Extreme phase differences are found only in the field; locusts maintained in the laboratory under conditions of crowding and in isolation only approach the full suite of features of the gregarious and solitarious phases, respectively. The features differing between the gregarious and the solitarious phases are termed 'phase characteristics'.

Locust phase polyphenism is continuous in two senses: (1) all kind of intermediates can be found between the two extreme phases and (2) induction of a phase change is not stage-specific. The latter means that phase characteristics can be shifted to either direction and the direction of the shift is reversible in any stadium, all in response to changes in the density of the population in the field or in the laboratory (Pener, 1991, pp. 5–6). Even the eggs are subject to phase transformation; experimental treatment within a short while after oviposition induced some gregarious phase characteristics in hatchlings from eggs laid by solitarious females of *S. gregaria* (McCaffery *et al.*, 1998; Simpson *et al.*, 1999) (see Section 16).

The term 'locust' was defined by the senior author as short-horned grasshoppers (Orthoptera: Acrididae) that ''meet two criteria: (1) they form at some (rather irregular) periods dense groups, comprising huge numbers, bands of hoppers and/or swarms of winged adults which migrate; and (2) they are polymorphic in the sense that individuals living separately differ in many characteristics from those living in groups'' (Pener, 1983, p. 379, 1991, p. 6). In this definition, 'polymorphic' should be altered to 'polyphenic', but otherwise this definition seems to be valid to date. The first criterion does not need explanation; it is known from biblical times (the plague inflicted on Pharaoh and the people of Egypt, Exodus, Chapter 10, verses 4–19; the prophecy of Joel, Chapters 1 and 2). As for the second criterion, phase characteristics or the amplitude of their change or both differ in different locust species. Some authors define locusts by only one of these two criteria. For example, Pierozzi and Lecoq (1998, p. 25, abstract and p. 27) define locusts as acridid species that exhibit phase polyphenism (see later discussion on *Rhammatocerus schistocercoides*). In contrast, some Australian authors define locusts as swarming acridids (see, e.g., Elder, 1997, p. 63, first sentence of the discussion).

L. migratoria and *S. gregaria* exhibit the most extreme, externally very distinct differences between the solitarious and gregarious phases, whereas *C. terminifera* shows externally only slightly distinguishable phase differences. According to Uvarov (1977, p. 356), gregarious males of *C. terminifera* are larger than solitarious males, but in the females, such difference is scarcely observable. In other locust species (Table 1), phase characteristics are in between these extreme cases. It is important to realize that different locust species belong to several different subfamilies of the family of Acrididae (Table 1, subfamilies according to Uvarov, 1966). Therefore, in this family, phase polyphenism presumably evolved several times, by convergent, or partially convergent, evolution. In fact, many acridid species show some tendency to aggregation and swarming (see also Song, 2005). Such species may be considered as less typical locusts or aggregating/swarming grasshoppers (Table 1, species marked by an asterisk). For example, *Melanoplus differentialis* Uhler was claimed to exhibit some phase polyphenism (Dingle and Haskell, 1967); crowded insects were smaller and darker than isolated ones, though morphometric ratios E/F an F/C (E = length of tegmina, F = length of hind femora, C = maximum width of head) did not seem to be different. More recently, Fielding and DeFoliart (2005) reported that high density induced darkening in another species of the same genus, *Melanoplus sanguinipes* (Fabricius). Also, in this species, 'good fliers', supposed to be migrants, showed a higher hyperlipaemic response to adipokinetic hormone I (Locmi AKH I) than 'bad fliers' considered to be non-migrants (Min *et al.*, 2004). Although the effect of density on adipokinetic response was not investigated in *M. sanguinipes*, it is known that this response in crowded *L. m. migratorioides* adults is much higher than in isolated conspecifics (Ayali and Pener, 1992), and young crowded adults of *S. gregaria* show a higher flight-induced hyperlipaemic response than isolated young adults (Schneider and Dorn, 1994). Moreover, the distinction between good fliers and bad fliers is also somewhat similar to locust phase polyphenism. Studying the neural correlates to flight-related phase polyphenism

TABLE 1 Acrididae that show elements of density-dependent polyphenism

Species Name	Common Name	Subfamily
Schistocerca gregaria	Desert locust	Cyrtacanthacridinae
Schistocerca piceifrons	Central American locust	Cyrtacanthacridinae
Schistocerca cancellata	South American locust	Cyrtacanthacridinae
Schistocerca interrita	Peru locust	Cyrtacanthacridinae
Nomadacris septemfasciata	Red locust	Cyrtacanthacridinae
Nomadacris succincta[a,b]	Bombay locust	Cyrtacanthacridinae
Austracris guttulosa[a]	Spur-throated locust	Cyrtacanthacridinae
Anacridium melanorhodon	Sahelian tree locust	Cyrtacanthacridinae
Locusta migratoria	Migratory locust	Oedipodinae
Locustana pardalina	Brown locust	Oedipodinae
Chortoicetes terminifera	Australian plague locust	Oedipodinae
Aiolopus simulatrix[a]	Sudan plague locust	Oedipodinae
Austroicetes cruciata[a]	Small plague grasshopper	Oedipodinae
Oedaleus senegalensis	Senegalese grasshopper	Oedipodinae
Gastrimargus musicus[a]	Yellow-winged locust	Oedipodinae
Calliptamus italicus	Italian locust	Calliptaminae
Dociostaurus marrocanus	Moroccan locust	Gomphocerinae
Gomphocerus sibiricus	Siberian locust	Gomphocerinae
Rhammatocerus schistocercoides[a]	Mato Grosso grasshopper	Gomphocerinae
Ceracris kiangsu	Yellow-spined bamboo locust	Acridinae
Melanoplus spretus (extinct)	Rocky Mountain locust	Catantopinae
Melanoplus differentialis[a]	Differential grasshopper	Catantopinae
Melanoplus sanguinipes[a]	Migratory grasshopper	Catantopinae

Note: Subfamilies are according to Uvarov (1966).
[a]Partial phase change; as also occurs in other species not listed, suggesting that the entire family is predisposed to evolving phase polyphenism.
[b]Also known as *Patanga succincta*.

in *S. gregaria*, Fuchs *et al.* (2003) and Ayali *et al.* (2004) found a lower threshold for wind-induced flight initiation in crowded than in isolated adults.

Another example is the Brazilian Mato Grosso acridid, *Rhammatocerus schistocercoides* (Rehn), that exhibits conspicuous gregarious behaviour, forming hopper bands (Lecoq *et al.*, 1999) and swarms of adults that migrate, though the displacement of the swarms is rather limited, up to 2.5 km per day (Lecoq and Pierozzi, 1996a). The species shows no morphological, morphometrical or colour-related phase differences (Pierozzi and Lecoq, 1998). As a result, these authors concluded that this insect is a grasshopper and not a locust (see earlier for their definition of locusts). However, in a later publication, *R. schistocercoides* was nevertheless connoted as a locust (Lecoq, 2000), presumably just reflecting the difficulty of a sharp distinction between locusts and grasshoppers. No experiments were reported on possible physiological or behavioural phase differences in this species. On the contrary, *A. guttulosa* is considered to be a locust by some Australian authors, despite that it exhibits only swarming of prereproductive adults and no overtly discernible phase polyphenism (see Section 8.5 for some additional details). Heifetz and Applebaum (1995) found some density-dependent physiological and behavioural changes, resembling locust phase characteristics, in the grasshopper, *Aiolopus thalassinus* (Fabricius), though density did not affect colouration and morphometrics, and no marching or swarming of this species have been recorded. However, another species of the genus, the Sudan plague locust, *Ailopus simulatrix* (Walker), possibly migrates (Steedman, 1990, p. 118). It seems that the whole family of Acrididae has some tendency of aggregation and swarming, as outlined by Uvarov (1977, pp. 142–150) and termed by him as 'antecedents of gregarious behaviour' (see also Song, 2005; Song and Wenzel, 2007). In conclusion, it is difficult to make a strict demarcation between locusts and grasshoppers. Density-dependent physiological and behavioural studies are needed to demonstrate that an aggregating and swarming acridid species shows at least some aspects of phase polyphenism.

Mapping information about phase characteristics onto independently derived phylogenetic trees for the acridids offers the opportunity to explore issues regarding the evolution of phase characteristics. Such issues include to what extent have phase characteristics arisen independently within and across acridid lineages; how are different phase characteristics coupled (if at all); and what are the environmental correlates that have accompanied the acquisition or loss of phase characteristics? (Simpson and Sword, 2009).

Several aspects of locust phase polyphenism were reviewed recently. They are as follows: Simpson *et al.* (1999) on phase-related behaviour and factors affecting behavioural phase state in *S. gregaria* (see Sections 11, 14 and 15); Hassanali and Torto (1999), Ferenz and Seidelmann (2003) and Hassanali *et al.* (2005a) on locust pheromones with some contradictory conclusions (see Sections 8.1 and 12); Dorn *et al.* (2000) and Breuer *et al.* (2003) on endocrine aspects of locust phase polyphenism (see Section 9); De Loof *et al.* (2006) on molecular markers of locust phase characteristics (see Section 10). Recently, Tanaka (2005) reviewed the effects of the dark-colour-inducing neurohormone (DCIN, also termed [His^7]-corazonin; see Sections 7.2.2 and 9.3.4) on certain phase characteristics such as colouration (Section 7.2.2), morphometrics (Section 5.2) and number of antennal sensilla (Section 5.3.1). In another, more detailed, review on the same subject, Tanaka (2006) extended the discussion in a wider context of locust phase polyphenism. Additionally, there are two recent reviews dealing among other topics with the adaptive significance and evolution of phase polyphenism (Simpson *et al.*, 2005; Simpson and Sword, 2009) and two articles on phylogenetic perspectives of the evolution of locust phase polyphenism, based on Cyrtacanthacridinae (Song, 2005), as well as on cladistic analysis of this subfamily (Song and Wenzel, 2007).

The review of Hartfelder and Emlen (2005) devotes a section to locust phase polyphenism, but these authors have been much influenced by Dorn *et al.* (2000) who claimed that JH has a major causative role in locust phase polyphenism. Hartfelder and Emlen (2005) strongly supported the claim of Dorn *et al.* (2000), despite many findings that disagree with this claim (see Section 9.1). In contrast, Dingle (2002), devoting a section to locust phase polyphenism, accepts that although JH promotes some solitarious phase characteristics, such as green colour and higher fecundity, ''it is not the primary physiological factor responsible for the solitary syndrome'' (Dingle, 2002, p. 260).

When the present review had been completed and when we were checking the list of references, an additional review appeared by Verlinden *et al.* (2009), devoted to endocrinological aspects of reproduction and phase change in locusts.

3.2 SOME OTHER INSECTS

Phase polyphenism is not limited to locusts. Pener (1991, p. 8) listed some references on density-dependent polyphenism in other, non-acridid, insects, all from the Orthopteroid complex.

The case of the Mormon cricket, *Anabrus simplex* Haldeman, which belongs to the katydids or bush-crickets (Tettigoniidae) should also be mentioned. This brachypterous species may show phase polyphenism (Gwynne, 2001, pp. 85–87 and Table 4.3); the gregarious insects have dark colour and migrate by marching, whereas the solitarious insects are more sedentary and have green or brown colouration (Lorch and Gwynne, 2000; Gwynne, 2001, pp. 64–65 and Plate 9). The coulee cricket, *Peranabrus scrabricollis* Thomas, again a flightless tettigoniid, also can form marching bands that migrate (Lorch and Gwynne, 2000). Recently, however, Sword (2005) questioned the true nature of phase polyphenism in the Mormon cricket. He quantified the effects of long-term differences in rearing density versus short-term presence or absence of conspecifics on behaviour. Short-term presence of conspecifics played a greater role in inducing movements of migratory band-forming Mormon crickets than did endogenous behavioural phase changes mediated by high local rearing density. Also, Sword (personal communication) was unable to obtain green solitarious colouration by individual separation of field-collected first-instar nymphs that were progeny of gregarious parents. Certainly, further investigations are needed to clarify the effect of density on possible physiological and other differences between 'gregarious' and 'solitarious' Mormon crickets. It is interesting to note that a recent analysis indicated substantial genetic divergence between Mormon cricket populations from the eastern and western slopes of the American Rocky Mountains. The former are solitary, whereas the latter are mainly, but not always, gregarious migrants. This finding suggests that the difference in appearance between the two forms may be genetically based rather than environmentally induced (Bailey *et al.*, 2005).

Finally, certain moth larvae exhibit phase polyphenism. The colour of such larvae depends on population density; they are dark under conditions of crowding, but cryptic, green or light coloured, under isolation. This polyphenism is similar in certain features to locust phase polyphenism, though the major component of its endocrine control differs from that of the locusts. Dark colouration in these lepidopteran larvae is induced by the melanization and reddish colouration hormone (MRCH) and related neurohormones of the pyrokinin (PK) and pheromone biosynthesis activating neuropeptide (PBAN) (see later) family, whereas dark colouration in locusts is induced by the DCIN (DCIN, also termed [His^7]-corazonin) (see Sections 7.2.2 and 9.3.4). The MRCH in adult female moths has a completely different function; it acts as a PBAN. There is no recent review on the phase polyphenism in larvae (though there are several reviews on PBAN). However, treatment of the subject by Applebaum and Heifetz (1999), Dingle (2002, pp. 262–263), in the introduction to a recent research article (Ben-Aziz *et al.*, 2005) and by another research report (Lee and Wilson, 2006), as well as the references cited in these publications, provides a fairly comprehensive picture.

4 Locust phase characteristics

Phase characteristics of locusts not only depend on the species (see Section 3.1) but also on developmental stadium, sex and even geographic strains within the same species. There are many stadium-dependent differences in phase characteristics. For example, hatchlings of gregarious *L. m. migratorioides* are blackish, whereas conspecific solitarious hatchlings are light grey. However, gregarious hoppers in the fifth (last) nymphal instar have a dirty orange background colour and black patterns, whereas solitarious hoppers are either green or homochrome. Similarly, gregarious hatchlings of *S. gregaria* are usually blackish, whereas solitarious hatchlings are mostly light greenish. In the last nymphal instar, gregarious *S. gregaria* hoppers have a bright yellow background colour and black patterns, whereas solitarious hoppers are green or greenish, or at low humidity, beige-brown without marked black patterns (Hunter-Jones, 1962). For a detailed account and relevant references on phase- and stadium-dependent colour polyphenism in *L. migratoria* and *S. gregaria*, as well as in other locusts, the reader may refer the work by Pener's (1991, pp. 12–16) review as well as Section 7 later in the present review.

Stadium-dependent differences in phase characteristics are even more marked between hoppers and adults. For example, the yellow background colour and black patterns characteristic of last-instar gregarious *S. gregaria* nymphs disappear in the adults; the latter are pinkish after the last moult, then become pinkish-beige, beige and eventually yellow. On the contrary, conspecific solitarious adults are mostly beige and never become yellow (Pener, 1991, pp. 18–19, and references therein); for a recent interpretation of this phase difference, see Sas *et al.* (2007).

Sex-dependent differences in phase characteristics are also considerable. In adults of *L. migratoria*, *S. gregaria* and *N. septemfasciata*, solitarious females are larger than conspecific gregarious females, but in adult males of these species, the situation is reversed. Therefore, the relative difference in body size between the phases is sex-dependent; the difference in size between the females and the males is smaller in the gregarious than in the solitarious phase (for body size of locusts, see also Section 5.1).

Another example of stadium-, sex- and even age-dependent difference in phase characteristics is the production of the pheromone, phenylacetonitrile (PAN) (also termed as benzyl cyanide), by sexually mature gregarious adult males of *S. gregaria* (see Sections 8.1, 12.1.3, 12.4 and 12.5). Nymphs, solitarious adults of both sexes, sexually immature gregarious adult males and females and mature gregarious adult females do not produce this substance; moreover, isolation of crowd-reared males soon results in cessation of PAN production (Deng *et al.*, 1996; Seidelmann *et al.*, 2000).

There are even strain-dependent differences in phase characteristics of the same species. The best known example is *Schistocerca gregaria gregaria* (usually connoted only as *Schistocerca gregaria*) that exhibits major phase differences, contrasting to *Schistocerca gregaria flaviventris* (Burmeister), the South African strain (or subspecies) that shows much smaller phase differences in every investigated aspect (Uvarov, 1966, pp. 363–364 and pp. 374–375; Uvarov, 1977, p. 522). Other cases of strain-dependent differences in the phase characteristics of *L. migratoria* were reviewed by Pener and Yerushalmi (1998). In laboratory experiments, Schmidt and Albütz (1996) found differences in morphometrics between two populations of *L. migratoria*, one originating from Greece and the other from Nigeria. The same authors (Schmidt and Albütz, 1999) reported differences in development time between crowded and isolated *S. gregaria* of various geographical origin. More recently, morphometrical differences (see Section 5.2) between crowded (gregarious) and isolated (solitarious) adults of the Okinawa (Japan) strain were found to be much smaller than in a West African strain of *L. migratoria* (Yerushalmi *et al.*, 2001). Also, adults of the former showed a higher response to the DCIN of locusts (see Section 7.2.2) than the latter (Grach *et al.*, 2004). Zhang and Kang (2005) found genetic divergence among different geographical populations of *L. migratoria* in China. Recently, Chapuis *et al.* (2008a) found that a historically outbreaking population of *L. migratoria* from Madagascar (presumably *L. m. capito*) exhibited parentally inherited density-dependent phase changes to a greater degree than a historically non-outbreaking population from France (presumably *Locusta migratoria cinerascens*). See also Sections 11.2 and 16.3 for strain-dependent differences in phase characteristics and Section 17.6 for population genetics of various geographic strains of *L. migratoria*.

In the following sections, we review recent data on phase characteristics, and factors affecting them, mostly published after compilation of the reviews by Pener (1991) and Pener and Yerushalmi (1998). Older publications, overlooked or not related to the subjects treated in the preceding reviews, are also included.

5 Morphology

5.1 SIZE

It was already mentioned that the size of adult locusts depends on species, sex and phase (see Section 4). However, in different locust species, phase-dependent changes in size are strongly dissimilar. In *S. gregaria*, *N. septemfasciata* and *L. migratoria*, solitarious females are larger than conspecific gregarious females, but solitarious males are slightly smaller than gregarious ones (Uvarov, 1966, p. 3, Table 2; p. 343, Fig. 199). In contrast, each sex of gregarious adults is larger than conspecific solitarious adults of the same sex in *D. maroccanus*, *L. pardalina* and *C. terminifera* (Uvarov, 1966, p. 3, Table 2; Uvarov, 1977, pp. 471–472 and Fig. 257; for *D. maroccanus*, see also Latchininsky and Launois-Luong, 1992, pp. 50–51). The difference between *L. migratoria* and *L. pardalina* is especially noteworthy because these are closely related species from the same subfamily (Oedipodinae). *C. terminifera* also belongs to the same subfamily and, as in the case of *L. pardalina*, Uvarov related the difference in size between gregarious and solitarious locusts of *C. terminifera* to phase. However, it cannot be discounted that the difference is more related to the availability of food and subsequent nutritional factors than to phase. According to Hunter (1989), in the arid areas of Australia, occasional adequate rainfall induces both hatching of *C. terminifera* eggs and germination with subsequent growth of ephemeral grasses. If the rainfall is sufficient to keep the grass green during nymphal development of the locusts, the adults are relatively large, they swarm and migrate. However, if the rainfall is insufficient to keep the grass green during late nymphal development, the resulting adults are relatively small, persist locally, but do not swarm and migrate, nor they have sufficient lipid reserves for migratory flights (Hunter, 1989). Such small non-migrating adults may easily be regarded as representatives of the solitarious phase.

5.2 MORPHOMETRY

Phase-related morphometry of locusts is based on morphometric ratios or on multivariate analysis or both. The morphometric ratios most widely used are F/C (length of the hind femur:maximum width of the head) and E/F (length of the tegmen:length of the hind femur).The letter 'E' is an abbreviation for the word 'elytra'; however, usage of this word for the fore-wing of locusts is incorrect. The proper term for the fore-wing of a locust is 'tegmen'. Often the height or length or width of the pronotum, or any combination of these dimensions are also measured, especially in *L. migratoria*, in which solitarious adults usually have a convexly arched pronotum, forming a median crest. Conspecific gregarious adults have a rather straight or even slightly concave pronotum (Pener, 1991, p. 9, Fig. 1).

The E/F ratio is higher and the F/C ratio is lower in gregarious than in solitarious adults of *S. gregaria*, *L. migratoria*, *N. septemfasciata* and *L. pardalina* (see Uvarov, 1966, pp. 344–351, and references therein; see also Nolte, 1967). Dirsh (1953) studied phase-related morphometry in adult males and females of *S. gregaria*, including the F/C and E/F ratios. Dirsh's measurements were made on dry specimens, originating from various gregarious and solitarious populations. Based on the ratios obtained, Dirsh (1953) constructed a scale of gregarization in percent values, ranging from 0% (solitary [presently connoted 'solitarious']) to 100% (gregarious), for the F/C and separately for the E/F ratio for males and females. Considering the heterogeneity of the populations and the dry material of the museum, this scale should not be used as an absolute standard of the phase state. Nevertheless, the results, especially the F/C ratio, provide a reasonable rough guide to distinguish between gregarious and solitarious *S. gregaria*. Uvarov (1966, pp. 344–351, 367–368 and 370–374) and Pener (1991, pp. 9–11) discussed phase-related morphometric ratios, including the effect of abiotic factors, such as temperature, on morphometrics.

Changes in density, leading to phase transformation, are accompanied by a corresponding shift in the F/C and E/F ratios. However, the absolute values of these ratios for solitarious and gregarious locusts and the amplitude of the shift strongly depend not only on the species and sex but also on the subspecies/geographic race, especially in *L. migratoria* (Uvarov, 1966, pp. 370–374; Heifetz *et al.*, 1994; Schmidt and Albütz, 1996; Yerushalmi *et al.*, 2001). This means that these ratios should be considered as exact indicators of the phase state only when they are obtained from a definite population. In *L. migratoria*, even the relative height of the pronotal crest of solitarious adults may differ in different races (Schmidt and Albütz, 1996).

More recently, Bouaïchi and Simpson (2003) investigated density-dependent accumulation of phase characteristics over consecutive generations (see Section 16), including the F/C ratio, in a geographically defined population of *S. gregaria* in Morocco. Their study started with measurements of field-captured male and female parents from a high-density subpopulation and, separately, from a low-density subpopulation. Additional F/C ratios were obtained from adult offspring of the high-density subpopulation, maintained under crowding in cages, and from field-captured adult offspring from the same subpopulation. In a parallel way, adult offspring from the low-density subpopulation, maintained under isolation, and field-captured adult offspring from the same subpopulation were also measured. The results showed that crowding of the offspring of the high-density subpopulation shifted the F/C ratio to values comparable to those recorded by Dirsh (1953) for gregarious *S. gregaria*. Isolation of the offspring from the low-density subpopulation did not result in a large shift of this ratio, presumably, because its values were high (i.e. more solitarious) in their low-density subpopulation parents.

Abbassi *et al.* (2003b) reported the F/C and E/F ratios of *S. gregaria* in a 1995 outbreak in South Morocco. The measurements were carried out at two geographical areas, each with various localities. Eventually, these authors defined the locusts as 'gregarious', 'transiens', or 'solitarious' according to Dirsh's scale of percentage of gregarization. Although the data are valuable, Dirsh's scale should not be used as an absolute basic standard of morphometrical phase state (see earlier).

In laboratory experiments, Deng *et al.* (1996) investigated changes of the F/C and E/F ratios after altering the density from crowding to isolation and vice versa, through several consecutive generations, of adult males and females of *S. gregaria*. As expected, the F/C ratio was more sensitive to changes in rearing density than the E/F ratio. These authors also explored the changes of these ratios in adults after transferring first-instar isolated hoppers to rearing in groups of two or four per cage. A significant shift of the F/C ratio was observed with four locusts per cage, but the values obtained were still higher (more solitarious) than those of the crowded controls.

Franc *et al.* (2005) recently reported field data on morphometric ratios of *N. septemfasciata* in Madagascar, concluding that solitarious, transiens, and gregarious phases are present in this island. In addition, these authors compared the morphometric ratios (E/F and F/C) obtained in their study for solitarious and gregarious males and females to the respective ratios reported by other authors for *N. septemfasciata* from various localities in Africa. Through this comparison, Franc *et al.* (2005) provide a good list of references on the morphometric ratios in this species.

Data on the morphometry of different populations of *D. maroccanus*, as reported by several authors, are summarized by Latchininsky and Launois-Luong (1992, pp. 48–49). The data show that the E/F ratio is higher in the gregarious phase, like in other locusts.

Morphometric ratios of a Greek and a Nigerian race of *L. migratoria* were studied in the laboratory by Schmidt and Albütz (1996), leading to the conclusion that thc morphometry of different subspecies/geographical races should be considered separately and confirming Uvarov's (1966) conclusion. Schmidt (2001) also investigated the effect of crowding on various phase characteristics of a Greek race of *L. migratoria* (presumably, *L. m. cinerascens*), including morphometry. He confirmed that low density induces a high F/C ratio. In contrast, Heifetz *et al.* (1994) reported that the F/C ratio is unaffected or only slightly affected by density in the Israeli strain of *L. migratoria*.

Yerushalmi *et al.* (2001) observed that an albino strain of *L. migratoria*, originating by a mutation from normally coloured locusts in Okinawa, Japan, and maintained in the laboratory (for description of the strain, see Tanaka, 1993; Hasegawa and Tanaka, 1994), exhibits rather solitarious values and only minor density-dependent changes of the F/C and E/F ratios in comparison to a normally coloured West African strain. Even crowded albinos showed higher F/C ratios (indicating a more solitarious characteristic) than isolated normal phenotypes. To reveal whether the albinism, or the origin of the strain (Okinawa versus West Africa), is responsible for this observation, Yerushalmi *et al.*

(2001) produced congenic albinos and normal phenotypes by repeated crossings and investigated their morphometrics with an increasing, up to 99.6%, West African gene pool. It was possible to obtain such congenic albinos and normally coloured phenotypes because the albinism in the Okinawa strain is controlled by a single recessive Mendelian unit (Hasegawa and Tanaka, 1994). The data obtained by Yerushalmi *et al.* (2001) were analysed by the classical morphometric ratios, F/C and E/F, and by canonical discriminant (multivariate) analysis based on measurements of F, C and E, as well as of M (minimum width of the pronotum) and H (maximum height of the pronotum). The results of both kinds of analysis showed that the morphometric values of the albinos with 99.6% West African gene pool are much closer to those of the congenic normal phenotypes than to the albinos with 100% Okinawa gene pool. It was concluded, therefore, that the more solitarious values and the smaller amplitude of morphometric phase change in the Okinawa albinos are related primarily to the geographic strain. However, the pigmentation (albino versus normal colouration) also affected morphometric phase characteristics; albinos with 99.6% West African gene pool showed more solitarious values and more limited morphometric phase change than congenic normal phenotypes. The effect of the pigmentation was considerably smaller than that of the strain. These results refuted Nolte's (1967, 1968) claim that albinos constitute an extreme solitary phase. Yerushalmi *et al.* (2001) clearly demonstrated that albino locusts, with 99.6% West African gene pool, exhibit considerable morphometric phase transformation, although it is less extensive than that of congenic normal phenotypes. Even albinos with 100% Okinawa gene pool exhibited a limited morphometric phase transformation.

By assessing the F/C ratio and by canonical discriminant analysis, Hoste *et al.* (2002a) confirmed that Okinawa albinos have rather solitarious morphometrics, even when crowded, and a smaller amplitude of density-dependent morphometric changes, than a West African strain. These authors also observed that the E/F ratio does not change with rearing density in the Okinawa albino strain. In albino females, both E and F of the locusts kept under isolation were significantly longer than of those kept under crowding, whereas in albino males, neither E nor F was affected by the density. In contrast, males of the normally coloured West African strain did show density-dependent changes of the E/F ratio. Unfortunately, the data presented for normally coloured West African females are erroneous; 8.20 mm for E and 3.00 mm for F are impossible values (Hoste *et al.*, 2002a, p. 800, Table 5, Females, normal, crowded locusts). Values should have read 48.20 and 23.00, respectively (Hoste, personal communication). In the same article, density-dependent behavioural changes are also reported (see Section 11.2).

The albinism in the Okinawa strain of *L. migratoria* is caused by the absence of a hormonal factor that is responsible for darkening (Tanaka, 1993; Tanaka and Pener, 1994). This factor, extracted from the corpora cardiaca (CC) of normal phenotypes of crowded *L. migratoria* and *S. gregaria*, was purified and identified as the 'dark-colour-inducing neurohormone (DCIN)' of locusts, also named as 'dark pigmentotropin', '[His^7]-corazonin' (Tawfik *et al.*, 1999a) and 'Scacorazonin' (Tanaka, 2006). The DCIN is responsible for the dark solitarious colour in *L. migratoria* and for the dark patterns in gregarious *L. migratoria* and *S. gregaria* (see Section 7.2.2 for colouration and Section 9.3 for neurohormones involved in locust phase polyphenism). Surprisingly, injection of DCIN into second- and third-instar isolated hoppers of Okinawa albinos and of a normally coloured strain from Ibaraki (Japan, north of Tokyo) of *L. migratoria* not only induced darkening but shifted the F/C and E/F ratios towards the direction of gregarious values in the subsequent adults (Tanaka *et al.*, 2002a). In this study, the Okinawa albino hoppers received 2×50 pmol and the Ibaraki normally coloured hoppers received 2×1 nmol of DCIN, indicating that normally coloured locusts, which have endogenous DCIN, are less sensitive to the exogenous hormone than the albinos lacking endogenous DCIN (see also Section 7.2.2). In the normally coloured strain, that was injected with the higher dose, even the convexity of the pronotum was reduced, again a shift towards the gregarious phase. Importantly, injection of DCIN did not affect the duration of nymphal development or the number of nymphal instars. However, although DCIN reduced the F/C ratio of isolated albinos to a level comparable to that of crowded albinos, the values of this ratio still remained as high as those obtained for isolated normal phenotypes of the Ibaraki strain. Tanaka *et al.* (2002a) eventually concluded that the solitarious morphometrics associated with the Okinawa albinos cannot be explained solely by the absence of the DCIN. This conclusion agrees with that of Yerushalmi *et al.* (2001) (see earlier). It may be added that Okinawa and Ibaraki are separated by about 10° of latitude, and locusts from these two localities do not necessarily belong to the same population.

In another study, Hoste *et al.* (2002b) investigated the effect of injected DCIN on morphometrics in *S. gregaria*. Isolated fourth-instar nymphs were injected three times, at 2-day intervals, with 1 nmol of DCIN in 2 μl of soya oil (DCIN is effective when injected in oil, see Section 7.2.2). Controls were injected similarly, but with oil only. Measurements of the subsequent adults were taken and analysis of the data was based on E, F, C, PL (length of pronotum), PH (height of pronotum) and DE (distance between the eyes), F/C and E/F ratios, as well as on canonical discriminant analysis. The results showed that DCIN shifted the morphometrics towards crowded values. In fact, however, the differences between DCIN-injected and oil-injected locusts were significant only for F in both sexes, E, PL, and PH in females, as well as for the F/C ratio in males, whereas the differences in the F/C ratio in females and the E/F ratio in both sexes were not significant (Hoste *et al.*, 2002b, Table 4). The DCIN-induced shift in morphometrics towards crowded values, as found by Hoste *et al.* (2002b), was widely cited by Breuer *et al.* (2003), perhaps over-emphasizing the importance of the effect in locust phase transition. It may be added that Hoste *et al.* (2002b) found that DCIN did not induce behavioural phase change in *S. gregaria* (see Section 11).

In a later publication, Maeno *et al.* (2004) paid attention to the well-known fact that solitarious *S. gregaria* may have an extra moult, that is an extra nymphal instar, whereas gregarious *S. gregaria* do not have such an extra instar. The number of instars is easily recognized by the number of eye stripes which is equal to N+1, where N = number of nymphal instars and the number 1 is the adult. This strict correlation between the number of nymphal instars+1 and the number of eye stripes is characteristic to the subfamily Cyrtacanthacridinae that includes *Schistocerca*, *Nomadacris*, *Anacridium* and other genera, as well as to some acridids from other subfamilies (for further details, see Albrecht, 1955, pp. 124–125; Albrecht, 1967, p. 32; Pener, 1991, p. 11 and 26–27; Uvarov, 1966, p. 198, 278 and 337 as well as further references in these publications). Maeno *et al.* (2004) consequently repeated the experiments of Hoste *et al.* (2002b) on the effect of DCIN on *S. gregaria* morphometrics, but separating the data for isolated (solitarious) adults that underwent five nymphal instars from those that underwent six. Maeno *et al.* (2004) found that both F/C and E/F ratios were significantly different between adults with five and six nymphal instars, justifying separate analyses. DCIN injected into isolated nymphs significantly shifted the F/C ratio towards lower (more gregarious) values in both sexes that underwent either five or six nymphal instars and in males that underwent five nymphal instars, all in comparison with the respective oil-injected controls. There were only a few isolated males that underwent six nymphal instars; therefore, no statistical analysis was presented for the F/C ratio in these males. It is known that a higher proportion of females than males exhibits the extra instar (Injeyan and Tobe, 1981a). The E/F ratios showed relatively small variations, but this ratio was significantly higher (more gregarious) in DCIN-injected males that underwent five nymphal instars than in the respective oil-injected controls.

Importantly, injection of DCIN did not seem to change the proportion of nymphs undergoing five or six nymphal instars, neither in females nor in males (cf. Maeno *et al.*, 2004, Table 1). It may be inferred, therefore, that these inter-related phase characteristics, number of nymphal instars and number of eye stripes, are not affected by DCIN.

In the same study, Maeno *et al.* (2004) also investigated the effect of timing of the injection of the DCIN on the shift of the morphometrics in *S. gregaria* and in the Okinawa albino strain of *L. migratoria*. They found that in both cases, the earlier the injection of the DCIN into the isolated nymphs, the larger is the shift towards gregarious morphometric ratios. Although DCIN undoubtedly affects morphometrics, it cannot be the sole responsible factor for these phase characteristics because Okinawa albinos do show limited density-dependent morphometric phase change (Yerushalmi *et al.*, 2001; Hoste *et al.*, 2002a), and albinos of *L. migratoria* with 99.6% West African gene pool (Yerushalmi *et al.*, 2001) show considerable density-dependent morphometric phase change in spite of the fact that these strains lack DCIN. A hypothetic explanation of the effect of DCIN on morphometrics and on an additional morphological phase characteristics (number of antennal sensilla) is offered by Maeno and Tanaka (2004) (see Section 5.3.1).

Recently, a strange but interesting finding was reported by Tanaka (2007), who injected CC extracts, obtained from crowded last-instar Okinawa albino nymphs of *L. migratoria*, into conspecific second-instar isolated Okinawa albino nymphs. The injection, five CC equivalent per recipients, did not induce darkening as expected, but shifted the morphometric ratios F/C and E/F of the subsequent adults towards gregarious values, like the effect observed by Tanaka *et al.* (2002a) after injection of DCIN into *L. migratoria* hoppers. However, Okinawa albinos of *L. migratoria* do not have DCIN (Schoofs *et al.*, 2000; Baggerman *et al.*, 2001). If so, how does injection of CC extracts from crowded to isolated Okinawa albinos affect morphometric ratios? Tanaka (2007) offers two theoretically possible explanations: (1) the albino CC factor is a mutated DCIN that lost its effect on dark-colour induction, but still affects morphometrics like the non-mutated DCIN; or (2) there is another factor in the CC of crowded albino locusts that affects morphometrics, but it is released only in the presence of DCIN and injection of CC extracts into isolated albinos provides both this other factor and the releasing DCIN. According to Tanaka (2007), the former explanation is more probable and this explanation is also mentioned in his recent review (Tanaka, 2006). A former finding of Baggerman *et al.* (2001) may support this explanation. They refer to a DCIN immunoreactive material in some fibres of the outer periphery of the CC of Okinawa adult albinos. However, mass spectrometric evidence showed that this staining is not due to DCIN (also termed [His^7]-corazonin). Surprisingly, the same immunopositive material is absent in the CC of normally coloured locusts, suggesting that it may be a defective DCIN molecule that still has affinity to the antibodies, but lost its dark colour–inducing activity. It may well be the substance that induces morphometric changes in Okinawa albino recipients of the CC of Okinawa albinos. A final conclusion can be made only after identification of the CC factor that affects morphometrics, but does not induce darkening.

5.3 SENSILLA

This section deals only with those sensilla that change in number in response to locust density. Such sensilla are found on the antennae, frons and outer side of the hind (metathoracic) femora. Only material related to the surface morphology of these sensilla is presented here; their neurobiology in relation to function and physiology is discussed in Sections 11.6, 12.3 and 14.

5.3.1 *Antennal sensilla*

According to Chapman (2002), there are four types of sensilla on the antenna of acridids: trichoid, large basiconic, small basiconic and coeloconic sensilla. These terms are used by most locust workers, but the nomenclature of the antennal sensilla is far from being uniform. For example, Ochieng *et al.* (1998) use the term chetic sensilla (‘*sensillum chaeticum*’), which is equivalent to trichoid sensilla in Chapman’s (2002) terminology. Moreover, Chapman’s small basiconic sensilla is named as trichoid sensilla (‘*sensillum trichodeum*’) by Ochieng *et al.* (1998), as well as by Ochieng’ and Hansson (1999). Thus, the same term, trichoid sensilla, is used by different authors for different sensilla. Chapman’s nomenclature is followed later. Most authors count the annuli of the flagellum from the distal (apical) annulus that has the number 1. The reason for this concept is that the number of annuli increases from instar to instar and the new annuli are added at the pedicel, at the proximal end of the flagellum (Chapman, 2002, Fig. 1).

Greenwood and Chapman (1984) investigated the number of each type of sensilla on the antenna in crowded (gregarious) and isolated (solitarious) fifth (last)-instar hoppers and adult males and females of *L. migratoria*. As expected, the number of sensilla increased from the fifth nymphal instar to adults in both sexes and under both conditions of density. Phase did not affect the number of annuli which was the same in crowded and isolated adults: 23 and 24 in males and females, respectively. However, coeloconic, small basiconic and especially large basiconic sensilla were more numerous in isolated than in crowded locusts. Detailed investigation of annuli 1, 4, 8, 12 and 16 revealed significantly higher number of large and small basiconic sensilla in the isolated than in the crowded adults in all of these annuli. Coeloconic sensilla were also significantly more numerous in isolated adults on annuli 4, 8 and 12, but not on annuli 1 and 16. The number of trichoid sensilla did not differ greatly between crowded and isolated locusts.

A little caution should be exerted with regard to the findings of Greenwood and Chapman (1984). These authors kept the isolated locusts in 2-l containers, probably under high humidity, because the majority developed green colouration. It is well known that high humidity promotes green colour in solitarious *L. migratoria* (see Pener, 1991, pp. 12–14 and Section 7.2.1 of the present review). However, some of the coeloconic sensilla are hygroreceptors in *L. migratoria* and in other insects (see Altner’s, 1977, review and references therein). This was recognized also by Greenwood and Chapman (1984), but they did not distinguish between these sensilla according to function. Moreover, in a recent general review on insect sensilla, Hartenstein (2005, p. 382, legend of Fig. 3) regards both basiconic (Chapman’s large basiconic) and coeloconic sensilla as hygro/thermosensory organs. This generalization does not seem to be valid for acridid antennae; nevertheless, the higher humidity experienced by the isolated locusts of Greenwood and Chapman (1984) may have resulted in an increased number of these sensilla. It is known that environmental factors (kind of food and odour, detailed later) affect the number of antennal sensilla in acridids.

Heifetz *et al.* (1994) investigated density-dependent differences in the number of sensilla on antennal annulus number 8 of adult males and females of a West African strain and the Israeli strain of *L. migratoria*. They employed the terms of Chapman (2002), except that 'large basiconic' sensilla were termed 'basiconic' sensilla. Heifetz *et al.* (1994) found no significant differences in the number of trichoid sensilla between crowded and isolated locusts or between the two strains. In both sexes and both strains, coeloconic sensilla were significantly more numerous in isolated than in crowded locusts. Basiconic and small basiconic sensilla were also significantly more numerous in both sexes of isolated locusts in the West African strain, but not in the Israeli strain. These results fully confirmed those of Greenwood and Chapman (1984) for the West African strain (presumably *L. m. migratorioides*) and partially confirmed for the Israeli strain. It may be recalled that Heifetz *et al.* (1994) found that locust rearing density does not affect the F/C ratio in the Israeli strain (see Section 5.2) and the amplitude of phase change is more limited in this strain.

Heifetz and Applebaum (1995) studied rearing density-dependent differences in the number of sensilla on annulus number 8 of the antenna in adult males and females of the grasshopper, *Aiolopus thalassinus*, in comparison to a West African strain of *L. migratoria*. The latter was used as a kind of reference. The nomenclature was that of Heifetz *et al.* (1994). In both sexes of *L. migratoria*, significantly more small basiconic and coeloconic sensilla were found in isolated than in crowded adults. In *Aiolopus thalassinus*, a significant density-dependent difference was found in the coeloconic, but not in the small basiconic sensilla. The coeloconic sensilla were more numerous in the isolated grasshoppers. The authors do not present data on (large) basiconic and trichoid sensilla. As noted earlier in relation to the study of Greenwood and Chapman (1984), it cannot be ruled out that the isolated insects of Heifetz *et al.* (1994) and Heifetz and Applebaum (1995) experienced higher humidity than their crowded insects. If so, it is possible that the humidity affected the number of coeloconic sensilla.

Phase-dependent differences in the antennal sensilla of *S. gregaria* were studied by Ochieng *et al.* (1998). Unfortunately, the source of the crowded (gregarious) locusts differed from that of the isolated (solitarious) ones; consequently, strain-dependent differences might have influenced the results. Although the humidity in the rooms of the crowded and isolated locusts was the same (45% R. H.), isolated locusts were kept in Perspex jars that probably provided a higher humidity than that in the cages of crowded locusts. Ochieng *et al.* (1998) investigated the sensilla on the ventral surface of the antenna in first-, third- and fifth-instar hoppers and in adults. In the first-instar hoppers and in the adults, the sensilla were counted on each antennal annulus, whereas in the third- and fifth-instar hoppers, the sensilla were counted only on annuli number 2, 8 and 14. They found no sex-related differences in the kind and number of sensilla and pooled the data for both sexes in each age group. Chetic sensilla (equivalent to Chapman's, 2002, trichoid sensilla) were significantly more numerous in isolated than in crowded first-instar hoppers, but no significant difference was found in the adults. Isolated adults possessed significantly more basiconic, coeloconic and trichoid (equivalent to Chapman's, 2002, small basiconic) sensilla than crowded adults (Ochieng *et al.*, 1998, Fig. 6). The authors noted that there are two kinds of coeloconic sensilla, but did not distinguish between them in the counting. The number of basiconic sensilla was the highest on annuli 9–15 and that of the coeloconic sensilla on annuli 8–14; density-dependent differences were also the highest on these annuli. Sensillar distribution in the second annulus of the third- and fifth-instar nymphs showed no density-dependent difference. On the 8th and 14th annuli of the fifth-instar nymphs, basiconic and coeloconic sensilla were more numerous in isolated than in crowded locusts. In surprising contrast, in the third instar, these sensilla were less numerous on the same (8 and 14) annuli in the isolated than in the crowded hoppers. It seems that Ochieng *et al.* (1998) did not distinguish between isolated locusts that underwent five or six nymphal instars. Either their fifth-instar nymphs were actually a mixture of the fifth and the sixth stadium (in both instances, these are last-instar nymphs) or none of their isolated locusts underwent six nymphal instars.

Chapman and Lee (1991) investigated the effect of crowding, food and exposure to various odours on coeloconic, basiconic and trichoid sensilla in the antenna of the grasshopper, *S. americana*. No distinction was made between small and large basiconic sensilla. They found a higher number of coeloconic sensilla in isolated grasshoppers than in crowded ones on annuli 6–9, but no density-dependent differences were seen in these sensilla on annuli 10–14. In another test, investigating only annuli 4, 8, 12 and 16, they found a greater number of coeloconic sensilla on annuli 4 and 8 in isolated than in crowded grasshoppers, but not on annuli 12 and 16. Basiconic sensilla were slightly more numerous in isolated grasshoppers on annuli 4, 8 and 12, but not significantly so. However, on annulus 16, there were significantly fewer basiconic sensilla in the isolated hoppers. Rearing density did not affect the number of trichoid sensilla. In addition to these density-related effects, Chapman and Lee (1991) also found effects of the food and exposure to different odours. Grasshoppers fed on a single plant, lettuce, from hatching until the adult stadium, showed a higher number of coeloconic sensilla on annuli 6–9 than those fed on lettuce with addition of various other plants. The single plant diet also resulted in fewer trichoid sensilla on annuli 10–14. Exposure to odour of a mixture of commonly occurring plant volatiles in paraffin oil also affected annulus-dependent differences in the number of coeloconic and basiconic sensilla.

In another study on *S. americana*, Bernays and Chapman (1998) investigating the effect of an artificial diet as opposed to lettuce and the effect of various chemicals added to the artificial diet also found that these conditions greatly affect annulus-dependent number of antennal sensilla, differently in different kinds of sensilla.

Rogers and Simpson (1997) showed that the food, wheat versus various synthetic diets, experienced by nymphs of *L. migratoria*, affected the number of sensilla on the 11th annulus of the adult's antenna. In this study, the annuli were counted from the base of the antenna. These authors demonstrated that the influence of the chemosensory environment on the development of sensilla numbers was localized to the particular sensory field experiencing stimulation. For example, when locusts were reared over the final two nymphal stadia on wheat, they developed a larger number of sensilla on the 11th annulus of the antenna and on the maxillary palps than did insects reared on nutritionally adequate but chemically bland synthetic food. When chemostimulation of the antennae was augmented by providing wheat odour to nymphs reared on synthetic food, antennal sensilla numbers (especially multiporous olfactory sensilla) increased, while there was no increase in numbers of palp sensilla.

Disregarding minor species- and strain-dependent differences, the emerging picture indicates a higher number of coeloconic and basiconic sensilla on the antennae in isolated than in crowded locusts in two species (*L. migratoria* and *S. gregaria*) and a higher number of coeloconic sensilla in isolated than in crowded specimens of two non-swarming grasshoppers (*A. thalassinus* and *Schistocerca americana*). However, the kind of food and the odour environment also have considerable effects on the number of antennal sensilla, and some of these effects are annulus-dependent. Therefore, care should be exercised in relation to this phase characteristics; it is much influenced by factors other than density. Also, any possible effect of the humidity on the number of antennal sensilla, especially of the coeloconic sensilla, should be explored.

A functional hypothesis attempting to explain the reason for the higher number of coeloconic and basiconic sensilla in solitarious locusts is offered by Greenwood and Chapman (1984). According to these authors, gregarious locusts require less physiological sensitivity than solitarious individuals, and consequently, they have fewer antennal sensilla. The data from non-locust acridids show a similar trend in regard to coeloconic sensilla. However, these grasshoppers do not have distinct phases and it is unclear why they need more sensilla under isolation.

Yamamoto-Kihara *et al.* (2004) studied the effect of the DCIN (see Section 7.2.2) on the number of antennal sensilla on annulus number 8 in isolated female adults of *L. migratoria*. Nymphs were injected twice, on day 3 of the second stadium and on day 3 of the third stadium, with 1 nmol of DCIN, in 1 μl of rape seed oil, each time. Controls received only rape seed oil injections. DCIN did not affect the number of trichoid sensilla (*sensu* Chapman, 2002), nor that of basiconic A and basiconic B sensilla (Chapman's, 2002, large and small basiconic sensilla, respectively). However, coeloconic sensilla were significantly fewer in the DCIN-injected locusts than in the controls. Fewer coeloconic sensilla are a gregarious characteristic (see earlier), and it may be recalled that injection of DCIN shifted the morphometrics of isolated locust towards the direction of the gregarious phase (see Section 5.2).

Maeno and Tanaka (2004) injected 2 nmol of DCIN, in 1 μl of rape seed oil, into 1-day-old third-instar nymphs of isolated *S. gregaria* and counted the different types of sensilla on annuli 2, 8, and 14 of the antenna in the subsequent adult females that underwent five nymphal stadia. Rape seed oil–injected isolated locusts served as controls. In addition, for the sake of comparison, counting of the sensilla on annulus 8 of un-injected crowded locusts was carried out. Type A basiconic sensilla (Chapman's, 2002, large basiconic sensilla) and coeloconic sensilla were significantly less numerous in the DCIN-injected insects than in the oil-injected controls on all three annuli (2, 8 and 14); in fact, the number of these two types of sensilla in DCIN-injected insects did not differ significantly on annulus 8 from that of the crowded un-injected locusts. On all three annuli, DCIN did not affect the number of type B basiconic sensilla (Chapman's, 2002, small basiconic sensilla), and on annulus 8, there were no significant differences between DCIN-injected, oil-injected and crowded locusts. Trichoid sensilla were slightly, but significantly, more numerous in DCIN-injected locusts than in oil-injected controls on annulus 2, but on annulus 14, no significant difference was found. On annulus 8, crowded locusts possessed more trichoid sensilla than oil-injected controls, but these sensilla in DCIN-injected locusts did not differ significantly from oil-injected ones or from crowded locusts (Maeno and Tanaka, 2004, Fig. 3). These results again indicate that DCIN promotes a gregarious characteristic in regard to the number of antennal sensilla.

In the same study, Maeno and Tanaka (2004) also investigated the effect of timing of the DCIN injection; they injected second-, third- and fourth-instar nymphs and concluded that the earlier the injection the greater the effect on the total number of sensilla on annulus 8.

Maeno and Tanaka (2004) attempted to explain the promotion of gregarious characteristics by the DCIN. Briefly, they suggest that crowding stimuli perceived by the antennae lead to increasing concentration of the endogenous DCIN, which in turn reduces the number of certain antennal sensilla, shifts morphometrics towards more gregarious values (see Section 5.2) and induces black patterns in the integument of gregarious nymphs (see Section 7.2.2). This explanation has not been repeated in Tanaka's (2006) recent review on 'corazonin and locust phase polyphenism'.

5.3.2 *Frontal hairs*

The frons-vertex region of the head of locusts bears mechanical aerodynamic hairy sensilla; wind blown on these hairs induces flight in adult locusts (Weis-Fogh, 1949, 1956; Haskell, 1960; Sviderskii, 1969). Fuchs *et al.* (2003) demonstrated that the number of these hairs is significantly higher in isolated than in crowded adult females of *S. gregaria*, in spite of the finding that the stimuli conveyed to the flight motor centres are weaker in the isolated locusts. The difference was also significant in each of the specific receptor fields, hair field 1 and hair field 2, which are known to mediate the response to the wind.

Fuchs *et al.* (2003) cite Sviderskii (1969) who had shown that the number of frontal hairs increases in consecutive nymphal stadia. Fuchs *et al.* (2003), being aware of the fact that isolated *S. gregaria* (especially the females, cf. Injeyan and Tobe, 1981a) may have an extra nymphal stadium, mention the possibility that the higher number of nymphal stadia results in a higher number of frontal hairs in the isolated nymphs. This assumption needs experimental testing and, possibly, comparison with *L. migratoria*, which has the same number of nymphal stadia in the gregarious and the solitarious phase.

5.3.3 *Mechanosensory trichoid sensilla on the outer side of the hind femur*

Roessingh *et al.* (1998) and Hägele and Simpson (2000) demonstrated that repeated touching of isolated nymphs of *S. gregaria* elicits rapid and full gregarious behaviour. It was subsequently found that the outer surface of the femur of the hind legs was the most effective site for eliciting behavioural gregarization upon repeated mechanical stimulation (Simpson *et al.*, 2001). Rogers *et al.* (2003) further investigating the neurobiological aspects of this phenomenon (see Section 14) discovered that isolated last-instar nymphs of *S. gregaria* have 29% more mechanosensory hairy trichoid sensilla on the outer surface of each hind femur than do crowded nymphs (135 ± 6.1 and 105 ± 5.4, respectively). In these nymphs, no density-dependent difference was found in the number of such sensilla on the hind tibiae and tarsi, nor on the front and middle legs. Isolated adults had 26% more of these sensilla on their hind femora than crowded adults (177 ± 4.6 and 140 ± 6.3, respectively). In contrast to last-instar nymphs, isolated adults had significantly fewer tactile hair sensilla on the hind tarsi, front femora and front tarsi than did crowded adults. The number of such sensilla seemed to be phase-independent on the hind tibiae of the adults. More detailed investigation showed that the additional hairy sensilla of the isolated locusts were not evenly distributed over the hind femur. On the outer surface of the hind femur of last-instar nymphs, there were 60% more sensilla on the most distal dorsal region and 36% more sensilla on the most proximal ventral region than on the equivalent parts of the femur of crowded nymphs, but similar numbers on the distal ventral and proximal dorsal surfaces. A similar distribution of these sensilla was found on the hind femora of the adults. Isolated locusts had significantly fewer bimodal basiconic sensilla on the distal region of the hind femur than crowded locusts. According to Rogers *et al.* (2003), these basiconic sensilla contain a single mechanosensory and several chemosensory afferent neurons.

The outer surface of the hind femur is most likely to be touched by other locusts. The higher number of sensilla on this surface may imply that if a solitarious locust becomes mixed with a group or band of other locusts, it is advantageous to alter the solitarious behavioural state as rapidly as possible to the gregarious behavioural phase state.

6 Anatomy

The number of ovarioles is higher in solitarious than in gregarious females of *L. migratoria*, *S. gregaria* and *N. septemfasciata* (see Uvarov, 1966, pp. 353–355; Pener, 1991, pp. 22–23; and references therein). However, it should be stressed repeatedly that the average weight of an egg pod of *L. migratoria* is more or less equal in crowded and isolated locusts (Albrecht *et al.*, 1958). Similarly, the average vitellin content of ovaries containing mature oocytes does not differ between crowded and isolated *S. gregaria* females (Injeyan and Tobe, 1981b). The eggs of the solitarious locusts are smaller and lighter, but more numerous per egg pod, than those of the gregarious locusts. The phase-dependent difference in the number of ovarioles is already evident in hatchlings. This presumably means that the phase state of the parent generation is transmitted to the progeny (see also Section 16).

Farrow and Longstaff (1986, p. 212, Table 1) reported some field data on the number of ovarioles in plague versus recession populations of locusts. In the Australian race of *L. migratoria*, the average number of ovarioles was 103.8 and 59.1 in recession and plague populations, respectively. Based on estimations, partial data and references, Farrow and Longstaff (1986, p. 213) concluded that the number of ovarioles in gregarious females from swarms (plague population) is about 30% fewer than in solitarious (recession population) females in *S. gregaria* and *N. septemfasciata*. In strong contrast, well-based field data demonstrated that the number of ovarioles in plague and recession populations of the Australian plague locust, *C. terminifera*, is practically equal.

Quesada-Moraga and Santiago-Álvarez (2001a) reported another trend, higher number of ovarioles in gregarious than in solitarious females of *Dociostaurus maroccanus*. However, the actual density was not described in this article; instead, the E/F ratio of the locusts (connoted T/F ratio according to the terminology of these authors) was supposed to reflect the phase state. Indeed, according to the data, the higher the E/F ratio (a gregarious characteristic), the higher the number of the ovarioles. However, even if this correlation is significant, it is not necessarily an independent phase characteristics. Gregarious *D. maroccanus* are considerably larger than solitarious individuals (see Section 5.1), and it is known that larger species of acridids have a higher number of ovarioles (Uvarov, 1966, p. 144). Therefore, the higher number of ovarioles in gregarious *D. maroccanus* females may just reflect their larger size.

Quesada-Moraga and Santiago-Álvarez (2001a) found no clear relation between the E/F (connoted T/F) ratio and the number of testicular follicles in *D. maroccanus* males.

7 Colouration and pigments

7.1 GENERAL BACKGROUND

Colour differences between gregarious and solitarious locusts are probably the best known and most easily recognized visible phase characteristic. It is not surprising, therefore, that a plethora of literature accumulated on the effects of phase, environmental factors, as well as physiological factors, on acridid and especially on locust colouration. The subject was thoroughly discussed in the reviews by Uvarov (1966), Albrecht (1967), Rowell (1971), Fuzeau-Braesch (1972, 1985) and Pener (1991, pp. 12–21). To save space, the old references cited in these reviews will not always be cited later, though historical perspectives or different contexts or both in the present review may necessitate the citation of some of the old literature.

There are three kinds of colour polyphenism in acridids: (1) phase colour polyphenism; (2) 'green-brown' colour polyphenism (though the so-called brown form is not necessarily brown, but it is not green); and (3) homochromy. The latter means adaptation of the body colour to the colour of the underlying background. Depending on the species, an acridid species may not show any colour polyphenism, or may exhibit only one kind, or a combination of two, or even all three. The colour polyphenism of nymphs and adults of the same species may be similar or different.

All three types of acridid colour polyphenism are based on 'morphological colour changes' (cf. Raabe, 1982, 1983, 1989; Fuzeau-Braesch, 1985), meaning that the visible colour changes because of alterations in synthesis, degeneration, oxidation–reduction, etc., of one or more pigments. This contrasts to 'physiological colour changes' that occur quickly, sometimes within second(s), due to concentration, dispersal or reorganization of the pigment granules within the epidermal cells. Physiological colour changes are not known from locusts, but a non-locust grasshopper species, *Kosciuscola tristis* Sjöstedt, from Australia shows temperature-dependent diurnal colour changes; the day (warm) colour is bluish-green and the night (cold) colour is dark. This colour change is caused by movement of dark pigment granules in the epidermal cells from the top to the basal part (day colour) and from the basal part to the top (night colour) (Key and Day, 1954a,b). It seems that adult males exhibit the most marked colour change, studied by Filshie *et al.* (1975). These authors found that small granules of uric acid and a pteridin, having a mean diameter of 0.17 μm, are responsible for the blue (warm) colour by the physical phenomenon of Tyndall scattering of the light. These granules are white in an isolated state. The dark (cold) colour is caused by spherical brown granules with a mean diameter of 1.0 μm. There are many microtubules directed from the basal to the apical part of the epidermal cells; possibly, these are associated with the movement of the granules.

The relations between visible colour and pigments are often blurred. For example, a blue plus a yellow pigment results in green visible colouration. Oxidation–reduction, number and site of double bond(s), in the pigment molecule, binding, etc., may change the colour in the same type of pigment (cf. Kayser, 1985; see also Section 7.4).

7.2 HOPPERS AND ASSOCIATED HOPPER-ADULT FEATURES

7.2.1 *Previous knowledge and relevant recent findings*

As outlined in Section 7.1, previous knowledge is based on earlier reviews and on the references listed in these reviews.

Late-instar hoppers of *L. migratoria* show all three kinds of colour polyphenism, namely, phase colour polyphenism, green-brown colour polyphenism and homochromy (see Section 7.1) hierarchically superimposed upon one another (Fig. 1A). The colouration of crowded (gregarious) hoppers is contrasting, dirty orange with black patterns, with no or little differences among individuals in the same crowd. However, even this relatively rigid colouration is subject to abiotic environmental effects; high temperatures reduce the areas of the black patterns and low temperatures increase them. The colour of the isolated (solitarious) hoppers is non-contrasting, cryptic, with a uniform shade dominating over the whole body; nevertheless, some areas of the body may be darker than the dominating shade. The sites of these darker areas correspond to those of the black patterns in the gregarious hoppers, though in the solitarious hoppers, they are less extensive in size and less intense in darkness. Changes in the density of the population result in the so-called 'transiens' forms that show almost endless intermediate colouration during a stadium or two.

Isolated (solitarious) hoppers of *L. migratoria* exhibit the green-brown polyphenism, the second kind in the hierarchy. Under high humidity, the hoppers are green, whereas under low humidity, they are not green and show homochromy, which is the third kind of colour polyphenism in this species. Homochromy stands at the lowest level of the hierarchy. Homochrome colouration ranges from off-white, through straw yellow, beige, buff, brown, reddish iron-rust colour, light or dark grey to black and intermediate shades between these colours, to match the underlying background in the field, or the inside (especially the floor) of the container in which an isolated hopper is kept in the laboratory.

Green-brown colour polyphenism and homochromy are subject to individual variations. Most acridids exhibiting one or both of these two kinds of polyphenism show a dominating shade, but a minority may show deviations from the dominating shade. Also, there are differences among individuals in the response to the same environmental factor. For example, in *L. migratoria* hoppers, at intermediate humidities, some individuals are green, light green, olive green or greenish, whereas some others show homochromy.

Like *L. migratoria*, hoppers of *Locustana pardalina* show all three kinds of colour polyphenism. *S. gregaria* and *N. septemfasciata* hoppers exhibit only phase and green-brown polyphenism, but not homochromy. The gregarious hoppers of these two species have contrasting colouration. In *S. gregaria*, black

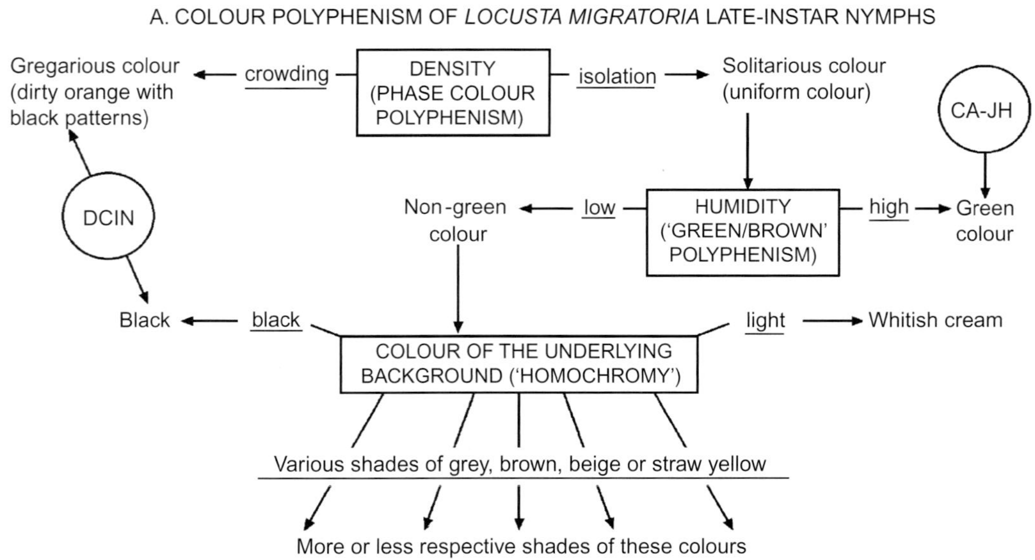

FIG. 1A Diagrammatic presentation of the three different kinds of superimposed colour polyphenism in final-instar hoppers of *Locusta migratoria*. The environmental factors that exert the major effects on each type of colour polyphenism (in parentheses) are boxed. The appropriate changes in the environmental factors are underlined, whereas the actual colour(s) exhibited by the hoppers are not. Endocrine effects on colour are encircled. CA, corpora allata; DCIN, dark-colour-inducing neurohormone (also termed [His^7]-corazonin); JH, juvenile hormone. Fig. 1A is updated from Pener (1991).

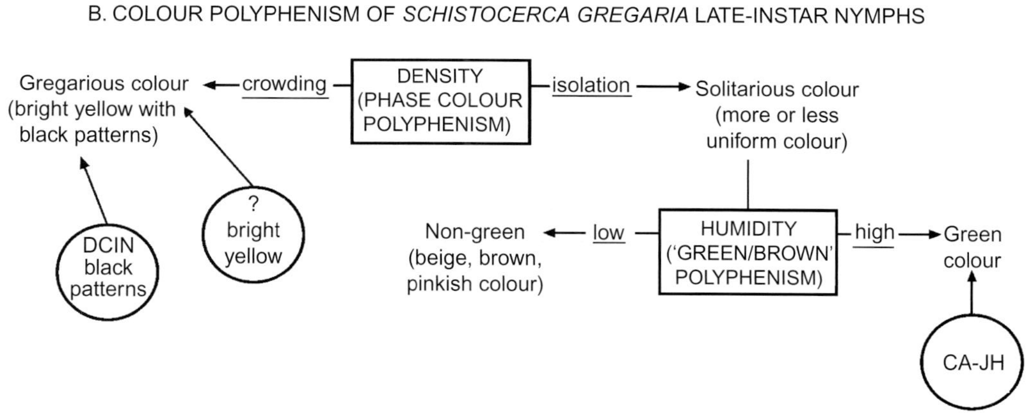

FIG. 1B An equivalent schematic for final-instar hoppers of *Schistocerca gregaria*. For abbreviations see caption of Fig. 1A and for further details, see text.

FIG. 2 Major forms of colour polyphenism in final-instar nymphs of *Schistocerca gregaria*. The gregarious form is uppermost, while green and beige solitarious forms are shown below. The latter two solitarious forms are induced by high and low humidity, respectively. © 2009 Meir Paul Pener.

patterns are present on an off-white-light rose, or yellow background body colour in early and late hopper stadia, respectively (Figs. 1B and 2; see also Stower's, 1959, colour plates). The areas covered by black patterns in *S. gregaria* depend on the temperature in a way similar to that in *L. migratoria* (see earlier). Investigating biotic effects on the black patterning, Lester *et al.* (2005) kept previously isolated *S. gregaria* hoppers under various conditions from the second or third stadium and noted the extent of these patterns in the last nymphal stadium on the head, pronotum and abdomen. The test insects were kept isolated, or exposed to conspecific or heterospecific (*L. migratoria*) olfactory (smell) stimulus, olfactory plus visual (smell and sight) stimuli, or to actual contact with a group that provided combined stimuli of smell, sight and touch. A significant effect of the stimulus type and of the species was found on the black patterning of the head and the pronotum. Conspecific cues induced more extensive black patterns than heterospecific cues. Black patterning was the least extensive when smell alone was provided and it was the most marked when the test hoppers were reared within a group. The nature of the stimuli significantly affected the extent of abdominal black markings, which were the strongest when the test hoppers were reared in contact with other hoppers. Whether or not stimuli came from conspecific or heterospecific hoppers had no significant effect on abdominal black patterning, however. The same article also deals with these biotic effects on the yellow background colour, but this is discussed in Section 7.2.3.

In *N. septemfasciata*, the black patterns are on a brownish yellow background colour (for colour plates, see Faure, 1932, pp. 336–337 and Plate 16 and Popov, 1989, pp. 82–83). Nymphs of *Dociostaurus maroccanus* show little phase colour polyphenism, and according to Latchininsky and Launois-Luong (1992, pp. 42–44, and references therein), the colour is not a sufficient indicator of the phase state of this locust.

It was already mentioned that some extant species of the genus *Melanoplus* may be considered as less typical locusts or aggregating/swarming grasshoppers (see Section 3.1). Two species of this genus was reported to exhibit density-dependent colour polyphenism; crowded *M. differentialis* and *M. sanguinipes* are darker than conspecific isolated specimens (Dingle and Haskell, 1967; Fielding and Defoliart, 2005). It is a question of semantics whether this phenomenon should be considered as colour phase polyphenism.

The effects of environmental factors are not necessarily uniform in all acridids. There may be differences, even in the same genus. For example, in *Schistocerca americana*, which is a grasshopper rather than a locust (cf. Sword, 2003), green-brown colour polyphenism of the hoppers is affected by temperature, whereas humidity, background colour and rearing density have little influence; also the effect of temperature on the black patterns is more marked than that of the rearing density (Tanaka, 2004a). Y. Tanaka (2008) studying the effect of density, humidity, photoperiod, surrounding background colour and temperature on the green-brown colour polyphenism of the pyrgomorphid grasshopper, *Atractomorpha lata*, found that only temperature exerted a significant effect; higher temperatures increased the frequency of the brown form. In *Conocephalus maculatus* (Le Guillon) belonging to the Tettigoniidae, another Orthopteran family, green-brown colour polymorphism (rather than green-brown colour polyphenism) seems to be controlled mainly by direct genetic factors, and it is not or little affected by environmental factors (Oda and Ishii, 2001).

Phase colour polyphenism also exists in hatchlings and it is somewhat different from that in the later hopper stadia. In *L. migratoria*, the hatchlings from eggs laid by isolated females are mostly light grey, whereas those originating from crowded parents are usually much darker. In *S. gregaria*, isolated females produce mostly light pale green hatchlings, whereas the hatchlings from eggs of crowded parents are dark, partially or almost completely black. Bouaïchi and Simpson (2003) tested the effect of parental density and egg pod density on hatchlings' colour in low-density and high-density subpopulations of *S. gregaria* in the field in Morocco. Egg pod density had no effect, but confirming previous concepts, the high-density subpopulation produced preponderantly dark hatchlings, whereas the low-density subpopulation produced a majority of green hatchlings.

Interestingly, in comparison with the controls, there was an increase in the frequency of green hatchlings from eggs laid by crowded *S. gregaria* parents that had been infected as fifth-instar hoppers with the pathogen fungus, *M. anisopliae* var. *acridum*, and developed a behavioural fever; similar results were obtained when young adults were submitted to high temperatures simulating the temperatures of the behavioural fever (Elliot *et al.*, 2003). The mechanism of this effect is yet unclear, although Elliot *et al.* (2003) offer some speculations. Unfortunately, the authors did not weigh the eggs. It cannot be excluded that the stress exerted by the fever led to a reduction of the weight of the eggs and subsequently increased the proportion of the green hatchlings; it was already mentioned that solitarious females lay more eggs per egg pod than gregarious females, but the eggs of the former are lighter than those of the latter (see Section 6) and the lighter eggs of the solitarious females give rise to green hatchlings (see earlier and also later).

Food may also affect hatchling colouration. Jackson *et al.* (1978) found that crowded *S. gregaria* fed on *Dipterygium glaucum* Oecn produced more green hatchlings than those fed on other plants. Again, the eggs were not weighed in these experiments, but the fecundity of the parents fed on *D. glaucum* did not seem to be inferior to the fecundity of those fed on *Pennisetum typhoides* (Burm).

Recently, Tanaka and Maeno (2006) reinvestigated the maternal and crowding effects on the body colour polyphenism in hatchlings of *S. gregaria*. They scored the colour of the hatchlings from 1, uniformly green, through 2, 3 and 4, increasing black markings, to 5, entirely black. They found a positive correlation between the weight of the hatchlings and the dark-colour scores; hatchlings with darker body colouration were heavier, indicating that hatchling body colour was dependent on hatchling body weight. Gregarious females lay heavier eggs; therefore, the hatchlings from these eggs are heavier and the heavier hatchlings are darker, all confirming previous findings that the hatchling progeny of gregarious parents are usually darker than those of solitarious females. In the same study, Tanaka and Maeno (2006) did not find that early egg separation results in development of green rather than black hatchlings. As a result, they questioned the effect of a recently suggested gregarious egg pod foam pheromonal factor (McCaffery *et al.*, 1998; Simpson *et al.*, 1999; see Section 16) on the colour of the hatchlings. However, as outlined by Simpson and Miller (2007), among other concerns, egg mortality in the experiments of Tanaka and Maeno (2006) was very high; given that dark hatchlings are larger and more robust than green ones, it is likely that the smaller eggs destined to produce green hatchlings suffered a higher mortality, shifting the results towards predominantly black hatchlings. Responding to the criticism by Simpson and Miller (2007), Tanaka and Maeno (2008) repeated their experiments, achieving low egg mortality, and reached the same conclusion as in their earlier (Tanaka and Maeno, 2006) paper. The issue is still open, but it may be stressed that the laboratory stock of the locusts of Tanaka and Maeno (2006, 2008) was different from that stock on which the egg pod foam pheromonal factor was found to affect hatchling colour (cf. McCaffery *et al.*, 1998; Simpson *et al.*, 1999; see Section 16.3).

Tanaka and Maeno (2006) also reported that the colour of the hatchlings did not change during the first stadium and the effect of rearing density on the colour became overt only in the second stadium, in accord with the observations of Injeyan *et al.* (1979). Islam *et al.* (1994a) reported an effect of density experienced by hatchlings on colour during the first stadium. Inspection of the data in that paper indicates that two of the six treatment pairs contributed most strongly to this conclusion. The equivalent treatment pairs to those used by Tanaka and Maeno (2006) (i.e. isolating or crowding hatchlings from gregarious parents that had oviposited under crowded conditions, or isolating or crowding hatchlings from solitarious parents in which the female was isolated from the time of mating until oviposition) had little effect on hatchling colour during the first stadium. However, there was evidence of marked darkening within the first stadium in response to crowding when solitary-reared females were mated with solitary-reared males then crowded between mating and oviposition. When these hatchlings were reared alone, they were predominantly green, but when crowded upon hatching, they were significantly darker, suggesting rapid post-hatching melanization. There was also a difference in colour induced during the first stadium when crowd-reared mothers were mated with crowd-reared males and then isolated until oviposition. When the resulting hatchlings were crowded, they were mainly black, whereas those reared in isolation had a mean of 30%–60% black markings.

Recently, Maeno and Tanaka (2008c) found that the first egg pod laid by crowded females of *S. gregaria* produced many green hatchlings, whereas the hatchlings from egg pods laid later were predominantly black. Isolated females of *S. gregaria* showed an opposite trend; the eggs of their first egg pod produced a relatively high proportion of black hatchlings. The difference in hatchlings' colour between the first and the subsequent egg pods is clearly seen in Fig. 4C and 4F of Maeno and Tanaka (2008c) for egg pods laid by isolated and crowded females, respectively. However, these figures (Fig. 4C and 4F) do not demonstrate clearly differences in the proportion of the green or of the black hatchlings from the second egg pod onwards (however, see Fig. 5D and 5H). In later egg pods, laid consecutively by the same females, such differences do not seem to exist (Fig. 5C and 5G). Transfer of females from isolation to crowding increased the proportion of the black hatchlings, whereas an opposite transfer increased that of the green hatchlings (Maeno and Tanaka, 2008c, Fig. 5D and 5H).

In another recent article, Maeno and Tanaka (2007) found a positive correlation between hatchlings' colour and colour of last-instar hoppers of *S. gregaria*. Under crowding, the black patterns of the last-instar hoppers were more intense in those that had been darker at hatching. Also, darker hatchlings showed darker yellow (orange) background body colour in the last hopper stadium, whereas lighter hatchlings give rise to a lighter yellow colour. Under isolation, the hoppers showed green-brown colour polyphenism, and the highest percentage of brownish last-instar nymphs was obtained from the darkest hatchlings.

In acridids that exhibit green-brown colour polyphenism, JH is responsible for induction of the green colour (Fig. 1A and 1B; for references, see Pener, 1991, pp. 16–17). Even in those acridids that do not show regular green-brown colour polyphenism, JH promotes the green colour of the haemolymph as observed a long time ago by Pfeiffer (1945), who implanted extra corpora allata (CA) (the endocrine glands that secrete the JH) into *Melanoplus differentialis*. Implantation of extra CA into hoppers of *Acrida turrita* (L.) resulted in green colour (P. Joly, 1952), and similar treatment led to similar results in three species of African grasshoppers (Rowell, 1967); one of these species even exhibited green adult colouration, despite that green adults have not been seen in natural populations. All these species are grasshoppers, not locusts, though *M. differentialis* may be considered as a less typical locust or an aggregating grasshopper (see Section 3.1 and Table 1).

In locust species that exhibit green-brown colour polyphenism, the green colour–inducing effect of implantation of extra CA, or administration of JH or JH analogues (JHAs), has been repeatedly confirmed without any exception. These treatments induce green colour even in crowded locust hoppers that show a reduction or disappearance of the gregarious colouration with the increasing green colour. Even an albino strain of crowded *L. migratoria*, which has no natural dark colouration (Tanaka, 1993), responded by green colour to JHAs (Hasegawa and Tanaka, 1994).

Employing chemical allatectomy by topical application of the anti-allatin substance, precocene, to 3-day-old fourth-instar isolated green or greenish hoppers of *L. migratoria*, Pener *et al.* (1992) showed a considerable reduction of the green colour in the next, morphologically normal fifth-instar hoppers. The green colour completely disappeared in the subsequent adults. In these experiments, the timing of the chemical allatectomy, leading to absence of JH, was crucially important. If precocene was applied earlier, the insects moulted to precocious adults (adultiforms) in the fifth stadium. If precocene was applied later, some of the hoppers were already in apolysis; in this state, the presence of the moulting fluid made questionable the penetration of the topically applied precocene into the body and subsequent chemical allatectomy. The results revealed that the chemically allatectomized fifth-instar hoppers that lost completely or almost completely their green colour showed solitarious homochrome colouration and no gregarious colouration. Chemical allatectomy did not affect the colour of the isolated non-green (homochrome) hoppers or the gregarious colouration of crowded hoppers. From these results, it was concluded (Pener *et al.*, 1992) that JH is not a primary factor that induces the solitarious phase and it does not play a central role in locust phase changes.

It was outlined earlier that although the green colour is a solitarious characteristic, it is not a necessary solitarious characteristic because of the existence of the 'brown' solitarious hoppers (Pener, 1991, p. 20). However, in a recent article, based on findings in the north-east corner of China, Tanaka and Zhu (2005) reported green adults of *L. migratoria* that might have come from a high-density population. If these green locusts were really in the gregarious phase and did not represent some transiens form, it may be inferred that the green colour is not an exclusive solitarious characteristic.

A DCIN originating from the brain neurosecretory cells (NSC) and the CC of locusts was quite recently purified and characterized (Tawfik *et al.*, 1999a). Its effect on the dark colouration is detailed in Section 7.2.2 and its chemical traits are discussed in Section 9.3.4. The promotion of some gregarious characteristics by the DCIN was already discussed in relation to morphometrics (Section 5.2) and number of antennal sensilla (Section 5.3.1).

7.2.2 *Effects of the dark-colour-inducing neurohormone*

Earlier studies indicated that the haemolymph of crowded *S. gregaria* hoppers induces dark patterns in conspecific isolated hoppers (Nickerson, 1956). It was also discovered that the corpora cardiaca (CC) (Staal, 1961) and the CC with the protocerebral neurosecretory cells (NSC) contain some factor(s) that increase(s) the dark patterns in crowded hoppers of *L. migratoria* (Girardie and Cazal, 1965; and others). The same factor(s) seemed to be responsible also for induction of dark/black colouration in isolated *L. migratoria* hoppers (Nicolas and Ismail, 1978).

This dark-colour-inducing effect did not seem to be restricted to locust colour polyphenism; presumably, the same factor(s) induced black homochrome response in the grasshopper *Oedipoda coerulescens* that exhibits marked homochromy, but neither phase nor green-brown polyphenism (Moreteau-Levita, 1972; Moreteau, 1975).

Although this dark-colour-inducing effect was known for about 30 years, purification and identification of the causative factor(s) were not attempted. *In vitro* tests were not available and *in vivo* tests required thousands of light (off-white) isolated hoppers for bioassays in which darkening induced by different fractions could be assessed. To obtain so many light coloured isolated hoppers was a practically impossible task.

This situation was changed by Tanaka's (1993) discovery of a few albino individuals of *L. migratoria* in a stock colony originating from Okinawa, Japan, from which he derived a laboratory colony of an albino strain. In this strain, the albinism is caused by deficiency of a neurohormone that induces darkening; implantation of CC from conspecific, normally coloured fifth-instar nymphs into fourth-instar crowded albino nymphs resulted in darkening. In the next (fifth) stadium, the recipients of the implants exhibited darker colouration, similar to the shades observed in normally coloured homochrome solitarious hoppers. In some instances, even gregarious hopper colouration (black patterns with dirty orange background body colour) was obtained. Similar procedure of implantation of CC from normally coloured crowded hopper into conspecific field-collected green solitarious hoppers yielded results similar to those observed in the albino recipients. Implantation of brain or of thoracic ganglia from normally coloured hoppers into crowded albino hoppers also induced darkening, which was, however, milder than that induced by implantation of CC. Implantation of suboesophageal ganglion was ineffective.

Tanaka (1993) also reported that implantation of CC or brain from various non-acridid insects led to darkening of the recipient crowded Okinawa albino hoppers. In a later article, Tanaka (2000a) found that implantation of CC and brain, taken from 52 different species, belonging to 10 insect orders, induces darkening in the recipient Okinawa albino nymphs of *L. migratoria*. Only implants of CC or brain from the Coleoptera (eight species tested) were inactive.

Following Tanaka's (1993) discovery, it was soon demonstrated by Hasegawa and Tanaka (1994) that a single recessive Mendelian unit is responsible for the albinism of the Okinawa albino strain of *L. migratoria*. These authors also showed that the Okinawa albinos are able to exhibit green colouration after administration of JHAs (see Section 7.2.1) and crowded sexually mature male adults become yellow, like normally coloured *L. migratoria* (see Section 7.3).

Albinism is well known from various acridid species. Faure (1932) described it in *L. pardalina*, Hunter-Jones (1957) in *S. gregaria* and Putnam (1958) in *Melanoplus sanguinipes* (under the name *Melanoplus bilituratus* Walker). It was reported several times from different geographical races of *L. migratoria* (Verdier, 1965; Nolte, 1969; Tanaka, 1993). Recently, another albino strain of *S. gregaria* was found by Rahman *et al.* (2003a), presumably different from the conspecific albino strain described by Hunter-Jones (1957).

Not necessarily the same causative factor is responsible for the albinism in different acridid species, or even in different geographical races of the same species. Nolte (1971) reported that different factors control albinism in two different albino mutants of *L. migratoria*. He crossed an albino mutant originating from a race in Tanzania with Verdier's (1965) albinos from Sardinia. The crossing resulted in normal phenotypes in F_1 and in a ratio of about 9:7, normal phenotypes:albinos, in the F_2 generation, as expected for a dihybrid crossing of two different recessive mutants.

Tanaka's (1993) discovery provided a convenient bioassay for purification and characterization of the dark-colour-inducing neurohormone (DCIN). Tanaka and Pener (1994, 1995), using the Okinawa albino hoppers of *L. migratoria* for bioassays, demonstrated that this neurohormone is a heat-stable neuropeptide, soluble in methanol or in saline, but shows activity only when injected in oil. Tanaka (1996) reported that the CC of the cricket, *Gryllus bimaculatus* De Geer, has a better dark colour–inducing activity in the Okinawa albinos than the CC of normally coloured *L. migratoria*. The factor inducing dark colour was discovered also in *S. gregaria* (Tanaka and Yagi, 1997); implantation of CC or injection in oil of dried methanol extracts of brain and CC from *G. bimaculatus*, normal phenotypes of crowded *L. migratoria* and *S. gregaria* hoppers, all induced darkening in green solitarious hoppers of *S. gregaria*. Darkening of Okinawa albino hoppers of *L. migratoria* was also obtained by injection in oil of dried methanol extracts of brain and CC from normal phenotypes of crowded *S. gregaria*.

Using the Okinawa albino strain of *L. migratoria* for bioassay, Tawfik *et al.* (1999a) finally purified the DCIN from 2950 pairs of CC from crowded adult *S. gregaria* and 4700 pairs of CC from normally coloured crowded adult *L. m. migratorioides*. High performance liquid chromatography (HPLC) purification, amino acid sequence determination and mass spectrometry (MS) led to identification of this neurohormone. In both species, it turned out to be the same, [His^7]-corazonin, a blocked undecapeptide, pGlu-Thr-Phe-Gln-Tyr-Ser-His-Gly-Trp-Thr-Asn-NH_2 (pGlu = pyroglutamate). [His^7]-corazonin was previously isolated and identified by Veenstra (1991) from the CC of *Schistocerca americana* without known function. [Arg^7]-corazonin had been isolated even earlier from the cockroach, *Periplaneta americana*; it accelerated the rate of heart beat in this insect (Veenstra, 1989). However, neither [Arg^7]-corazonin nor [His^7]-corazonin affected heart beat rate in *S. americana* (Veenstra, 1991). Today, several isoforms of corazonin are known from various insects (Predel *et al.*, 2007) with various possible functions (see Section 9.3.4), but insofar as known, only in acridids do they induce darkening. Therefore, [His^7]-corazonin found in *S. americana* (Veenstra, 1991), *S. gregaria* and *L. migratoria* (Tawfik *et al.*, 1999a) was connoted as DCIN by Tawfik *et al.* (1999a).

Tawfik *et al.* (1999a) also examined the dose–response relations of injection in oil of synthetic DCIN into early fourth-instar hoppers of the Okinawa albinos. These authors recorded darkening, by arbitrary colour grades, 3 days after the injection, still in the same stadium. The dose of 1 fmol already induced discernible darkening and with increasing doses darkening became more intense. The doses of 10–100 pmol induced maximum response, completely or almost completely black hoppers. It is noteworthy that the difference between discernible and maximum effects extends 4–5 orders of magnitude, reconfirmed also by Yerushalmi *et al.* (2002). Such a wide dose range is exceptional for a neurohormone. In addition, Tawfik *et al.* (1999a) showed that injection of DCIN dissolved in water, or in 10% aqueous ethanol, is highly ineffective; slight discernible darkening was observed after injection of 100 pmol, and Shalev *et al.* (2003) observed such slight darkening only after injection of 1 nmol of DCIN dissolved in water.

Earlier studies (Tanaka, 1993; Tanaka and Pener, 1994) showed that implantation of CC from normally coloured locusts into Okinawa albino hoppers not only induces darkening (similar to homochromy of normally coloured solitarious hoppers) but also sometimes induces gregarious colouration, dirty orange with black patterns. Tawfik *et al.* (1999a) clarified this point by finding that 0.1 nmol of DCIN injected in oil into third-instar Okinawa albino hoppers, not at the beginning of the stadium, but 3 days after the moult, induced full gregarious colour in the next (fourth) stadium. Moreover, in absence of *S. gregaria* albinos lacking DCIN (see later), Tawfik *et al.* (1999a) injected 1 nmol of DCIN in oil into third-instar normal phenotypes of green solitarious *S. gregaria* hoppers 1 day after the moult and obtained in the fourth stadium gregarious black patterns, even though the injected nymphs were kept continuously under isolation. However, it was stressed later by Tanaka (2001) that, although inducing the gregarious black patterns, DCIN does not induce the bright yellow background body colour (see Section 7.2.3) characteristic to last-instar gregarious nymphs of *S. gregaria*.

Another study by Tanaka (2000b) showed that the colour induced by injection of DCIN into Okinawa albino hoppers of *L. migratoria* depends on the dose and timing of the injection. He injected synthetic DCIN in oil into third-instar Okinawa albino hoppers and distinguished 12 types of colour variations of the response in the fourth stadium. Four of these were quite similar to homochrome colourations, the hoppers showed increasingly darker colours up to complete black. Another four types resembled gregarious colouration with increasing dark patterns. Three exhibited mixtures of homochrome solitarious and gregarious colouration and one showed uniform reddish iron-rust colour (see Fig. 1 in the work by Tanaka, 2000b). Following the moult of these hoppers to the fifth stadium, Tanaka (2000b, Fig. 2) distinguished six types of colouration. However, he noted that in both, the fourth and the fifth stadia, the colouration of these types was not sharply different, but rather exhibited a continuous range. In general terms, the results showed that the albino hoppers injected with a high dose, 100 pmol of DCIN in oil, at the beginning of the third stadium turned completely black after the next moult, whereas those injected in the middle of the third stadium developed black patterns with a dirty orange background body colour, similar to that of fourth-instar gregarious nymphs. Quite surprisingly, most of the hoppers showing gregarious colouration in the fourth stadium became black in the fifth stadium, two moults after the injection.

In the same study, Tanaka (2000b) also tested the effect of injection of DCIN in oil on normally coloured crowded hoppers of *L. migratoria*. Young third-instar hoppers injected with 1 nmol of DCIN became completely black in the fourth, as well as in the fifth stadia. Similar hoppers injected with 100 pmol became mostly black in the fourth stadium and completely black in the fifth stadium. Doses of 1 or 0.1 pmol had no effect on these crowded normal phenotypes.

In another study, Tanaka (2000c) investigated the effect of interaction between DCIN and JH. Third-instar Okinawa albino hoppers injected with 100 μg JH III, 2 days before the moult to the fourth stadium, turned green after the moult. Intensity of the green colour in the fourth stadium was dose-dependent within the range from 0.1 through 1, 10, 100 and up to 300 μg of JH III injected to 3.5-day-old third-instar hoppers. Tanaka (2000c) also reconfirmed earlier findings of Hasegawa and Tanaka (1994), showing that Okinawa albino hoppers that became green after injection of JH, or JHAs, differ from conspecific normally coloured solitarious green hoppers; in the former, the ventral side of the abdomen remains whitish, whereas in the latter, the same area is darker. The interaction between JH and DCIN, both injected into third-instar Okinawa albinos, was found to be complex. The colour observed in the fourth stadium was affected by both the dose and the day of the injection (age) in the third stadium.

In other experiments reported in the same article (Tanaka, 2000c), field-collected, presumably fourth-instar solitarious green hoppers of a normally coloured strain of *L. migratoria* were injected with 1 nmol of DCIN. After the next moult, these hoppers developed black patterns, but the rest of their body remained green. The intensity of the black patterns seemed to depend on the time elapsing between the injection and the moult. Similar tests with brown solitarious hoppers injected with DCIN 2 days before the moult resulted in hoppers with black patterns and orange background body colour, like the colouration of gregarious last-instar nymphs.

In hoppers of both *L. migratoria* and *S. gregaria*, high temperatures reduce the black patterns and low temperatures increase them (refer reviews by Uvarov, 1966; Pener, 1991; see also Section 7.2.1 of the present review). Tanaka (2003) investigated the interaction between rearing temperature and injected DCIN in crowded Okinawa albino hoppers and in normally coloured hoppers of *L. migratoria*. Injection of 1 pmol of DCIN in oil into 0-day-old fourth-instar albino nymphs, maintained at 25 or 30 °C, induced beige-brown colouration. Surprisingly, at 34 °C, slightly darker beige-brown colouration was observed, but at 38 °C, little darkening occurred and at 42 °C darkening was strongly inhibited (see Fig. 3A in the work by Tanaka, 2003). Injection of 10 pmol of DCIN into similar albino nymphs induced darkening even at high temperatures; dark grey colouration was observed at 30 and 34 °C and a somewhat lighter grey at 38 and 42 °C (see Fig. 3B in the work by Tanaka, 2003). Interestingly, nymphs maintained at 25 °C and injected with 10 pmol of DCIN showed reddish-dark brown colour. All treated nymphs in these experiments resembled various homochrome colourations, characteristic to normally coloured solitarious hoppers. Tanaka (2003) offered several explanations of these results, but drew no firm conclusion. In additional experiments (Tanaka, 2003), Okinawa albino nymphs that received 1 pmol of DCIN were transferred from 30 to 42 °C and vice versa. The transfer affected the colour any time after the injection. Nymphs transferred from the lower to the higher temperature became lighter than those maintained at the lower temperature, but not as light as untreated controls. The reverse transfer showed an opposite trend. Normally coloured nymphs kept at the highest temperature, 42 °C, exhibited no or little black patterns, but depending on the timing of the injection of 1 nmol of DCIN, they developed such patterns, or became completely black.

It was already mentioned (see earlier) that different factors can cause albinism in acridids. This fact is illuminated also by the findings of Yerushalmi *et al.* (2000), who implanted CC, or injected, in oil, dried methanol extracts of the CC, obtained from crowded normal phenotypes of *L. m. migratorioides*, or from Hunter-Jones' (1957) crowded albinos of *S. gregaria*, into fourth-instar Okinawa albino hoppers of *L. migratoria*. These treatments induced similar darkening in the Okinawa albinos, regardless of the donors of the CC. It was concluded, therefore, that Hunter-Jones' (1957) *S. gregaria* albinos possess DCIN, nevertheless they maintain albinism. Injection of synthetic DCIN into penultimate-instar albino hoppers of *S. gregaria* did not induce darkening, except injection of 50 nmol, a massive dose, that induced slight black patterns in the last stadium. Assuming that a fourth-instar *S. gregaria* hopper has about 50 μl of haemolymph, 50 nmol of DCIN is equivalent to a solution of 10^{-3} M, clearly not a physiological dose. However, 10 nmol per hopper, equivalent to about 2×10^{-4} M, still a very high non-physiological dose was completely ineffective.

Independently from Yerushalmi *et al.* (2000), Schoofs *et al.* (2000) published a few months later results of somewhat similar experiments, using Hunter-Jones' (1957) *S. gregaria* albinos (Schoofs *et al.* mention that they received the albino strain of *S. gregaria* from Professor S. Andersen of Denmark; however, the senior author of the present review was informed by Professor Andersen that he had received the strain from the same institute in London [Anti-Locust Research Centre] where the strain was originally reported by Hunter-Jones). Schoofs *et al.* (2000) found that dried methanol extracts of CC or brain obtained from Okinawa albinos, or from albinos of *S. gregaria*, or from normally coloured *S. gregaria*, did not induce dark colour in *S. gregaria* albinos. In contrast, brain and CC taken from *S. gregaria* albinos induced darkening in Okinawa albinos of *L. migratoria*. Injection of synthetic DCIN or [Arg^7]-corazonin up to 1 nmol into *S. gregaria* albinos did not induce any darkening. These results were practically similar to those of Yerushalmi *et al.* (2000). Additionally, using immunocytochemistry, Schoofs *et al.* (2000) found, in both normally coloured and albino *S. gregaria*, immunoreactive staining to the DCIN in two pairs of four lateral neurosecretory cells (NSC) in each half of the brain, in the nervi corporis cardiaci (NCC) II and in the CC, as well as in a single cell at the base of each optic lobe. Apparently, they did not test such immunoreactive staining in the NSC of the thoracic ganglia. Tanaka (1993) found that implantation of these ganglia from normally coloured *L. migratoria* into Okinawa albinos induces darkening. The presence of DCIN (also termed [His^7]-corazonin) in both normal and albino *S. gregaria* was shown also by reverse phase high performance liquid chromatography (RP-HPLC) and mass spectrometry (Schoofs *et al.*, 2000). Similarly to the findings of Schoofs *et al.* (2000) with Hunter-Jones' (1957) albino strain of *S. gregaria*, Rahman *et al.* (2003a) found DCIN also in hoppers of a presumably new albino strain of this species.

Baggerman *et al.* (2001) reconfirmed the above findings of Schoofs *et al.* (2000). In addition, Baggerman *et al.* (2001) demonstrated the presence of DCIN in green solitarious hoppers of a normally coloured *S. gregaria*. This finding presumably means that the DCIN is not released into the haemolymph, or it is not active (absence of receptors?) in the solitarious hoppers of this species. Probably, the DCIN in these solitarious hoppers is in 'stand by' to induce gregarious black patterns if crowding occurs. This situation seems to be similar in *L. migratoria*. Implantation of CC from field-collected normally coloured solitarious fifth-instar nymphs into fourth-instar Okinawa albino hoppers induced dark colouration (Tanaka, 1993, Table 1). Also, dried methanol extracts prepared from CC or brain of green or brown fifth-instar isolated hoppers of *L. migratoria* induced dark colouration when injected into Okinawa albino nymphs (see Table 4 of Tanaka and Pener, 1994). The 'stand by' DCIN in the CC of solitarious *L. migratoria* presumably has an additional function; it probably induces homochrome darkening if the substratum becomes dark/black ('fire darkening', see Burtt, 1951 and some earlier references therein; reviews by Uvarov, 1966, pp. 52–56; Rowell, 1971; Fuzeau-Braesch, 1985).

The fact that injection of dried CC or brain extracts or of DCIN in oil is highly effective whereas in aqueous solution it is much less so (Tanaka and Pener, 1994; Tawfik *et al.*, 1999a) needs some comments. The kind of oil (rape seed oil, peanut oil, soybean oil, olive oil, sesame oil) does not change significantly the effect of DCIN (Tanaka, unpublished observations, cited by Tanaka, 2000b). DCIN is not the only neuropeptide that is much more effective when injected in oil than in aqueous solution. Janssen *et al.* (1998) found that the threshold concentration of the trypsin modulating oostatic peptide in the flesh fly, *Neobellieria bullata*, is a thousand times lower when injected in oil than in Ringer's solution. A partial explanation was found for the highly reduced activity of DCIN in aqueous solution. In a nuclear magnetic resonance (NMR) study, Shalev *et al.* (2003) showed that both DCIN and [Arg^7]-corazonin aggregate when dissolved in water. These peptides dissolved in dimethylsulphoxide (DMSO) are dispersed and their injection in that solvent improves their dark colour–inducing effect by 1.5 or 2 orders of magnitude. However, injection of these peptides in DMSO is still less effective by about 3 orders of magnitude than their injection in oil (see Fig. 1 in the work by Shalev *et al.*, 2003). It has been assumed that the release of these peptides from the injected oil droplet is much slower than from the DMSO, thereby increasing the physiological response, if it depends on their prolonged presence.

Recently more DCIN-related neuropeptides were discovered in various, non-acridid, insects. Today, a small family of corazonins is known (Predel *et al.*, 2007; see also Section 9.3.4 of the present review). All of those tested on Okinawa albino nymphs of *L. migratoria* induced darkening.

The dark colour–inducing effect of DCIN is not restricted to locusts, though it seems to be limited to acridoids. Tanaka (2000a) injected 1 nmol of DCIN into nymphs of five different grasshopper species that exhibit green-brown, or green-brown and homochrome, colour polyphenism. All were affected by the DCIN; they became darker or exhibited dark patterns. However, injection of DCIN or [Arg^7]-corazonin into a katydid (Tettigoniidae), *Euconocephalus pallidus* (Redtenbacher) failed to induce darkening.

Yerushalmi and Pener (2001) tested the dose–response relations of injection of DCIN into nymphs of the grasshopper, *Oedipoda miniata* (Pallas), a species that exhibits homochromy, but neither phase nor green-brown colour polyphenism. In a laboratory colony of this grasshopper, the hoppers from the second stadium onwards exhibited light brown colouration with some darker spots. Penultimate-instar nymphs were injected with DCIN and the colour was noted 4 days after the injection, still in the same stadium, and again after their moult to the last nymphal stadium. Slight, but discernible darkening was observed with the dose of 1 pmol. Maximum darkening was seen in the penultimate nymphal stadium after injection of 10 nmol, but 1 nmol was sufficient to induce maximum darkening in the last nymphal stadium. Therefore, susceptibility to DCIN was lower in this grasshopper than in Okinawa albinos of

L. migratoria, and the range from discernible to maximum effect covered only 3–4 orders of magnitude.

Tanaka (2004b) studied the effect of DCIN in nymphs of the American grasshopper, *Schistocerca americana*. It may be recalled that [His^7]-corazonin was first isolated and identified from this species (Veenstra, 1991), though then without known function. Tanaka (2004b) found that injection of DCIN in oil resulted in a dose-dependent induction of black patterns, but did not affect the green-brown polyphenism known from this grasshopper. Recording the dark patterns of the pronotum, discernible darkening was found with the injection of 10 pmol and maximum darkening was seen with the injection of 1 or 10 nmol. This range is only slightly different from that found for nymphs of *O. miniata* by Yerushalmi and Pener (2001).

7.2.3 *The yellow background colour of* Schistocerca gregaria *hoppers and aposematic colouration*

The background body colour of late-instar gregarious nymphs of *S. gregaria* is bright yellow, though in certain parts of the body it may be orange (see later). This background colour is especially vivid in fifth (last)-instar hoppers at high temperatures, but it may be seen also in the fourth stadium. The background body colour of the second- and third-instar hoppers is off-white and slightly pinkish. All these stadia have black patterns on the background colour. In contrast, solitarious nymphs of this species exhibit a more or less uniform colouration, green or beige-brown often with a pinkish flush (see Section 7.2.1). The colouration of gregarious and solitarious (as well as some transient) nymphs of *S. gregaria* is illustrated, for each stadium, by Stower's (1959) colour plates.

The induction of the black patterns in isolated (solitarious) nymphs of *S. gregaria* by sensory stimuli originating from crowded nymphs was reported by Lester *et al.* (2005) and was discussed in Section 7.2.1. It was also discussed (Section 7.2.2) that DCIN, injected into isolated hoppers of *S. gregaria*, induces the gregarious black patterning, but does not induce the bright yellow background colour (Tanaka, 2001). Lester *et al.* (2005) studied the effect of different sensory stimuli on the induction of the bright yellow background colour. Similarly to the study of the black patterning (see Section 7.2.1), isolated *S. gregaria* hoppers from the second or third stadium were exposed to different gregarious stimuli and their background colour was scored, in the last nymphal stadium. The test insects were exposed to conspecific or heterospecific (*L. migratoria*) olfactory (smell) stimulus, or olfactory plus visual (smell and sight) stimuli, or to actual contact with a group that provided combined stimuli of smell, sight and touch. Nymphs maintained in isolation up to the last nymphal stadium served as controls for comparison. Neither heterospecific nor conspecific smell induced the yellow background colour, but combined smell and sight led to moderate yellowing. Physical contact with a conspecific crowded group resulted in maximum yellowing, which was similar to that of nymphs kept under crowding from hatching onwards. The yellowing induced by the physical contact with a conspecific group was significantly higher than that induced by the heterospecific group.

It is known that mechanical touching induces gregarious behaviour (Simpson *et al.*, 2001; for details see Sections 5.3.3 and 14). It seems that the test insects of Lester *et al.* (2005) experienced quite similar mechanical touching in the heterospecific and the conspecific groups, because their behaviour became gregarious, yielding low '*p*(solitary)' values (see Section 11.1 for quantifying behavioural phase state) without significant difference. Therefore, Lester *et al.* (2005) concluded that the results imply the presence of a species-specific contact chemical cue that is needed to induce full intensity of the bright yellow background body colour.

Maeno and Tanaka (2007) found a positive correlation between hatchling colour and the colour of the last-instar hoppers of *S. gregaria*. The findings of this study on black patterns are noted in Section 7.2.1. Concerning the yellow background body colour of crowded hoppers, these authors found that darker hatchlings become orange, whereas lighter hatchlings become yellow in the last nymphal stadium. However, the distinction between yellow and orange was based on the thoracic pleura and the first two pairs of legs; the photographs of the hoppers in their article show no difference in the background colour of the abdomen and the hind legs, which was also bright yellow in the 'orange' hoppers.

Aposematic colouration, also connoted warning colouration, is widespread in the Animal Kingdom and abundant among insects (refer review by Guilford, 1990; for recent concepts and theories, see Endler and Mappes, 2004; Broom *et al.*, 2006; Speed and Ruxton, 2007; and references in these articles). Aposematic species are unpleasant, unprofitable or dangerous to predators and advertise these traits, to visually hunting predators, by characteristic, conspicuous, usually highly contrasting body colour patterns. The aposematic colouration, commonly yellow, red or orange, combined with dark or black patterns or stripes, encourages predators to learn to avoid such colourations. However, there is debate and speculation about the evolution of aposematism. At an early state of evolution in a population, a few initially conspicuous individuals should attract predators more than cryptic (camouflaged) individuals. At the same time, if there are only a few aposematic specimens, predators will be less likely to learn to avoid them through repeated exposure. There are currently several hypotheses concerning the initial origin and spread of aposematism in a population (see reviews earlier), and even a gradual transition from cryptic to aposematic phenotypes over time cannot be ruled out (Lindström *et al.*, 1999). Some hypotheses invoke kin selection among related phenotypically similar individuals, but fast acting defences that allow a potential prey to survive an attack by a predator (non-suicidal stings, urticating hairs, etc.) can enable the evolution of aposematism by individual selection alone.

Sword (1999) found density-dependent colour polyphenism in the grasshopper, *Schistocerca lineata* Scudder (known for a while as *S. emarginata*, but Song, 2004, revised its status and reverted it to *S. lineata*). Typically, the nymphs are green under conditions of low density, but yellow and black under high density. Certain populations of hoppers of this species in Texas prefer feeding on the toxic plant, *Ptelea trifoliata*, despite the presence of other acceptable plants. This preference leads to congregation of the hoppers on *Ptelea*. Laboratory experiments demonstrated that most yellow-black hoppers reared on *Ptelea* were unpalatable to an insectivorous lizard species, but all hoppers reared on an alternative host plant, *Rubus trivialis*, were palatable even with yellow-black colouration. Sword (1999) concluded, therefore, that the yellow-black colouration of *S. lineata* (under the name *S. emarginata*) nymphs at high local densities may function as aposematism when the nymphs have been feeding on plants containing unpalatable chemicals. Sword (1999) demonstrated that unpalatability was mediated by the presence of plant material in the gut, rather than sequestration or *de novo* synthesis of toxic compounds (see also Sword, 2001). Sword (1999) theorized that similar gut-content-mediated aposematism may be extended to the desert locust, *S. gregaria*.

In a subsequent study, Sword (2002) found that density-dependent colour polyphenism of *S. lineata* nymphs is more marked in a *Ptelea*-associated unpalatable population than in a *Rubus*-associated palatable population. The area of the body covered by black patterns was larger in crowded *Ptelea*-associated population than in crowded *Rubus*-associated population. A similar trend was found in isolated nymphs, though with much reduced black patterns. In crowded nymphs, the background body colour was more uniformly yellow in the *Ptelea*-associated than in the *Rubus*-associated population. In the *Ptelea*-associated population, phenotypic variation was reduced in crowded nymphs as compared to that in isolated nymphs. In contrast, the *Rubus*-associated population exhibited much lower variation under isolation than under crowding. Sword (2002) concluded that the differential expression of the density-dependent colour polyphenism between palatable (*Rubus*-associated) and unpalatable (*Ptelea*-associated) *S. lineata* is in accordance with the expected costs and benefits of conspicuousness. Additionally, he concluded that density-dependent plasticity in colour polyphenism may provide an adaptive intermediate stage to the evolution of aposematism. Sword (2002) also mentioned that population-level variation in the phenotypic plasticity found in *S. lineata* may reflect an evolutionary trend towards a partial loss of density-dependent colour polyphenism in grasshoppers from palatable populations. Sword and Dopman (1999, p. 443) suggested that ''genetic differentiation between populations, utilizing *Rubus* and *Ptelea*'' may exist; they also stated that ''Depending on the degree of genetic isolation, these host-associated populations may represent host races … or distinct sibling species''. A subsequent phylogeographic analysis of *S. lineata* based on mitochondrial DNA evidence supported this scenario of genetic differentiation. *S. lineata* from the *Ptelea*- and *Rubus*-feeding populations comprise two different genetically distinct lineages associated with each host plant (Dopman *et al.*, 2002).

Sword's (1999) findings in *S. lineata* were tested experimentally in *S. gregaria* (Sword *et al.*, 2000). As already noted, hoppers of the desert locust exhibit phase-dependent colour polyphenism. Isolated (solitarious) hoppers show cryptic colouration and gregarious late-instar nymphs exhibit black patterns on a bright yellow background body colour (see Fig. 2), typical of aposematic colouration. Even earlier gregarious nymphal stadia have contrasting black and white, or black and pinkish colouration.

Sword *et al.* (2000) reported that an insectivorous lizard species showed a high percentage (67%) of rejection of gregariously coloured individuals after eating a single gregarious nymph of *S. gregaria* fed on the toxic plant *Hyoscyamus muticus* (Egyptian henbane). In contrast, after eating a single cryptic solitarious nymph fed on the same plant, rejection rate to a second solitarious nymph was significantly lower (14%), indicating that the lizards did not associate the cryptic solitarious colouration with locust toxicity. Rejection rate of a second gregarious nymph, after eating a single gregarious nymph fed on three other species of plants, was also low (6%, 13% and 27%).

More recently, Despland and Simpson (2005a) tested in *S. gregaria* the response to hyoscyamine of crowd-reared (gregarious) nymphs, isolated-reared (solitarious) nymphs and of newly crowded nymphs in transition from the solitarious to the gregarious phase state. The authors used an artificial diet, composed for *L. migratoria* (Simpson and Abisgold, 1985), adjusted to *S. gregaria* (Simpson *et al.*, 2002), with or without L-hyoscyamine. This alkaloid is the main defensive compound found in *H. muticus* and is a powerful neurotoxin to vertebrates – less so to insect herbivores. Isolated insects discriminated against consuming the diet with 2% dry weight of hyoscyamine (a concentration similar to that at which it occurs in Egyptian henbane), whereas crowded insects fed equally from a choice of food that did or did not contain hyoscyamine. Newly crowded nymphs accepted and even seemed to prefer the diet with hyoscyamine. Despland and Simpson (2005a) concluded that feeding behaviour is one of the traits that contribute to density-dependent cryptic and aposematic strategies in the desert locust. Solitarious nymphs avoid predator detection by crypsis and would not benefit greatly from consuming toxic plants. Gregarious nymphs migrate by marching and they switch frequently between the temporarily available food plants and could acquire anti-predator defence by aposematism and feeding occasionally on a variety of toxic plants. Newly crowded solitarious nymphs may benefit more from chemical defence than isolated nymphs because predator avoidance learning is enhanced when distasteful preys are aggregated (see references in the work by Despland and Simpson, 2005a). Although some predators do not seem to associate the cryptic colouration with toxicity, at least after a single experience (see earlier), they may do so after repeated experience. Furthermore, the ability to learn to avoid cryptic but toxic prey may differ in different predators and with different defensive plant compounds (Sword *et al.*, 2000).

In another study, Despland and Simpson (2005b) again tested the prediction that in *S. gregaria*, gregarious and newly crowded nymphs (the latter are behaviourally already gregarious, but still have cryptic solitarious colouration) feed more readily on a toxic compound than do solitarious nymphs. They again used an artificial diet containing 2% of L-hyoscyamine against the same diet without hyoscyamine as a control and measured the following variables: (1) frequency of meal initiation at first contact with the food; (2) mean duration of the first feeding bout; (3) total amount of food eaten during the trial and (4) mean proportion of the assay time spent feeding. The results showed that phenotype did not affect behaviour of the nymphs on the control food. However, solitarious nymphs tested alone showed shorter first feeding bout, ate smaller amounts and spent less of the total assay time feeding on hyoscyamine-containing food than on control food. The results of these experiments (Despland and Simpson, 2005b) and those of the former tests (Despland and Simpson, 2005a) provided empirical evidence to support their prediction. Additionally, Despland and Simpson (2005b) constructed a computer simulation showing how crypsis ceases to be effective as an anti-predator strategy when solitarious locusts become crowded, how chemical defence becomes favourable as conspicuousness increases with local density and how aposematic colouration becomes advantageous under these conditions.

The gregarious colouration of *S. gregaria* hoppers was claimed to mediate nymphal aggregation (Ellis and Pearce, 1962). In contrast, Gillett (1973), comparing albino and normally coloured *S. gregaria*, found that albinos grouped less among themselves than normally coloured locust among themselves. However, in mixed tests, albinos exposed to normally coloured locusts did not group more than albinos among themselves, and normally coloured locusts tested with albinos did not group less than normally coloured locust among themselves. Therefore, no effect of the gregarious colouration on grouping was found. Gillett (1973) speculated that the gregarious colouration may assist in maintenance of cohesion in marching bands by means of visual compensation or optomotor response. In an earlier study, Ellis (1953b) concluded that gregarious colouration and physical contact increase the intensity of marching in *L. m. migratorioides* hoppers. The design of these experiments did not exclude olfactory stimulus because they were made in an era when no or little attention was paid to olfactory (pheromonal) cues. Interestingly, even earlier, Chauvin (1941) claimed that locust gregariousness is induced by visual and tactile stimuli.

The claim that gregarious colouration mediates nymphal aggregation was refuted by Sword and Simpson (2000) studying the possible intraspecific role of phase colour polyphenism in last-instar *S. gregaria* nymphs. Specifically, they investigated the effect of the gregarious yellow-black colouration in comparison with the effect of solitarious green colouration on (1) direct induction of the initial process of behavioural phase change and (2) subsequent group formation among locusts through the immediate effect on behaviour. Sword and Simpson (2000) found no effect of the gregarious colouration as either a gregarizing stimulus to behaviourally solitarious nymphs or as a visual aggregating stimulus to gregarious nymphs. However, they strictly prevented olfactory or tactile contact between the stimulus (gregarious or solitarious colouration) and the test insect. This procedure has an advantage and a disadvantage. On one hand, it limits the effect to a purely visual stimulus. On the other hand, however, it disregards the possibility that, even if visual stimulus alone is ineffective, visual stimulus combined with other kind of stimuli may act in synergy.

Previous results indicated that visual stimulus alone, that is, the sight of crowded nymphs, has a gregarizing effect in some solitarious nymphs of *S. gregaria* (Hägele and Simpson, 2000, Fig. 1, experiment 2), more so when such visual stimulus is presented as long as 24 h (Roessingh *et al.*, 1998, Fig. 4). Olfactory stimulus, that is, smell of crowded nymphs, alone had no effect (Roessingh *et al.*, 1998; Simpson *et al.*, 1999; Lester *et al.*, 2005). However, sight and smell resulted in synergy; solitarious nymphs exposed for 4 h to the combined stimuli, showed gregarious behaviour in most of the test insects (Roessingh *et al.*, 1998, Fig 1; Simpson *et al.*, 1999, Fig. 7; Lester *et al.*, 2005, Fig. 2; see Section 14).

It seems, therefore, that a possible effect of visual stimulus in combination with other stimuli on induction of gregarious behaviour cannot be ruled out, and it is possible that the gregarious colouration plays some intraspecific role. This possibility does not weaken its interspecific role as concluded by Sword and Simpson (2000). Additional experimental work is needed to clarify this point and to determine whether the gregarizing influence of visual stimuli (with or without odour stimuli) comes from the colour pattern or movement of other locusts, or both.

Interestingly, the fact that desert locusts may be toxic is illuminated also by Jewish tradition. The rules of the Jewish religion allow eating the desert locust; it is the only insect that is 'kosher'. Yemenite and Algerian Jews indeed ate desert locusts but removed the head that was believed to be toxic. Obviously, removal of the head opens the foregut (stomodaeum) and often pulls it out, leading to removal of a possible toxic content. Removal of the head, or even the head with the digestive tract, of the locusts was practiced also by Arab populations in the Middle East (Amar, 2004, pp. 51–53). It seems that not only humans exercise this precaution. Observing the effect of natural enemies on a population of *S. gregaria*, Stower and Greathead (1969, p. 212) reported that a bird, the isabelline wheatear, *Oenanthe isabellina isabellina*, "was observed taking locusts and perching on the corner posts ... to eat them. Below the posts discarded heads and legs of locusts were found". Stower and Greathead (1969, p. 206) noted the history of the locust population, which was complex with overlapping generations. It started with the '*gregaria*' phase, followed by dissociation, then congregation, again dissociation, then slight congregation and eventually change towards the '*solitaria*' phase (the terms gregaria and solitaria in italics are used here as in the original paper). From comparison of the dates of observations and the timing of the preceding phase-related changes of the population, it seems that the birds were eating locusts from some transient population.

Experimental work is needed for clarifying intraspecific or interspecific or both roles of density-dependent colour polyphenism in other locust species. *L. migratoria*, *Locustana pardalina* (the brown locust) and *Nomadacris septemfasciata* (the red locust) exhibit phase-dependent colour polyphenism. The gregarious nymphs of these species have conspicuous colouration, which may be aposematic (Faure, 1932; Popov, 1989), whereas the solitarious nymphs are cryptic. Simpson and Sword (2009) provides some possible explanations concerning the aposematic colouration of gregarious *L. migratoria* nymphs. The first of these explanations is the simple fact that experiments investigating aposematism in this species have not yet been carried out. This explanation is valid also for *L. pardalina* and *N. septemfasciata*.

7.3 ADULTS

In *L. migratoria* and *Locustana pardalina*, green solitarious hoppers experiencing high humidity (see Section 7.2.1) may moult to green adults (Faure, 1932). A few days after fledging, solitarious adults of *L. migratoria* lose their ability to become green, and if green adults are transferred to dry condition, the green colour slowly, though incompletely, fades away (Albrecht, 1965). The green colour also fades slowly if green isolated adults are transferred to a crowd (Pener, 1976b). In *S. gregaria*, green or greenish adults may be obtained by injection of a massive dose of pyriproxyfen (a JHA) into last-instar crowd-reared or isolated-reared nymphs (Pener, unpublished observations), but under normal conditions, newly fledged adults of this species are not green.

Many locust species show colour changes during the adult's life (refer the review by Uvarov, 1966, pp. 297–299). In crowded *S. gregaria*, newly fledged adults are pinkish, or even dark pink at low temperatures. The pink colouration fades, slowly at low temperatures, and the adults become pinkish-beige, then beige-brown. Thereafter, adult colour changes are sex-dependent. With sexual maturation of crowded males, the colour changes to yellow, then to bright yellow. In crowded females, the colour changes from beige-brown to yellowish, but never becomes bright yellow (for a detailed description of these colour changes, see Chauvin, 1941; Norris, 1954, Pener, 1967). Isolated adults of *S. gregaria* do not show appreciable colour changes; both sexes are usually beige-brown, perhaps a little lighter than conspecific crowded adults in the beige brown stage.

Crowded adults of *L. migratoria* are greyish after fledging, the colour changes to greyish-brown and eventually crowded males of this species become yellow, then bright yellow after sexual maturation, similarly to crowded males of *S. gregaria*. In contrast to *S. gregaria* females, crowded *L. migratoria* females do not become yellowish, except for a very slight yellowish tint on the bases of the hind wings (for a detailed description of the colour changes of crowded *L. migratoria*, see Pener *et al.*, 1972). Different strains or laboratory stocks of *L. migratoria* may show differences in the rate and intensity of yellowing of crowded males (Pener, 1976b). Crowded adult males of the Okinawa albino strain of *L. migratoria* also show yellowing, albeit it seems to be less intense than that of normally coloured conspecific crowded adult males (Hasegawa and Tanaka, 1994).

Isolated adults of *L. migratoria* may remain green (see earlier), or beige-brown, but their ability to exhibit homochromy is more limited than that of isolated hoppers; nevertheless, they show the dark/black homochrome response (see later and colour plates in the work by Fuzeau-Braesch, 1985). Isolated adults of *L. migratoria* do not become yellow, except on the hind wings of both sexes. This yellow colour on the hind wings is much brighter than the slight yellowish colour on the hind wing bases of conspecific crowded adults (Pener, 1976b).

Yellowing with sexual maturation also takes place in adults of *L. pardalina*, and *N. septemfasciata*, as well as in some other species of locusts (refer review by Uvarov, 1966). Like in *S. gregaria*, the yellowing in the crowded males of *N. septemfasciata* is more complete than in crowded females (Michelmore and Allan, 1934). Solitarious adults of *N. septemfasciata* show no homochromy, but a limited homochrome response exists in isolated *L. pardalina* adults.

Sexual maturation is not necessarily associated with yellow colouration. For example, in the less typical South American locust, *Rhammatocerus schistocercoides* (see Section 3.1), the adults become partially or fully green with sexual maturation (Lecoq and Pierozzi, 1996b).

Yellowing of gregarious adults of *S. gregaria*, *N. septemfasciata* and *L. migratoria* depends entirely on the CA and their secretion, JH. Without any exception, allatectomy of last-instar hoppers or of young adults of these species prevents yellowing and reimplantation of active CA, or administration of JH, reinduces the yellow colour (for detailed references, see Pener, 1991, p. 19). Allatectomy of sexually mature yellow males of *S. gregaria* results in fading of the yellow colour (Loher, 1961; Pener, 1965). The yellowing that is restricted to the hind wings in both sexes of solitarious *L. migratoria* adults is also related to JH; allatectomy prevents or subdues its occurrence (Pener, 1976b). Although JH is absolutely necessary for yellowing of gregarious males of these species, it is not sufficient. Implantation of extra CA into isolated adult males of *L. migratoria* does not induce yellowing, but a simple transfer of isolated males into a newly formed crowd does, even without implantation of extra CA (Pener, 1976b).

Recent studies revealed an additional factor that is responsible for yellowing of gregarious adult males of *S. gregaria*. Wybrandt and Andersen (2001) purified a yellow protein from the abdominal integument of crowded sexually mature males of this species. The molecular mass of this protein was found to be

25 686 and its amino acid sequence was determined. The yellow colour is produced by carotene bound to the protein. The visible absorption spectrum of the protein shows a cluster of three maxima, at 425, 450 and 480 nm, as well as an ultraviolet (UV) maximum is at 275 nm. Wybrandt and Andersen (2001) found considerable amounts of the protein in the integument and minor amounts in the haemolymph and fat body of gregarious mature males, but did not find it in females or immature males. The authors did not investigate solitarious adults. They concluded (Wybrandt and Andersen, 2001, p. 1188) that ''it appears more likely that the protein is synthesized in the epidermal cells and that it receives its carotene by transfer from the haemolymphal lipophorins ...''.

Using sexually mature gregarious males of *S. gregaria*, Sas *et al.* (2007) cloned a part of the cDNA sequence of the yellow protein gene, comprising 266 bp. Its deduced amino acid sequence perfectly matched the amino acid sequence 98–186 of the yellow protein of Wybrandt and Andersen (2001). Sas *et al.* (2007) found abundant yellow protein transcripts only in the integument; all other tissues, namely, accessory gland complexes, testes, flight muscle, foregut, midgut, hindgut, brain and fat body, showed no positive signal. These authors also investigated age-dependent pattern of the yellow protein transcript. They found that in gregarious males, transcript level started to increase by day 10, showed a large increase by day 12, then remained quite constant at a high level. In gregarious females, solitarious males and females and in crowd-reared males that were isolated at the moult to the adult stadium, the yellow protein transcript remained low.

Hasegawa and Tanaka (1994) reported that in addition to the role of the CA/JH, a factor from the CC plays some role in the yellowing of gregarious males of *L. migratoria*. This finding led Sas *et al.* (2007) to test the effect of injection of crude methanol extracts of brain-CC complexes on yellowing. Fifty brain-CC complexes dissected from crowded males and separately, 50 such complexes dissected from crowded females of *S. gregaria*, were used to prepare the extracts. Crowd-reared males that have been isolated at the moult to adulthood and showed a low level of yellow protein transcript served as recipients. Two brain-CC equivalents were injected daily, between day 8 and 12, into each recipient. Injection of either male or female brain-CC extracts increased the transcript level of the yellow protein. This finding strongly suggests hormonal control of the synthesis of the yellow protein. The authors did not doubt the well-proven effect of JH on yellowing of adult gregarious males and forwarded the hypothesis that JH synergizes the activity of another hormonal inducer from the brain-CC.

Rahman *et al.* (2002a) found a 6.08-kD peptide in *S. gregaria* and revealed its amino acid sequence. This peptide is present in a much higher concentration in the haemolymph of crowded than in that of isolated locusts (Rahman *et al.*, 2002a, 2003b). It is found in the haemolymph of crowded fourth- and fifth-instar hoppers, as well as in male and female adults (Rahman *et al.*, 2008b). Rahman *et al.* (2008a) investigated, by immunocytochemical method, the location of this peptide in different tissues of crowded adults of *S. gregaria*; they found the strongest immunostaining in the follicle cells of the female's ovary and in the seminal vesicle of the male's accessory glands. Rahman *et al.* (2008b) also reported that injection of this phase-related peptide (they abbreviated it as PRP) induced production of mRNA of the yellow protein (described by Wybrandt and Andersen, 2001, see earlier) in pieces of abdominal cuticle at the age of 15 days. However, this phase-related peptide did not induce any actual yellowing. Comparison of the features of this phase-related peptide (see earlier) with those of the putative brain-CC peptidergic hormonal factor responsible for yellow protein mRNA transcription (Sas *et al.*, 2007) indicates that these two peptides are different. The article by Rahman *et al.* (2008b) does not mention or cite the article by Sas *et al.* (2007), despite that both articles are coming from the same laboratory and authored partially by the same persons. Further discussion in relation to the possible role of 6.08-kD peptide can be found in Sections 10.2.1 and 16.3

Sas *et al.* (2007, p. 43) claimed that their findings ''favor the hypothesis that *S. gregaria* has a male-specific hormonal control mechanism that operates in adult life''. This claim does not seem to take into consideration that, according to their own finding, brain-CC complexes of females also induced yellow protein transcript in the recipients (Sas *et al.*, 2007, Fig. 5). Therefore, females also seem to have the putative hormone, but it is not expressed by intense yellowing. Taking into consideration that the yellow protein is ''a cuticular protein that is synthesized in the epidermis'' (Sas *et al.*, 2007, p. 42), it seems that not the hormone but the response of the integument is sex-dependent. Perhaps, the hormone is not released or the transcript does not lead to translation in the females.

An old, almost forgotten, paper of Fogal (1968) strongly supports the differential response of the integument of gregarious males and females of *S. gregaria*. Fogal (1968) transplanted abdominal sternites from males into females and vice versa. When transplantation was made in the fourth or fifth (last) nymphal stadium, or in 1- to 2-day-old adults, the male patches in female recipients all showed yellow colour, approximately to the same extent as the male donors. Implanted female patches in male recipients showed no yellowing. Transplantation of patches from 5- to 9-day-old or 12- to 16-day-old males into females also induced yellowing of the implants, though the yellow colour was somewhat less intense than in the male donors.

It seems, therefore, that a complex chain of causative factors is responsible for the intense yellowing of gregarious males of *S. gregaria*; this chain includes JH, a putative hormone from the brain-CC and some factor(s) in the epidermis.

Presence of sexually mature males of crowded *S. gregaria* accelerates the maturation of conspecific immature adults of both sexes (Norris, 1954; refer recent reviews by Ferenz and Seidelmann, 2003; Hassanali *et al.*, 2005a). This effect has been repeatedly confirmed by all relevant studies. A 'maturation-accelerating pheromone', produced by sexually mature males, is responsible for the effect, and the production of the pheromone depends on the CA/JH (Loher, 1961; see also Section 8.1). Mahamat *et al.* (1993) investigated in *S. gregaria* the effect of the pheromone on maturation of immature males, using extracts obtained by washing yellow mature males with hexane, or by trapping volatiles from such males. These authors noted the time (age in days) elapsing between the moult to the adult stadium and appearance of the yellow colour in crowded males exposed to the extracts in comparison with crowded males not so exposed. Both hexane extracts and trapped mature male volatiles considerably accelerated the maturation as assessed by induction of the yellow colour. Mahamat *et al.* (1993) followed the yellowing of the young males up to Norris' (1954) stage III on her maturation scale of crowded males. She defined this stage as ''Marked fading of the brown colour from the tip of the abdomen. Traces of yellow usually present on the posterior abdominal tergites. Well marked yellow flush usually present on the hind wings. Mature'' (Norris, 1954, p. 14).

In a subsequent paper, Mahamat *et al.* (2000) reported similar experiments with *S. gregaria*, but instead of testing extracts, they prepared a synthetic blend of five major components of the mature males' emissions (see Mahamat *et al.*, 1993 for these components and see Sections 8.1 and 12.1.3), then tested the effect of the synthetic blend and blends in which one of the five compounds was missing. They recorded the days elapsing up to copulation and yellowing, but the latter again was followed only up to Norris' stage III. Therefore, in both studies (Mahamat *et al.*, 1993, 2000), it remained questionable whether the extracts, or the synthetic blends, would have been able to accelerate the induction of the fully bright yellow colour (stage V of Norris, 1954).

Mahamat *et al.* (2000) reported that maximum delay of yellowing was obtained with the blend in which phenylacetonitrile (PAN) was missing, indicating that this compound is the most important component of the maturation-accelerating pheromone. However, Schmidt and Albütz (2002) reported that in *S. gregaria*, neither PAN (under the name benzyl cyanide) nor volatiles of sexually mature males accelerated the appearance of the yellow colour in immature males. Pentane extracts, obtained by washing the cuticle of sexually mature yellow males, accelerated yellowing of immature males, comparably to the effect induced by presence of crowded mature yellow males.

In a recent review, Hassanali *et al.* (2005a) attempted to reconcile the different findings of Mahamat *et al.* (2000) and Schmidt and Albütz (2002) by stating ''that the volatile adult pheromone functions in the presence of other chemotactile signals available in groups of adults, mature or immature'' (Hassanali *et al.*, 2005a, p. 235). A synergistic effect of volatile and contact pheromones is certainly theoretically possible. However, it cannot reconcile the claim of Mahamat *et al.* (2000) that PAN is the most important compound that induces yellowing in contrast to the claim of Schmidt and Albütz (2002) that PAN (under the name benzyl cyanide) has no yellow colour inducing effect, though it ''serves as a maturation accelerating and synchronizing factor'' (Schmidt and Albütz, 2002, p. 139). Similarly, the claim by Mahamat *et al.*

(1993) that trapped volatiles accelerate yellowing and the claim by Schmidt and Albütz (2002) that volatiles do not accelerate yellowing cannot easily be reconciled. Certainly more research is needed to clarify the nature of the 'maturation-accelerating pheromone' and its mode of effect on yellowing of crowded adult males of *S. gregaria.*

Many acridids, including some locusts, such as solitarious *L. migratoria* and *Locustana pardalina* exhibit homochromy, which is often more limited in adults than in nymphs. In certain species, the adults are capable of exhibiting 'fire darkening', that is, they turn blackish or black on black or burnt ground (Burtt, 1951; Ergene, 1954; Levita, 1970; refer the review by Rowell, 1971, pp. 157–167, for adults, see p. 159).

Studying *L. migratoria*, Fuzeau-Braesch (1966, 1985, p. 573, Plate 1, photos 7–9) reported black homochromy in solitarious adults. In contrast, no homochrome darkening has been reported for conspecific gregarious adults. Fuzeau-Braesch (1985, p. 574) even stated that "gregarious adults are incapable of darkening if transferred onto a black background, as opposed to the solitarious kind that easily become homochromous: grouping inhibits homochromy". Despite denying dark homochromous response in gregarious *L. migratoria* adults, Fuzeau-Braesch (1985, p. 574) made a concession, stating that "Subjection to pure CO_2 for 3 min a day, which momentarily anaesthetizes the insects, removes the inhibition and allows gregarious locusts to turn homochromous black ...".

The dark colour–inducing effect of the DCIN (also termed [His^7]-corazonin) in locust and grasshopper nymphs has already been discussed in detail (Section 7.2.2). Its effect on adults was demonstrated by Grach *et al.* (2004), comparing on one hand the response to DCIN of isolated versus crowded *L. migratoria* and comparing on the other hand the response of different strains of this species. Locusts crowd-reared for many continuous generations and locusts isolation-reared for two or three generations were tested by injection of graded doses of DCIN in olive oil into adults. Darkening response was recorded by five arbitrary colour grades: 0 = off white, the colour of the Okinawa albino locusts (see Section 7.2.2); 1–3 = increasing darkening and 4 = blackish or black. Half grades were also used. As already described in relation to morphometric studies (Section 5.2), the albinism was 'transferred', by repeated crossings, from the Okinawa albino strain to a West African strain of *L. migratoria*. This was possible because albinism in the Okinawa strain is controlled by a single recessive Mendelian unit (Hasegawa and Tanaka, 1994). Eventually four lines, connoted four strains, were obtained: (1) albinos with 100% Okinawa genome; (2) albinos with 99.6% West African genome; (3), normally coloured locusts, originating from West Africa (normal phenotypes), with 100% West African genome and (4) normally coloured locusts with 99.6% West African genome. Isolated and, separately, crowded adults of the four strains were injected with DCIN 24–48 h after the moult to the adult stadium and their colour grade was recorded 10 days after the injection.

In the albinos having 100% Okinawa genome, discernible darkening was seen with 0.05 pmol of DCIN, and maximum effect (mean colour grade 3.9) was obtained with the dose of 500 pmol. Like in the case of conspecific hoppers (Section 7.2.2), the difference between discernible and maximum effect extended across 4–5 orders of magnitude. Importantly, no significant difference was found in the response of the crowd-reared and isolated-reared adults.

Albinos with 99.6% West African genome were less sensitive to DCIN than albinos with 100% Okinawa genome. Discernible darkening was seen with 5 pmol of DCIN, and maximum effect was obtained with 5000–50 000 pmol. Therefore, in spite of being albinos, these adults were about 100 times less sensitive to DCIN than the Okinawa albino adults. Moreover, maximum mean colour grades reached merely about 2.5–3.0, much less than the maximum effect in the Okinawa albinos (mean colour grade 3.9, see earlier). Again, no significant difference was found in the dose–response data between crowded and isolated adults.

The normal phenotypes with 100% and 99.6% West African genome were obviously moderately dark at the beginning of the tests. Their dark colour usually corresponded to colour grades 1.5–2.0 for crowded adults and 1.0–1.5 for isolated ones. Maximum responses obtained were mean colour grades of approximately 2.5–3.5, rather similar to that observed in albinos with 99.6% West African genome. As in the other strains, no significant difference was found between crowded and isolated adults in each strain.

The results of Grach *et al.* (2004) clearly reveal two facts. Firstly, the effect of DCIN in crowded (gregarious) adults is similar to that found in isolated (solitarious) adults; in other words, the response is phase-independent. This finding is surprising in light of Fuzeau-Braesch (1985) conclusion (see earlier) that gregarious adults of *L. migratoria* are unable to exhibit dark homochromy (except in the instance of daily repeated exposure to CO_2). It is known that DCIN is present in gregarious adults; in fact, it was purified from the CC of such adults (Tawfik *et al.*, 1999a). The findings of Grach *et al.* (2004) indicate that the target, probably the integument, of gregarious and solitarious adults responds similarly to the exogenous hormone, implying the presence of the relevant receptors in both phases. If Fuzeau-Braesch is correct and gregarious adults are indeed unable to show dark-black homochromy, it may be assumed that the DCIN is not released from the CC of gregarious adults in amounts sufficient to induce such homochromy. It is possible that the daily treatment with CO_2 (see earlier) induces release, or increases the rate of release, of the DCIN from the CC, leading to the capability of exhibiting dark homochromy by gregarious adults of *L. migratoria*.

The second major point demonstrated by Grach *et al.* (2004) is that, in *L. migratoria*, DCIN-induced darkening is strain-dependent. The large differences in the response between albinos with 100% Okinawa genome and albinos with 99.6% West African genome, both in relation to dose–response and maximum grade of darkening, clearly show that not the albinism but the origin of the strain (Okinawa versus West Africa) is the cause of these differences. It may be recalled that a similar conclusion has been drawn in relation to morphometry (see Section 5.2).

Grach *et al.* (2004) performed two additional sets of experiments. One was devoted to changes in darkening in relation to the time between the injection and recording of body colour. The other set explored the dependence of darkening on the age of the adults at the time of injection. In both sets of experiments, only crowded adults of Okinawa albinos were tested.

In the first set, graded doses of DCIN were injected into 24- to 48-h-old adults, and the colour was recorded 5, 10, 20 and 30 days after injection. Five days after injection, darkening was considerable, but not yet fully developed. Maximum darkening was observed 10 days after injection. On days 20 and 30, the dark colouration was fading, considerably with the lower doses and rather slightly with the higher doses.

In the other set of experiments, locusts were injected with graded doses of DCIN on day 1, or day 14, or day 28 after the moult to the adult stadium, and the colour grades were noted 10 days after the injection. Adults injected on day 1 were significantly more sensitive to DCIN than those injected on day 14 or day 28. With the highest doses (500 and 5000 pmol), mean maximum response of the day 1 injected adults almost reached colour grade 4. In day 14 and day 28 injected adults, the same doses resulted in mean maximum response of colour grades about 2–2.5. There was no significant difference in the response of the adults injected on day 14 and those injected on day 28.

Yerushalmi and Pener (2002) tested the effect of DCIN on adults of the grasshopper, *O. miniata*. This grasshopper exhibits homochromy, but neither phase nor green-brown colour polyphenism. Two sets of experiments were carried out. In the first, last-instar nymphs collected in the field and moulted to adults in the laboratory were used. In the second set, field-collected adults from the same locality, just before the end of their reproductive aestivation (cf. Broza and Pener,1969), were tested. In the young adults, 10 pmol of DCIN induced slight darkening that appeared between 7and 14 days after the injection. A higher dose of DCIN (100 pmol) induced substantial darkening by the same time. Young adults injected with 1 or 10 nmol of DCIN showed a similar darkening response to that obtained with 100 pmol, except that in those receiving 10 nmol some darkening appeared earlier, already apparent on day 7 after the injection. Old adults were 3.5 months older than the young ones and their response to the DCIN was quite limited; slight darkening was induced with 100 pmol and the response did not increase markedly even with injection of 10 nmol. Therefore, this non-locust grasshopper showed a similar trend to that of gregarious Okinawa albinos of *L. migratoria* (see earlier), namely a decreasing response to DCIN with ageing of the adults.

7.4 PIGMENTS

Relatively few recent studies have been devoted to locust pigments. Former reviews include those of Goodwin (1952), Nolte (1965), Uvarov (1966),

Rowell (1971) and Pener (1991). Fuzeau-Braesch (1985) presented a general review on insect colouration, with much data on locusts and their pigments.

Carotenoids, bile pigments, melanin and ommochromes are responsible for locust colouration. Pteridines and flavonoids may also contribute. As already outlined (Section 7.1), the relations between the actual, so-called visible colour and pigments are often blurred. Different pigments may produce the same visible colour; for example, both carotenoids and ommochromes may be responsible for similar reddish-pink visible colour. Blue plus yellow pigments result in green visible colour. The same kind of pigment may yield different visible colours, depending on oxidation–reduction, number and site of double bonds in the pigment molecule. Pigments may be free, but are often bound to a protein and the protein may affect the shade of the colour. The concentration and dispersal of the same pigment in the integument may yield different colours; melanin can produce orange, brown or black colour. Ommochromes and pteridines may be yellow or red. Biliverdins at different oxidation levels can produce different colours. The distribution of the pigments in the integument, epidermis and haemolymph is also important; pigments in the deeper layers are often masked by those in an upper layer. Pigments may appear in solution or as isolated granules and the latter may induce physical colours (see, e.g., the case of *Kosciuscola tristis*; Section 7.1). For the structure and properties of insect pigments, Kayser's (1985) chapter may be consulted. Although differences in the visible colouration between gregarious and solitarious locusts are among the best recognized phase characteristics, data on phase-dependent differences in pigments are rare, scattered and based mostly on old studies, often with controversial conclusions.

According to the old literature (Goodwin and Srisukh, 1951a; Goodwin, 1952), the green colour of the haemolymph and the integument of solitarious *L. migratoria* and *S. gregaria* (related by later authors to high humidity; see Section 7.2.1) are obtained from a mixture of two kinds of pigments, a blue bile pigment and yellow carotenoids. In contrast, Passama-Vuillaume (1965) claimed the presence of a green bile pigment alone in solitarious hoppers of *L. migratoria*. This controversy is still not yet solved. Rowell (1971) mentioned that the bile pigment may be sufficiently labile to be subject to colour change between green and blue. It may be added that Dadd (1961) obtained blue or greenish-blue locusts on a carotenoid-deficient diet. The bile pigment had been assumed to be mesobiliverdin, but later, it was claimed to be biliverdin IXα bound to protein (cf. Passama-Vuillaume and Barbier, 1966).

Recent studies show that a blue biliverdin and a yellow carotenoid are present in the haemolymph of hoppers and adults of both phases of *S. gregaria*. Mahamat *et al.* (1997a) reported that in the absorption spectrum of the haemolymph of gregarious (crowd-reared) locusts, there is a broad peak at 460 nm, corresponding to the absorption spectrum of yellow carotenoids. The absorption spectrum of solitarious (isolated-reared) locusts yielded additional peaks at 375 and 680 nm, characteristic of biliverdin. These authors found that the ratio 460/680 (reflecting the ratio, carotenoid/biliverdin) is phase-dependent. The difference between the two phases was highly significant; moreover, no overlap occurred. The mean ratios calculated for both sexes, nymphs, immature and mature adults, two different geographical strains and three kinds of diet, ranged from 3.69 to 4.78 and from 0.64 to 1.67 for the gregarious and solitarious phases, respectively. It is stressed that these ratios express the state of pigments in the haemolymph and do not necessarily express the pigments in the integument, or the colour of the integument. For example, in spite of showing relatively low ratios of 460/680 (meaning less carotenoid/more bile pigment in the haemolymph), solitarious adults of *S. gregaria* are usually not green.

Deng (2002) also assessed the absorption ratio, 460/680, in the haemolymph of gregarious (crowd-reared) and solitarious (isolated-reared) hoppers and adults of both sexes of *S. gregaria*. This author reported values of mean ratios for different stages within the range of 1.5–2.24 and 1.09–1.86 for gregarious and solitarious locusts, respectively, but despite the overlap, the difference was significant. These ratios for the gregarious locusts were much lower than those obtained by Mahamat *et al.* (1997a; see earlier), but for the solitarious locusts, the values in the two studies are quite comparable. Deng (2002) attributed the difference between his values and those of Mahamat *et al.* (1997a) to different generations of crowded locusts. However, such a large difference between the 21st (Mahamat *et al.*, 1997a) and the 28th (Deng, 2002) generations of crowding is unlikely. Both of these studies did not describe the visible colour of the integument, and it is unknown whether non-green solitarious hoppers were tested or not.

In another publication, Tawfik *et al.* (1997a) also assessed the absorption ratio, 460/680, in the haemolymph of crowded and isolated last-instar nymphs and in 10-, 20-, 30- and 40-day-old adults of *S. gregaria*, without differentiating the sexes. Rough calculation from the data presented in Fig. 4 of these authors shows that the mean absorption ratios for adults of different ages ranged from about 1.6 to 2.3 and from 1.1 to 1.3 in crowded and isolated locusts, respectively. These ratios for isolated locusts are within the range of mean ratios as reported by Mahamat *et al.* (1997a) and Deng (2002). However, the ratios for crowded locusts are comparable to those obtained by Deng (2002), but much lower then those recorded by Mahamat *et al.* (1997a). Tawfik *et al.* (1997a) stated that exposure to JH III of fifth-instar crowded hoppers of *S. gregaria* shifted the haemolymph pigment ratio towards that of the solitarious phase. Very unfortunately, however, the authors did not provide the actual values of the ratios found in the hoppers. Exposure of adults at any age to JH III did not affect significantly the 460/680 ratio.

Neither Mahamat *et al.* (1997a) nor Deng (2002) found sex-dependent differences in the 460/680 absorption ratios, and both articles support earlier publications in regard to presence of carotenoids and biliverdin in solitarious locusts (see earlier). Additionally, these authors, as well as Tawfik *et al.* (1997a), showed the presence of biliverdin in gregarious, *S. gregaria*, albeit in lower relative concentrations (less biliverdin) than those in the solitarious locusts.

Chino *et al.* (1983) purified and characterized a biliverdin-binding cyanoprotein in the haemolymph of crowd-reared adults of *L. migratoria*. These authors found that the molecular mass of the cyanoprotein is approximately 350 kD, reported its amino acid composition and disclosed its size and shape by electron microscopy. The apoprotein consisted of identical subunits, each with a molecular mass of 83 kD. Therefore, the cyanoprotein was assumed to be a tetramer. Each subunit contained two molecules of biliverdin, associated non-covalently with the apoprotein; thus, the cyanoprotein contained eight molecules of biliverdin. It also contained about 3.5% mannose, bound covalently with the apoprotein. It may be stressed that Chino *et al.* (1983) did not study isolated adults of *L. migratoria*. The absorption spectrum of the cyanoprotein showed maxima at 370 and 660 nm. The extracted blue pigment exhibited a maximum at 375 nm and a broad absorption at 640–660 nm. The difference between the latter and the maximum of 680 nm observed in *S. gregaria* by Mahamat *et al.* (1997a), Tawfik *et al.* (1997a) and Deng (2002) may be due to technical reasons or because of different locust species.

There are two types of carotenoids, carotenes (hydrocarbons) and xanthophylls, that are oxidized derivatives of the carotenes; insects, like all animals, must take them from their diet (Canavoso *et al.*, 2001, and references therein). According to the old literature, β-carotene is present in the haemolymph of locusts and two carotenoids are present in the integument of *S. gregaria* and *L. migratoria*, mostly β-carotene and some astaxanthin (Goodwin and Srisukh, 1949). The old literature (Goodwin, 1949, 1952) also reported that β-carotene is responsible for the yellow colouration of crowded sexually mature adult males of *S. gregaria* and *L. migratoria*. In a more recent study, Wybrandt and Andersen (2001) purified and determined the sequence of a yellow protein from crowded sexually mature males of *S. gregaria* (discussed in detail in Section 7.3). It may be added that the chromophore (carotenoid) is firmly, but not covalently, attached to the protein; denaturation of the protein by 6 M urea, or exposure to a pH of 2, do not dissociate the pigment from the protein, but treatment with acetonitrile, or ethylacetate, removes the pigment. Wybrandt and Andersen (2001) did not study the kind of the carotenoid, they only cited Goodwin's (1949) suggestion that the pigment is probably β-carotene. It was already discussed (Section 7.3) that Sas *et al.* (2007) did not find this yellow protein in isolated adults of *S. gregaria* and presented strong circumstantial evidence that the synthesis of this protein is controlled by some hormone originating from the brain-CC complex. Sas *et al.* (2007) did not investigate the pigment attached to the yellow protein.

Goodwin (1949) suggested that in each species, *S. gregaria* and *L. migratoria*, the difference in the yellow colour between solitarious and gregarious adults, as well as between sexually mature males and females, is not caused by differences in total carotene content, but rather due to differences in carotene distribution, especially near the cuticle. This old suggestion received some more recent support. It is recalled that in *S. gregaria*, Mahamat *et al.* (1997a) obtained no sex-dependent differences in the absorption spectra, and

Deng (2002) found no significant differences between males and females in the absorption ratio 460/680 (see earlier), albeit in the haemolymph. In an earlier study, J. Gross, M. P. Pener and M. Rothschild (unpublished) found that β-carotene content in the integument of allatectomized males, which never become yellow (see Section 7.3), increases similarly to that in normal yellow males of *S. gregaria*. This finding may indicate that in normal sexually mature gregarious males, the β-carotene is in a higher (more apical) layer of the integument than in the allatectomized males.

Insects produce melanin, though for its synthesis they may employ enzymes structurally and functionally different from those of the vertebrates (Sugumaran, 2002). Cherqui *et al.* (1998) indicated a eumelanin pathway in the haemolymph of last-instar nymphs of *L. migratoria*. Kayser and Palivan (2006) revealed differences between melanization and sclerotization in different insect orders, using electron spin resonance (ESR) spectroscopy of exuviae. From the Orthoptera, they investigated exuviae of *L. migratoria* and indicated that the ESR spectrum is probably due to the eumelanin type. The pale white exuviae of the Okinawa albino strain of *L. migratoria* (see Section 7.2.2 and refer the work by Hasegawa and Tanaka, 1994) did not show any ESR signal. The authors do not state, but presumably in the study of Kayser and Palivan (2006) and in that of Cherqui *et al.* (1998), crowded hoppers were used.

Melanin is responsible for the black patterning of gregarious hoppers of *S. gregaria* and *L. migratoria*. As already outlined (Section 7.2.1), high temperatures reduce the areas of the black patterns and low temperatures increase them. This means that the amount of melanin is increasing with decreasing temperatures and vice versa. The black patterns are induced by the DCIN (also termed [His^7]-corazonin), as shown by Tanaka's (2003) study on the inter-related effects of this hormone and temperature in *L. migratoria*, (for details, see Section 7.2.2). However, the induction of the dark colour may be a somewhat complex process. In the Okinawa albino hoppers of *L. migratoria* (see Section 7.2.2), injection of relatively high doses of DCIN induces uniformly dark blackish or black colour, resembling dark homochromy ('fire darkening') of normally coloured solitarious hoppers, or induces gregarious black patterns (Tawfik *et al.*, 1999a; Yerushalmi *et al.*, 2002). Although substantial darkening occurs already before the next moult, the exuviae shed at this moult do not have any dark colour. This means that the darkening observed within the stadium that received the DCIN is not due to exocuticular melanin. In contrast, the exuviae of the second moult after the injection of DCIN are clearly dark or have clear dark patterns (Tanaka and Pener, 1994; Yerushalmi *et al.*, 2002), indicating the presence of exocuticular melanin. It seems, therefore, that DCIN does induce melanin synthesis in the Okinawa albino hoppers. The question arises as to what is responsible for the non-exocuticular darkening within the stadium that receives the DCIN. No clear answer can be provided without proper biochemical research on the pigment(s). However, two assumptions may be feasible; deposition of melanin or dark ommochrome(s) in the epidermis and perhaps in the endocuticle. These two possibilities are not mutually exclusive, that is, both melanin and ommochrome may accumulate under the exocuticle. Also, the presence of dark patterns in the exuviae of the second moult after the injection of DCIN does not exclude a possible deeper presence of melanin or ommochrome or both. It may be mentioned that Fuzeau-Braesch (1966, 1985), investigating dark homochromy (fire darkening) in isolated adults of *L. migratoria* kept on dark ground, distinguished between 'primary' and 'secondary' pigmentation. Using radiolabelled tyrosine, she showed that primary melanization is based on tyrosine taken up after ecdysis and during the sclerotization of the cuticle. The secondary pigmentation, that is, the homochrome darkening induced by the black ground, is achieved by deposition of a dark ommochrome in the epidermis and by a secondary melanization that covers the epidermal ommochrome.

Ommochromes, produced by the tryptophan-ommochrome pathway in insects (Kayser, 1985), are noted in a review on spider colouration as yellow, red, brown and black pigments (Oxford and Gillespie, 1998, pp. 620–621). There is no reason to assume that insects differ from spiders in this respect. Ommochromes had been reported in the old literature under the name 'insectorubin', in both *S. gregaria* and *L. migratoria*, by Goodwin and Srisukh (1950) and Goodwin (1950, 1952). These authors observed that the isolated pigment is reddish-brown that quickly changes to dark brown. They also found that the amount of the pigment is increasing with decreasing temperatures and vice versa, similarly to melanin (see earlier). They did not find differences in the concentration of the pigment investigated in solitarious green and buff, as well as in gregarious *L. migratoria*. In *S. gregaria*, however, both solitarious buff and gregarious locusts contained greater amount of the pigment than green solitarious locusts.

Citing an unpublished report by L. R. Cohen on *S. gregaria*, Nolte (1965) stated that ommochrome in its reduced form is wine-red, but appears brown when bound to a protein. About 11% of the ommochrome occurred in a yellow, oxidized, state. Nolte (1965) confirmed the conclusions of Goodwin and Srisukh (1950) and Goodwin (1950, 1952) that (1) the amount of ommochrome decreases with an increase in temperature and (2) there is a smaller amount of ommochrome in green solitarious hoppers than in gregarious conspecifics. Both of these features are similar to the features of melanin.

Deploying radiolabelled tryptophan, Fuzeau-Braesch (1968) concluded that the ommochrome of gregarious *L. migratoria* and of the non-locust grasshopper, *Oedipoda coerulescens*, is xanthommatin, whereas Verdier's (1965) albino strain of *L. migratoria* has no ommochrome. She also found another pigment and considered it as another ommochrome, or a degradation product of xanthommatin. As for *O. coerulescens*, a species that exhibits homochromy, but neither phase, nor green-brown polyphenism, Helfert (1977, p. 277, Table 4) found that ommochrome content is highest in specimens showing black homochromy. Pinamonti *et al.* (1973 and references therein) studied the xanthommatin-forming enzyme system in crowded adults of *S. gregaria*. They found xanthommatin synthesis, *in vitro*, in homogenates and fractions of the eyes and the integument, but not in the fat body and the Malpighian tubules. These authors also tested xanthommatin synthesis in an albino strain of *S. gregaria* and found that its rate is one-tenth of that found in the normally coloured locusts. In a former study, Pinamonti and Petris (1966) found tryptophan pyrolase activity only in the fat body. Tryptophan pyrolase is the enzyme that cleaves the pyrole ring and transforms the tryptophan to *N*-formylkynurenine, the first step in xanthommatin biosynthesis. The second step is enzymatic transformation of the formylkynurenine to kynurenine, still in the fat body. Considering this activity of the fat body, but finding no xanthommatin in it, Pinamonti *et al.* (1973) concluded that kynurenine is transferred to the epidermal cells, where it is transformed to 3-hydroxykynurenine. All these compounds and processes constitute a part of the tryptophan-ommochrome pathway in insects (Kayser, 1985, p. 385, Fig. 10).

Bouthier (1975), studying fourth- and fifth-instar hoppers and adults of gregarious (crowd-reared) *L. migratoria*, found four kinds of ommochromes. The compound eyes contained all four, but only ommidin and cryptommidin were abundant. Two other ommochromes, connoted acridiommatin 1 and 2, were present in the integument, in the foregut and hindgut, as well as in the Malpighian tubules. These two pigments were also found in the faeces of young adults. The colour of the isolated pigments was purplish red or carmine-red. Bouthier (1975) did not provide the exact structure of the acridiommatins. These pigments, however, are related to xanthommatin; both xathommatin and acridiommatins are derived from xanthurenic acid (Kayser, 1985). Following the biosynthetic pathway of the ommochromes, Bouthier (1975) also revealed that acridiommatins' synthesis in the epidermal cells clearly correlates with the moulting cycle. The author also stated that there was a "very small amount of acridiommatins present in the *albino* integument during the larval instar" (Bouthier, 1975, p. 217). The albino strain investigated was that of Verdier (1965).

In a later article, Bouthier and Lhonoré (1984) studied the accumulation of several types of inclusions in the cytoplasma of the epidermal cells of crowded *L. migratoria*. They found that the epidermal cells in the longitudinal zones right and left of the tergal crest of the fifth-(last)-instar crowded nymphs is the richest in ommochrome granules, corresponding to the dark melanized patterns in the overlaying cuticle. Ommochrome content was lowest in the sternal area. The colour of the ommochrome granules was brown-red or brown-yellow. These authors also carried out electron microscopic investigations of the inclusions in pieces of integument of fourth and fifth abdominal segments of male locusts, taken daily from fifth-instar hoppers, at the ecdysis to adult and in 2- and 10-day-old adults. They found five types of cytoplasmic inclusions; types 1, 2 and 3 corresponded to successive steps of concentric pigment and mineral deposition. Ommochrome (acridiommatin 1 and 2) deposits reached maximum in type 3 granules and mineralized with calcium phosphate. Bouthier and Lhonoré (1984) also studied urate microcrystals (see later).

Ballan-Dufrançais (1978) investigated the inclusions in males and females of fourth- and fifth-instar hoppers, as well as of adults of various ages of *S. gregaria*. This author also found ommochrome granules associated with calcium. Ballan-Dufrançais (1978) also studied granules of purinic wastes, especially potassium urates, in the epidermis (see later).

It is astonishing that in regard to ommochrome-related differences between gregarious and solitarious locusts, we are confined to over half-a-century-old articles of Goodwin and Srisukh (1950) and Goodwin (1950, 1952) and to the summary of Nolte (1965) that is based mostly on unpublished reports. Presence of ommochromes in solitarious dark homochrome adults had been shown (Fuzeau-Braesch, 1966); the ommochromes were found in the epidermis beneath exuvial (exocuticular) melanin. However, a modern comparison of the ommochromes in gregarious and solitarious locusts is very unfortunately not available. In a recent study, two differentially expressed ommochrome-binding protein–like genes were found in the larval fat body of a lepidopteran insect, *Ostridia nubialis*, the European corn borer (Coates *et al.*, 2005), but studies on the molecular biology of ommochromes in locusts are completely missing.

As already discussed (Section 7.2.2), DCIN induces dark colouration, or gregarious black patterns, in hoppers of Tanaka's (Hasegawa and Tanaka, 1994) Okinawa albino strain of *L. migratoria*. However, in some instances, DCIN injected into third-instar hoppers induces reddish colour (Tanaka, 2000b, Fig. 1, photo D; confirmed by unpublished results of Yerushalmi and Pener). It is tempting to speculate that this colour is related to ommochrome(s). As already outlined (see earlier), ommochromes in a reduced state have such reddish colour. However, the pigment should be extracted and identified to support or contradict this speculation.

Pteridines are defined as substances that are derived from the basic pyrazino-pyrimidine ring. The pteridines that are colourless absorb in the UV between 340 and 370 nm and may, or may not, be regarded as pigments. Pteridines that yield visible colours are yellow, orange or red (Ziegler and Harmsen, 1969; Kayser, 1985; and references in these reviews).

Goodwin and Srisukh (1951b) reported the presence of pterin in the eyes of both *S. gregaria* and *L. migratoria*. They did not identify the pigment, but noted that it more closely resembles xanthopterin than any other pterin (reported up to 1951). The authors did not compare solitarious and gregarious locusts. Goodwin (1952) added that small amounts of the same pigment are also present in the integument.

Nolte (1965) found a pigment, called pteridine by him, in the integument and the eyes of *S. gregaria*, *L. migratoria* and *Locustana pardalina*. The pteridine showed blue-green fluorescence in alkaline solution and yellowish colour in acid solution. No great differences were found in its concentration between hoppers and adults nor between *S. gregaria* and *L. migratoria*. Nolte (1965) noted that pteridine is a background pigment that does not contribute to the visible colour. Based on an unpublished report by E. Schneiderman, Nolte (1965) concluded that the amount of pteridine in the integument was greater in solitarious than in gregarious locusts. In the eyes of *S. gregaria*, more of the fluorescent pigment was found than in the eyes of the other two species, and green solitarious locusts of all three species seemed to have more of the pigment than gregarious locusts.

Harmsen (1966) reported that xanthopterin and sepiapterin are present in mature yellow males of *S. gregaria*, noting, however, that the yellow pigmentation is mainly due to a chloroform soluble substance. It is not stated, but the yellow colour of the males surely reveals that crowded locusts were tested.

Bouthier (1966) investigated pterins in Verdier's (1965) albino strain of *L. m. cinerascens* in comparison with normally coloured conspecifics. He cited the publications of Goodwin and Srisukh (1951b) and Nolte (1965), as well as some other relevant studies. Bouthier (1966) found more pterins in the eyes of the albinos than in the eyes of the normal phenotypes. In contrast, somewhat less pterins were found in the abdomen of the albinos than in that of normal phenotypes. The pterin that showed blue fluorescence was abundant in the eyes and in the integument; it was assumed to be isoxanthopterin. Another pterin that exhibited green fluorescence, abundant in the integument of normal phenotypes and in the eyes of the albinos, seemed to be xanthopterin. Finally, the pterin that showed yellow fluorescence was presumably sepiapterin. Two other unidentified pterins were also reported by Bouthier (1966).

More recently, Němec *et al.* (2003) reinvestigated the presence of pteridine-like pigments in hoppers and adults in crowded *L. m. migratorioides* by thin layer chromatography (TLC). The authors sampled the haemolymph, the eyes and the pronotum (as a representative sample of the integument), extracted the pteridines and ran them in TLC. They attempted to identify the pteridines according to Rf values and fluorescent colours that each displayed under 350 nm UV illumination, as compared to 12 standard pteridines with known Rf values and fluorescent colour. The extract of the haemolymph of untreated nymphs and adults showed six different spots, that of the integument (noted as cuticle by the authors) yielded five spots. No qualitative difference was found between hoppers and adults. The extract of the eyes showed six spots in the hoppers, as well as in the adults, out of which five were the same, but the sixth seemed to be different between hoppers and adults. The Rf of a yellow spot in the eye extracts seemed to correspond to that of xanthopterin, though the xanthopterin standard gave yellow colour (see Table 1 of Němec *et al.*, 2003), whereas the colour of the pteridine from the eyes, with the same Rf value, was marked as white by the authors (see Table 2 of the same article).

Němec *et al.* (2003) also explored the effect of various substances on the pteridines; metyrapone was injected, methoprene (a JHA), precocene II and neem oil were topically applied. Metyrapone and neem oil exerted no effect on an externally visible change in the pigmentation of the eyes and the integument; with these substances, only the haemolymph was tested. The pteridine pattern of the metyrapone-injected locusts was very similar to the controls, but one yellow fluorescent spot with an Rf of 0.36 was absent. The pteridine pattern of the haemolymph of the neem oil–treated locusts showed major differences from that of the controls. The authors state for the neem oil–treated locusts that ''Not less than 6 spots were visible in the haemolymph of the larvae (Table 3), while 2 others were restricted to adults'' (Němec *et al.*, 2003, p. 21). However, inspection of Table 3 reveals nine spots for the nymphs, of which two have exactly the same Rf value as the controls, and in three others, the difference between the controls and the neem oil–treated nymphs is only 0.01 Rf. In the haemolymph of the neem oil–treated adults, five spots were found; three similar to those in the nymphs and two specific to the adults.

Methoprene, as all effective JHAs, shifts the colour of the gregarious hoppers of *L. migratoria* to green, characteristic of solitarious locusts. However, it should be kept in mind that induction of green colouration does not mean induction of the solitarious phase (see Sections 7.2.1 and 9.1). Němec *et al.* (2003) stated that methoprene treatment induced a pteridine pigment pattern of the haemolymph that differed from the controls by three extra spots. However, comparison of the second column (''Haem. larva control injected'') of Table 3 and the same column in Table 4 (''Haem. larva methopr-treated'') indicates that none of the Rf spots are the same, and if they are close, the fluorescent colours are different. In the larval cuticle, seven pteridine spots were found. According to the authors, the patterns from the eyes of the methoprene-treated locusts contained eight spots. However, the respective columns of Table 4 show five pteridine spots in the nymphs, seven pteridine spots in the adults, out of which three are the same in nymphs and adults. It is impossible to obtain the number eight from these data (2 [nymph-specific]+3 [the same in the nymphs and adults]+4 [adult-specific] = 9).

Precocenes cause atrophy of the CA (Pener *et al.*, 1978), the source of the JH, though in some instances, precocenes induce a temporary effect of JH excess (Fridman-Cohen and Pener, 1980). Precocene treatment induced changes in the distribution of the Rf spots; especially noteworthy was the difference between the eyes of the nymphs and the adults.

It remains to be seen how much these pteridines contribute to the visual colour of the locusts' integument, though some effect on the colour of the eyes seems to be feasible. It may be recalled that Nolte (1965) claimed that the pteridine found by him is a 'background pigment' that does not contribute to the visual colour (see earlier). Also, some of the spots reported by Němec *et al.* (2003) may be degradation products; spots attributed to degradation were also found in some of the 'standard pteridines' (see Němec *et al.*, 2003, p. 20, Table 1).

Flavonoids are O-heterocyclic compounds, based on the 2-phenylchroman skeleton, called flavan (Kayser, 1985). As already mentioned (Section 7.3), the colour of the hind wings of solitarious adults of *L. migratoria* is bright yellow. Receiving such wings from the senior author of the present review, Prof. T. L. Hopkins extracted the pigment and subjected it to HPLC with UV spectroscopy. The pigment turned out to be a flavonoid C-glycoside, which is resistant to acid hydrolysis, and therefore, it was impossible to isolate and identify the flavonoid. Therefore, the results remained unpublished. The grass fed to the locusts also contained the same flavonoid C-glycoside, and the locusts presumably obtained the substance from the diet because insects cannot synthesize flavonoids.

Flavonoids were discovered in the hind wings of a non-locust grasshopper, *Dissosteira carolina* (L.) (Oedipodinae), by Hopkins and Ahmad (1991) and were found to be quercetin and quercetin-β-3-O-glucoside, with minor amounts of luteolin.

Inorganic materials and purinic wastes, presumably mostly in the form of potassium urates, also contribute to the visible colour of *S. gregaria* (Ballan-Dufrançais, 1978) and *L. migratoria* (Bouthier and Lhonoré, 1984). According to the former study, inclosures of purinic wastes and ommochromes seem to be mutually exclusive in the hopper stadia, but in the latter study, they are claimed to be present together. According to Ballan-Dufrançais (1978), if urates are only present, the visible colour appears as white spots or lines.

This section is closed by emphasizing the almost complete absence of modern recent studies on locust pigments, especially in regard to pigment-related phase-dependent differences.

8 Reproduction

8.1 MATURATION-ACCELERATING AND MATURATION-RETARDING EFFECTS

Maturation time may be defined as the period elapsing between moult to the adult stadium (connoted fledging) and vitellogenetic development of the oocytes (or first oviposition) in females and physiological as well as behavioural readiness to copulate (or first copulation) in the males. Laboratory experiments demonstrated that in *S. gregaria*, crowded (gregarious) adults mature earlier than conspecific isolated-reared (solitarious) adults (Norris, 1954). In contrast, maturation time of *L. m. migratorioides* is shorter in solitarious than in gregarious adults (Norris, 1950; Pener, 1976a). Norris (1964) concluded that there are two counteracting pheromonal effects in the adults of each of these two species; presence of mature males accelerates the maturation of immature adults, but presence of young adults retards the maturation of other young adults. In *S. gregaria*, the accelerating effect dominates over the retarding one, whereas in *L. m. migratorioides*, the retarding effect is more prominent. Therefore, under crowding, changeover from retarding to accelerating effect occurs relatively later in *L. m. migratorioides* than in *S. gregaria*. Projection of the laboratory results to field conditions indicates that in both species, these counteracting effects result in a rather synchronous sexual maturation of the gregarious adults, constituting an important advantage by largely separating migration (in the air) from sexual activities (on the ground) of the swarm.

The maturation-accelerating effect exerted by mature males on immature adults of both sexes of *S. gregaria* is a well-established fact that was reported repeatedly (Norris, 1954; Loher, 1961; Richards and El Mangouri, 1968; Amerasinghe, 1978a). More recently, Mahamat *et al.* (1997b) reconfirmed the accelerating effect on females; exposed to the presence of conspecific mature males, females showed earlier presence of vitellogenin in the haemolymph than females kept without males or exposed to immature males. Presumably, the effect is pheromonal and the term 'maturation-accelerating pheromone' is frequently used in the literature. The emission of the pheromone is controlled by the JH that is produced by the CA (see later). Research on the pheromone was discussed in Section 7.3, in relation to yellowing of the sexually maturing males. Briefly, Mahamat *et al.* (1993) found that volatiles of mature gregarious males of *S. gregaria* exert a maturation-accelerating effect on conspecific adults. They found anisole, benzaldehyde, veratrole, 4-vinylveratrole and phenylacetonitrile (PAN) as the major volatiles. PAN was dominant, constituting almost 80% of these emissions. In addition, guaiacol, phenol, benzylnitrile, benzylalcohol and 2-benzoyloxy PAN were present in smaller amounts.

In a subsequent article, Mahamat *et al.* (2000) reported that a blend of five compounds, anisole, veratrole, benzaldehyde, PAN and 4-vinylveratrole, exerted a maturation-accelerating effect on young males, similar to that exerted by living crowded mature males. Maturation time was assessed by the days elapsing until stage III yellowing (see Section 7.3) and until copulation. Removal of anisole from the blend had no significant effect, but removal of any one of the other four compounds significantly reduced the maturation-accelerating effect of the remaining blend. Maximum reduction was observed in the absence of PAN from the blend. Mahamat *et al.* (2000) concluded that PAN appears to be critical to maturation-accelerating activity, though the other three compounds also make a contribution.

To provide a more complete picture, it is worth mentioning that Torto *et al.* (1994) concluded that four compounds, namely, PAN, benzaldehyde, guaiacol and phenol, act as adult 'aggregation' pheromones in *S. gregaria*. This conclusion was reinforced by Njagi *et al.* (1996), noting that solitarious adults are also attracted. These conclusions were repeated in the reviews by Hassanali and Torto (1999) and Hassanali *et al.* (2005a). The pheromonal effect on aggregation will be discussed in Sections 12.4 and 12.5, but the claim of a parsimonious double effect of PAN (maturation acceleration and aggregation promotion) should be noted here.

In contrast to Mahamat *et al.* (2000), Schmidt and Albütz (2002) reported that neither volatiles nor PAN (under the name benzyl cyanide) alone accelerated maturation of immature males of *S. gregaria*, but pentane extracts of the cuticle surface obtained from mature yellow males did so. Hassanali *et al.* (2005a, p. 235) suggested that this controversy could be could be reconciled if "volatile adult pheromone functions in the presence of other chemotactile signals available in groups of adults, mature or immature". However, in the first article firmly demonstrating chemical acceleration of the maturation process in crowded *S. gregaria*, Loher (1961) found that maturation is accelerated by volatiles, though he succeeded in concentrating these volatiles in oil. Also, two articles of Mahamat *et al.* (1993, 2000) reported that volatiles alone (visual and tactile stimuli excluded) accelerated maturation. It may be recalled that Mahamat *et al.* (1993, 2000) recorded yellowing only up to Norris' (1954) stage III, which indicates a feeble and much localized yellowing (see Section 7.3 for a more detailed discussion). Interestingly, Amerasinghe (1978a) reported that both volatiles and contact with mature male cuticular extracts accelerated maturation of young males as assessed by the onset of sexual behaviour. However, neither volatiles nor contact with cuticular extracts induced yellowing. Amerasinghe (1978a) obtained acceleration of yellowing only when the immature test males were kept with mature males. It is possible, therefore, that the different conclusions drawn by different authors are due to the mode of assessment of the maturation of the males (onset of sexual behaviour or yellowing or onset of PAN emission or any combination of these methods). Perhaps, the foregoing statement by Hassanali *et al.* (2005a, see earlier) may be modified to state that volatile adult male pheromone accelerates maturation, but chemotactile signals may amplify the maturation-accelerating effect, possibly differentially in different characteristics used for assessment of maturation. In conclusion, even after about half a century of research, the issue is not completely clarified.

Njagi *et al.* (1996) reported that isolated-reared (solitarious) mature males of *S. gregaria* do not produce PAN and veratrole; they produce little phenol, guaiacol and anisole and more or less the same amount of benzaldehyde as produced by gregarious males. Benzaldehyde was claimed to be an effective component in maturation acceleration (see earlier; Mahamat *et al.*, 2000) and also as a component of the aggregation pheromone blend (Torto *et al.*, 1994; see other relevant references earlier). If so, is there a physiological function of the benzaldehyde produced by solitarious mature males?

Changes of density rapidly affect the production of volatiles by mature males in *S. gregaria*. Deng *et al.* (1996) found that isolation of crowded mature males quickly (about 4 days) resulted in cessation of PAN production, whereas newly crowded, previously isolated, mature males started to produce PAN within 4 days. Similarly, Seidelmann *et al.* (2000) reported that PAN (under the name benzyl cyanide) is produced exclusively by mature gregarious males of *S. gregaria*. Isolation of previously grouped mature males reduced PAN emission to practically zero within 7 days, but regrouping of these males reinduced its emission, reaching a maximum within 4–5 days. These authors again isolated the regrouped mature males that led to a reduction of PAN emission, similar to that observed after the first isolation. Even a 'crowd' of two mature males resulted in a slight emission of PAN; in the case of three, four, or five males, PAN emission gradually increased. PAN production is not elicited in maturing males kept crowded with only females (Seidelmann *et al.*, 2000, Fig. 5) and is substantially delayed, along with other measures of maturation, when nymphs are present in groups of adult males (Assad *et al.*, 1997a, discussed later).

In contrast to the claim that PAN acts as an aggregation/cohesion pheromone (see earlier), Seidelmann *et al.* (2000), using an olfactometer bioassay in *S. gregaria*, found that PAN has a strong repellent effect on gregarious mature males and a weak repellent effect on immature males. Further publications from the same laboratory (Seidelmann and Ferenz, 2002; Ferenz and Seidelmann, 2003) reinforced the conclusion that PAN repels other sexually mature males. According to these authors, mature gregarious males emit PAN that serves as a volatile olfactory cue to repel other males during mounting or to prevent homosexual encounters or both. They named the PAN as a 'courtship inhibition pheromone'. Perhaps, the term 'rival male repelling pheromone' would be more accurate because the substance does not inhibit the courtship of the male that is mounting a female. A subsequent publication showed that wings and legs are the emission sites of the PAN, whereas veratrole is emitted all over the body (Seidelmann *et al.*, 2003). The glands secreting PAN have yet to be identified. Possibly, vacuolated epidermal cells are responsible, although these have yet to be demonstrated in the distal part of the wings of sexually mature males. The repelling effect is not all encompassing; Seidelmann (2006) reported that it is ignored by female-deprived crowded mature males that under such a condition mount one another.

It is difficult to reconcile the repeated claim by Torto *et al.* (1994), Njagi *et al.* (1996), Hassanali and Torto (1999) and Hassanali *et al.* (2005a) that PAN is a major component of an adult aggregation pheromone blend with that by Seidelmann and Ferenz (2002) and Ferenz and Seidelmann (2003) claiming that PAN acts as a repellent of other mature males. For some arguments about this controversial subject, see Hassanali *et al.* (2005a) and Seidelmann *et al.* (2005). Recently, Rono *et al.* (2008) reported that, in crowded *S. gregaria* adults, PAN at low concentrations exerts an aggregating effect, whereas at high concentrations, it repels rival males. The issue is discussed in detail in Section 12.5.

In light of the claim about the maturation-accelerating effect of volatiles and PAN (Mahamat *et al.*, 1993, 2000) and the claim by Schmidt and Albütz (2002) that cuticular surface extracts, but neither mature male volatiles nor PAN, accelerate maturation of immature males, the exact nature of the 'maturation-accelerating pheromone' is open to further investigations. The issue is even more complicated, because Norris (1954) found that mature females exert a weak stimulus capable of inducing accelerated maturation, but females do not produce PAN, benzaldehyde and veratrole (Torto *et al.*, 1994). Also, Richards and El Mangoury (1968), as well as Amerasinghe (1978a), observed that isolated mature males accelerate the maturation of young males, but isolated mature males do not produce PAN (Deng *et al.*,1996; Seidelmann *et al.*, 2000; see earlier). However, the isolated mature male ceased to be isolated when tested with a young male, and Seidelmann *et al.* (2000) have shown that a 'crowd' of two males resulted in a slight emission of PAN (see earlier). Therefore, it cannot be ruled out that the presence of the young male did induce some shift towards the gregarious phase of the formerly isolated mature male, leading to a subsequent emission of the maturation-accelerating pheromone.

Production of the maturation-accelerating effect in crowded *S. gregaria* is controlled by the CA and their secretion, the JH. Allatectomized males do not accelerate maturation of other adults (Loher, 1961; Amerasinghe, 1978b); in contrast, they retard maturation of young adults (see later). In fact, allatectomized adults of *S. gregaria* never become mature, males do not show yellowing (see Section 7.3) and do not exhibit mating behaviour (see Section 8.2). Tawfik *et al.* (2000) showed that pheromone emission by crowded adult males started between 10 and 15 days after fledging, coinciding with a conspicuous increase of the amount of JH III in the haemolymph.

The maturation-retarding effect exerted by young males on other young males in *S. gregaria* has received less attention than the maturation-accelerating effect. Nevertheless, the retarding effect was clearly recognized by Norris (1954, 1964), who tested crowded young males, keeping them with other young males, renewing the latter at weekly intervals. Thus, the test males were continuously exposed to young males. Richards and El Mangoury (1968; Richards maiden name is Norris, so Richards M. J. and Norris M. J. denote the same author) confirmed the effect; additionally, they found that fifth-(last)-instar hoppers exert a retarding influence on maturation, similar to that exerted by immature adults. Assad *et al.* (1997a) reconfirmed that maturation of young adults is retarded when they are reared together with conspecific crowded last-instar hoppers. In one experiment, the young adults were exposed to visual, tactile and chemical signals originating from male and female hoppers. In another experiment, the visual and tactile signals were excluded, but the results were similar, indicating that the effect is due to chemical signals. Nymphal faeces were ineffective; therefore, the authors turned to test captured volatiles emitted by the hoppers and a synthetic blend identified as nymphal aggregation pheromone components (cf. Hassanali and Torto, 1999; Hassanali *et al.*, 2005a; and references therein; see also Section 12.1.1). Both the captured volatiles and the synthetic blend were reported to retard maturation. Unfortunately, Assad *et al.* (1997a) confined their study to hopper-related effects and did not investigate the maturation-retarding effect of young males as originally described by Norris (1954). This situation is responsible for some gap in knowledge. Obeng-Ofori *et al.* (1994b) found guaiacol and phenol as the major volatiles of fifth-instar hoppers and young adults. Both these compounds are in the faeces. However, nymphal faeces were found to be ineffective by Assad *et al.* (1997a). The question arises what is (are) the factor(s) in young adults that retard(s) maturation of other young adults? It seems that the presence, composition or alteration of the so-called nymphal aggregation pheromone blend have not been investigated adequately in young adults.

It was already mentioned that juvenile hormone (JH) is responsible for the maturation-accelerating effect in crowded adults of *S. gregaria* (see earlier). Allatectomy results in absence of JH and allatectomized adults of both sexes retarded maturation of young adult males (Norris and Pener, 1965). Adults that had been allatectomized within a few days after fledging, but were tested for the retarding effect after 4 weeks, still retarded maturation of young adult males. It seems, therefore, that allatectomy, that is absence of JH, prolongs the period during which an inhibitory influence on maturation is exerted. The factor(s) causing the maturation inhibiting effect is not known; presumably, it is similar to the factor(s) present in normal young adults that cause(s) the same effect, which, however, is (are) again not completely clear (see earlier).

Other factors also affect maturation time of locusts. In crowded *S. gregaria*, fed on yellow senescent vegetation, maturation was delayed in comparison to conspecifics fed on fresh green vegetation. However, addition of gibberellin, or eugenol, to the diet of senescence vegetation shortened maturation time that becomes similar to that of the locusts fed on green vegetation (Ellis *et al.*, 1965; Carlisle and Ellis, 1965). Carlisle *et al.* (1965) reported that a single contact (no feeding) with terpenoids derived from *Commiphora myrrhae* accelerated maturation of both sexes of *S. gregaria* reared in pairs. In addition, Carlisle *et al.* (1965) presented field evidence, from the Somali Peninsula, that maturation of *S. gregaria* is initiated when *Commiphora* ssp. come into leaf and flower. More recently, Assad *et al.* (1997b) reinvestigated the effect of *Commiphora quadricincta* extracts on maturation time of crowded *S. gregaria* adults. They used extracts of plant samples, collected at the Red Sea coastal areas in Sudan, in October, 1994, before the winter rains, and in March, 1995, after the winter rains. Considerably fewer kinds of essential oils were found in the extracts taken after the rains than in those taken before the rains. Both oil samples significantly accelerated maturation of young adults, but those taken before the rains were more potent and their accelerating effect was similar to that of mature males.

Jackson *et al.* (1978) investigated effects of different plant diets, fed to crowded *S. gregaria*, on the maturation time of the adults and found considerable differences. However, the authors stressed that some plants were senescent, which delays maturation (see earlier). Quickest maturation was found when the locusts were fed on *Chrozophora oblongifolia*.

Factors affecting maturation time have been much less studied in *L. m. migratorioides* (*sensu* Uvarov, 1966; that is from Africa) than in *S. gregaria*. As already mentioned (see earlier), isolated *L. m. migratorioides* mature more rapidly than crowded consubspecifics (Norris, 1950). In a later article, Norris (1964) demonstrated that presence of mature males accelerates the maturation of young males. The maturation-retarding effect of young males was deduced from the fact that crowded adults of *L. m. migratorioides* mature more slowly than isolated ones.

8.2 MATING BEHAVIOUR

Quantitative assessment of the intensity of mating behaviour enables the sexual activity of conspecific males of different phases, or males submitted to different conditions, to be compared. Such quantitative assessment may be based on variables such as the average percentage of time spent on sexual behaviour (for a description of this method, see Pener, 1967), the average number of sexual attempts (to mount) per male and the percentage of sexually active males (Amerasinghe, 1978b). Male mating behaviour in several locust species includes mate guarding. The male mounts a female to prevent access and mounting of another male before the female becomes receptive (precopulatory mate guarding), or to prevent sperm competition, arising from a second insemination by another male, which may decrease the ratio of the progeny sired by the first insemination (postcopulatory mate guarding). Some aspects of male guarding in locusts are discussed by Parker and Smith (1975) and more recently by Zhu and Tanaka (2002). A male guarding a female is sexually active and therefore contributes positively to the quantitative assessment of the intensity of mating behaviour.

Pener (1976a) reported that mating behaviour of crowded males of *L. m. migratorioides* is more intense than that of consubspecific isolated males, as assessed by weekly average time spent on sexual behaviour (three observations per week, 2 h each observation). Also, in the frame of this experimental design, mature crowded males produced more spermatophores than mature isolated males (Pener, 1976a, Fig. 5B). These findings seem to reflect different sexual strategies of gregarious and solitarious locusts in the field. Males in a swarm have many females around, but also have a heavy competition from rival males. Therefore, a gregarious male should have more intense sexual behaviour to possess a female as soon as possible and approach and mount at the first occasion, leading also to longer precopulatory mounting. However, the male may have limited time to complete copulation and fertilize the female when the swarm is on the ground. This situation may explain the higher number of spermatophores produced by the crowded males within the 2-h period of observation. Such pressures do not seem to affect solitarious males for which finding of a mate (see Section 8.3) may constitute the major pressure and the main factor of competition.

More recently, Tanaka and Zhu (2003) reconfirmed phase-dependent differences in the mating strategy of *L. migratoria*. They found that isolated adults of the Okinawa albino strain (cf. Tanaka, 1993; Hasegawa and Tanaka, 1994) exhibit a shorter period of precopulatory mounting, and isolated adults of the Okinawa albino strain, as well as a normally coloured strain from Ibaraki (Japan), spend more time on actual copulation than crowded adults. These findings are again in accord with the solitarious and gregarious sexual strategy; precopulatory guarding seems to be less affected by presence of rival males in solitarious than in gregarious locusts. Tanaka and Zhu (2003) also reported that the proportion of the offspring sired by a second male, mating with the same female, was higher in isolated than in crowded adults. The phase-related advantage of this difference is not clear, but cannot be ruled out that it reflects less use of anti-sperm competition measures by solitarious than gregarious males, which would be in keeping with a lower probability of rival males being present. The advantage of a positive relationship between the time spent on precopulatory mounting and the time spent on copulation in crowded (Zhu and Tanaka, 2002; Tanaka and Zhu, 2003) but not in isolated locusts (Tanaka and Zhu, 2003) is also unclear.

The effect of the CA and the JH on male mating behaviour of *L. m. migratorioides* is phase-dependent (Pener, 1976a). Allatectomy of isolated males results in a delayed onset of mating behaviour, but after this delay, the intensity of mating behaviour (assessed by the average percentage of time spent on sexual behaviour) and rate of spermatophore production by the allatectomized males do not differ from those exhibited by unoperated control males. In contrast, allatectomy greatly reduces the intensity of mating behaviour and rate of spermatophore production in consubspecific crowded males (Wajc and Pener, 1969; Pener *et al.*, 1972; Pener, 1976a). Implantation of three pairs of CA into allatectomized males increased the intensity of the sexual behaviour to a level comparable to that shown by normal (non-allatectomized) males (Pener *et al.*, 1972). Implantation of three pairs of extra CA, into recipients having their own CA, induced an earlier onset of the mating behaviour in crowded males but did not increase the eventual intensity of the mating behaviour. Similar implantation of three pairs of extra CA (originating from crowded males) into isolated males has no effect on the onset, neither on the intensity of the mating behaviour (Pener, 1976a).

In the grasshopper *Melanoplus sanguinipes*, which may be considered as an intermediate species between grasshoppers and locusts, Cheeseman and Gillott (1990) described an effect of the CA on male sexual behaviour somewhat similar to that found by Pener (1976a) in *L. migratoria*. The age of onset of copulation was considerably delayed in the allatectomized than in the control males, but later, the proportion of copulating allatectomized and control males became similar. Also, in a 3-h observation period, the time to initiate copulation in 50% of the males was 60 min for allatectomized males, but only 10 min for control males. In addition, these authors found that at the age when 50% of the males showed first copulation, the accessory glands of the allatectomized males were less developed than those of the controls. Topical application of JH III to allatectomized males accelerated accessory gland development, but did not elicit more intense mating behaviour.

Mating behaviour of crowded (gregarious) males of *S. gregaria* completely depends on the CA and JH, as has been demonstrated by many authors (Loher, 1961; Pener, 1965, 1967; Odhiambo, 1966; Cantacuzène, 1967, 1968; Amerasinghe, 1978b; Pener and Lazarovici, 1979). Allatectomized males do not exhibit mating behaviour and do not show yellowing (see Section 7.3); in fact, they never become mature. Reimplantation of CA into allatectomized crowded males induced temporarily mating behaviour (Loher, 1961), though at a subnormal level (Pener, 1965). Late allatectomy, performed on sexually active crowded males, 4 weeks after fledging (Loher, 1961), or even 39–43 days after fledging (Pener, 1967), resulted in cessation of mating behaviour within 2–4 weeks after the operation.

Amerasinghe (1978b) found that injection of synthetic JH I induced sexual activity in crowded allatectomized males of *S. gregaria*. Five days after the injection, the intensity of mating behaviour of the allatectomized males temporarily approached that of normal (non-allatectomized) males. Injection of JH III into crowded allatectomized males induced feeble intensity of mating behaviour; its level was much lower than that induced by JH I. Both JH I and JH III promoted yellowing in crowded allatectomized males, but JH I was more effective than JH III. JH III is the natural JH in Orthoptera, including locusts (Huibregtse-Minderhoud *et al.*, 1980; Bergot *et al.*, 1981; Loher *et al.*, 1983; Pener *et al.*, 1986), and its milder effect can be explained by more rapid metabolic degradation than that of JH I. Both JH I and JH III were ineffective at inducing yellowing in isolated males, but unfortunately, Amerasinghe (1978b) did not report data from an experiment aiming to investigate the intensity of mating behaviour in normal, allatectomized and allatectomized then CA implanted, isolated males. Nevertheless, Amerasinghe (1978b, pp. 605–606) stated, in a section of his article devoted to yellowing, that after allatectomy then injection of JH "many of these isolated males showed sexual activity for a brief period (about 1 week) after the hormone treatment, particularly in the case of J.H.I.". It may be inferred from this single sentence that allatectomized isolated males, lacking JH, are sexually inactive, like allatectomized crowded males.

Pener and Lazarovici (1979) also tested the effect of exogenous JH I and JH III on mating behaviour and yellowing of crowded allatectomized males of *S. gregaria* (under the temporary name *S. americana gregaria*), employing injection or topical application. Injection of either JH I or JH III induced mild intensity of mating behaviour in crowded allatectomized males, as assessed by percentage of time spent on sexual behaviour, but topical applications were ineffective. Intensity as high as that showed by unoperated or sham operated controls was obtained with a 1:1 mixture of JH I and JH III, given in eight 20-μg injections of the mixture at 2-day intervals (cumulative dose 160 μg). In this experiment, the tested males showed mating behaviour for up to 4 weeks (including the first 14 days during which the injections were administered). In contrast to the ineffectiveness of topical application in inducing mating behaviour, this mode of treatment induced yellowing in the allatectomized males. Taking into consideration that the yellow protein (Wybrandt and Andersen, 2001) and JH are both needed for yellowing, as well as the conclusion of Sas *et al.* (2007) that the yellow protein is synthesized in the epidermis (see Section 7.3), it is easy to explain the effect of the topically applied JH on yellowing. By this mode of administration, the hormone was placed almost directly onto the target organ. This finding indicates that different target organs are responsible for the effect of JH on mating behaviour and yellowing. Unfortunately, Pener and Lazarovici (1979) did not test mating behaviour, or the possible effect of allatectomy and JH administration in isolated males.

The complete control of mating behaviour by the CA/JH is not limited to *S. gregaria* among the acridids. Pener (1968) demonstrated that crowded allatectomized males of the red locust, *Nomadacris septemfasciata*, do not show mating behaviour. Implantation of four pairs of CA, originating from adult, sexually mature males of *L. m. migratorioides*, into allatectomized *N. septemfasciata* males, induced mating behaviour, reaching temporarily the intensity of that shown by unoperated males. Again, no isolated males were tested. The CA/JH also completely control sexual behaviour in crowded males of *Anacridium aegyptium* (Greenfield and Pener, 1992), a species which is rather a grasshopper than a locust. All three species, *S. gregaria*, *N. septemfasciata* and *A. aegyptium*, belong to the subfamily of Cyrtacanthacridinae and all show ability of delayed maturation or adult reproductive diapause (see later).

The effect of the CA/JH on male mating behaviour varies among the acridids. In some species, the control is all encompassing, in some others, it is partial (see earlier), and in still others, it does not seem to exist. These species-dependent differences may be related to presence or absence of reproductive diapause or to the subfamily to which the species belong. A long discussion of this subject is unrelated to locust phases, except of the details discussed earlier; the reader may consult relevant reviews (Barth and Lester, 1973; Pener, 1986, 1997).

Data on mating behaviour of female locusts are scarce. Tanaka and Zhu (2003) found a significant effect of crowded females on the length of precopulatory mounting and copulation time in *L. migratoria*, but no similar effect was found in isolated locusts. Strong and Amerasinghe (1977) concluded that crowded allatectomized females of *S. gregaria* exhibit a permanent defence towards courting males, but no data are available for conspecific isolated females. Like in the males, the effect of the CA/JH on female mating behaviour in acridids depends on the species (refer reviews by Pener, 1986, 1997), but the subject is again unrelated to locust phases.

A pair of glands, named Comstock–Kellog glands, is located in the genital chamber of several subfamilies of female acridids. These glands are eversible organs and have glandular epithelium. They were observed in an eversed state before and during copulation, but their function is controversial (Uvarov, 1966, 1977; Whitman, 1990; and references in these reviews). The glands are present in *S. gregaria* (Thomas, 1965), like in other acridids belonging to the subfamily of Cyrtacanthacridinae. Recently, Njagi and Torto (2002), investigating crowded *S. gregaria*, extracted the glands and reported that the extract contained pentanoic acid, which was present in detectable amounts only in the glands of 14- to 16-day-old females. This age coincided with sexual maturation of the females. These authors also reported presence of hexanoic acid, heptanoic acid, octanoic acid, nonanoic acid and decanoic acid in the gland extracts. According to Njagi and Torto (2002), pentanoic acid significantly stimulated some behavioural patterns, considered by the authors as precopulatory behaviour, in conspecific crowded males. However, no more decisive experiments were made (e.g. analysis of male mating behaviour towards Comstock–Kellog glands excised females with that towards operated and then pentanoic acid–treated females). Again, no experimental data are available for isolated females.

8.3 MATE LOCATION

As outlined in Section 8.2, gregarious locusts in swarms have no difficulty to locate a mate. Both sexes have many individuals of the opposite sex around, and the pressure on the male is to overcome competition from rival males. The situation is different in solitarious locusts because the distribution of the individuals is scattered and they have to locate a mate to reproduce.

Although acoustic signals play an important role in locating and courting a mate in many acridids (Uvarov, 1966, pp. 176–188, 1977, p. 221, and references therein), the acoustic repertoire of *S. gregaria* is primitive, probably expressing only disturbance (Loher, 1959), and does not seem to have importance in mate location or recognition. The sound produced by *L. migratoria* is more developed (for a description of the sound-producing mechanism, see Albrecht, 1953, p. 28), but it is emitted by sexually mature males, mostly when disturbed while engaged in some kind of sexual activity. However, the sound does not seem to play a role in meeting of the sexes, and it is not a necessary part of male courtship and copulation in crowded locusts (Bar-Ilan *et al.*, 1969). Therefore, it is improbable that solitarious locusts of these two species utilize acoustic signals for mate location and recognition.

Inayatullah *et al.* (1994), using two kinds of wind tunnels, tested the effect of olfactory cues on attraction between the sexes in isolated *S. gregaria*. They reported that isolated males were attracted towards the females, even when the female was hidden and her wings were glued together to exclude visual and acoustic cues. In contrast, isolated females did not respond to hidden males. The authors concluded that solitarious females of this species emit a sex pheromone that attracts conspecific males. The chemical identity of the pheromone has not been identified.

Ould Ely *et al.* (2006) extended the studies on mate location in isolated *S. gregaria*. They used a flatbed wind tunnel and exposed isolated males and females, as well as newly crowded adults (isolated-reared as hoppers), to olfactory or olfactory plus visual cues from the opposite sex. Ould Ely *et al.* (2006) recorded five behavioural variables. They reported that isolated males showed some preference to olfactory signals from isolated females, but the difference from the controls (no signal) was not significant. Also, there was no significant difference between the results obtained in the photophase and the scotophase. The same results were reported when isolated females were exposed to olfactory cues from isolated males. Newly crowded males, which had been isolated throughout the hopper stadia, showed a significantly higher response to olfactory cues from isolated females than either continuously isolated or continuously crowded males. Isolated females exposed to olfactory cues from isolated males seemed to detect the presence of the males, but their response was weak and not significant in three out of the five behavioural variables. The response of newly crowded females did not differ significantly from that of continuously isolated or continuously crowded females. Combined olfactory and visual signals resulted in a significantly higher response of isolated males towards isolated females than that observed for control males that were tested in absence of these signals. Newly crowded males exposed to combined visual and olfactory stimuli from isolated females exhibited a higher response than either those exposed to olfactory stimuli alone or those not exposed to any stimulus. Isolated females also responded more strongly to combined visual and olfactory signals from isolated males, but their response was weaker than that of the isolated males to isolated females.

Ould Ely *et al.* (2006) interpreted the higher response of the newly crowded males to cues from solitarious females as a mode of 'recruitment' of solitarious locusts into a gregarizing (congregans) population and extended this conclusion also for the females. However, the opposite could equally be argued: by being attracted away from the aggregation towards solitarious females, such behaviour by recently crowded locusts would serve to disperse the group. There is no evidence that having mated the solitarious female would join the group, and having experienced reduced population density during mating, the male too might be expected to begin to resolitarize behaviourally (Bouaïchi *et al.*, 1995) and hence would not rejoin the group.

There are some contradictions between the results of Inayatullah *et al.* (1994) and Ould Ely *et al.* (2006). The former authors found a significant response of isolated males to olfactory signals from isolated females. The latter authors also found some difference in this response, but it was not significant. Also, in the experiments of Ould Ely *et al.* (2006), olfactory signals from isolated females did not affect significantly isolated males, but combined olfactory plus visual signals did, at least in comparison with the controls (no signal). In light of the results of Ould Ely *et al.* (2006), the possibility that visual signals play a major role in location of a solitarious mate cannot be excluded. However, a possible effect of visual signals alone was not reported.

It seems, therefore, that more research is needed to clarify the mechanism of mate finding in solitarious *S. gregaria*. Also, the results of Inayatullah *et al.* (1994) and Ould Ely *et al.* (2006) were obtained in wind tunnels of 135 and 180 cm length. This distance may not be sufficient for long-range mate location in solitarious adult locusts of this species.

8.4 OVIPOSITION AGGREGATING EFFECT

In acridids that lay egg pods into the soil, the site of oviposition depends on environmental factors such as soil moisture, soil compaction and vegetation cover.

However, such environmental factors, either alone or in combinations, are inadequate to explain completely the high degree of grouping of egg pods, caused by gregarious behaviour, resulting in oviposition aggregation. In the older literature, this behaviour is best investigated in gregarious *S. gregaria* in the field (Stower *et al.*, 1958; Popov, 1958) and in the laboratory (Norris, 1963, 1970). However, group laying is not limited to *S. gregaria*, it is observed in *L. migratoria*, *Dociostaurus maroccanus*, *Calliptamus italicus*, possibly in *N. septemfasciata*, and even in some non-locust grasshoppers (Uvarov, 1977, pp. 254–257, and references therein). Grouping of conspecific egg pods may well contribute to clumpy distribution and close proximity of the hatchlings (see Sections 16.1 and 17).

In her first article on the oviposition aggregating effect in *S. gregaria* in the laboratory, Norris (1963) used 16 kinds of tethered decoys (mature or immature, males or females, dead or live adults, contact or lack of contact with the decoys, paper dummy locusts, heterospecific *L. migratoria*, etc.) and concluded that there is no long distance attraction to the group and the effect is rather based on cohesion. Norris (1963) also concluded that the stimuli involved in group oviposition have visual, chemical, tactile and possibly acoustic components.

In a later article, Norris (1970) tested aggregation of ovipositing females of *S. gregaria* in darkness and concluded that the effect is mainly chemotactile (today also connoted as contact pheromonal). She concluded that the pheromone is produced also by locusts reared in isolation, although probably in lesser amounts than by those reared crowded. The pheromones produced by *S. gregaria* and *L. migratoria* are to some extent interspecific, but each species responds preferentially to its own pheromone.

Lauga and Hatte (1977, 1978), using a Y-shaped olfactometer, found that the sand into which gregarious females of *L. m. migratorioides* repeatedly laid attract conspecific mature adults of both phases from a distance of 0.5 m. The sand reused many times without washing or sterilizing acted as an attractant of both sexes and caused egg-laying preference of the females. The same authors also found that the reused sand affected the fecundity of isolated females; also, in the first five or six egg pods, it induced an increase in the mean weight of eggs, a clear gregarious characteristic. In contrast, Norris (1970) investigating the response to sand contaminated by previous egg laying of *S. gregaria* found no attraction to the contaminated sand by conspecifics.

The results and conclusions of Norris (1963, 1970) and Lauga and Hatte (1977, 1978) were discussed in several reviews in relation to locust pheromones (Loher, 1990; Whitman, 1990; Byers, 1991). However, no experimental progress was made to the oviposition aggregating effect in locusts until the mid-1990s. Ferenz *et al.* (1994) stated, without presenting any experimental data, that egg-laying *S. gregaria* females were not attracted to soil reused from previous ovipositions, or to egg pods, confirming Norris' (1970) conclusion. However, in contrast to Norris' results, tethered living or dead conspecific decoys were 'not very attractive'. The interpretation of this statement is left to the reader. However, Ferenz *et al.* (1994) stated that females in the process of egg laying often arrested fully mature probing females, which then started oviposition.

In contrast to Norris' (1970) results and the statement of Ferenz *et al.* (1994), Saini *et al.* (1995) related the group oviposition pheromonal effect in *S. gregaria* to the froth (or foam) of the egg pods and sand contaminated with such froth. Subsequently, Rai *et al.* (1997) identified veratrole and acetophenone as two major behaviourally active substances of the group oviposition pheromone in *S. gregaria*. Curiously, however, as outlined in the review by Pener and Yerushalmi (1998), in former studies from the same laboratory (International Centre of Insect Physiology and Ecology [ICIPE]), Torto *et al.* (1994) found no veratrole in volatiles of crowded adult *S. gregaria* females and neither veratrole Torto *et al.* (1994) nor female volatiles of conspecific adults (Obeng-Ofori *et al.*, 1994a) attracted young or older females and males. Also, Norris (1970) found that freshly killed last-instar hoppers and, separately, last exuviae of *S. gregaria* also attracted ovipositing females, but no acetophenone nor veratrole was reported as nymphal volatiles (Obeng-Ofori *et al.*, 1994b; Torto *et al.*, 1996; refer review by Hassanali and Torto, 1999), though veratrole is a component of volatile emissions of adult crowded males of *S. gregaria* (Torto *et al.*, 1994; Hassanali and Torto, 1999; Seidelmann *et al.*, 2003; Hassanali *et al.*, 2005a) and Fuzeau-Braesch *et al.* (1988) found it in vapour condensed from the atmosphere of cages with crowded fifth-instar hoppers of *S. gregaria* and, separately, *L. migratoria*.

Later publications by the ICIPE laboratory reported that gravid females of *S. gregaria* exhibited oviposition responses to sand into which conspecifics had laid previously; the effect of this sand, without froth or eggs, was compared to clean sand control (Torto *et al.*, 1999a). Analysis of trapped volatiles showed the presence of three electrophysiologically active compounds, unsaturated ketones, identified as (*Z*)-6-octen-2-one, (*E*,*E*)-3,5-octadien-2-one and the geometric isomer of the latter, (*E*,*Z*)-3,5-octadien-2-one (Torto *et al.*, 1999a,b). These substances originated from the eggs (Torto *et al.*, 1999b).

Bashir *et al.* (2000) showed in combined field surveys and field cage experiments that solitarious *S. gregaria* females laid mainly in the vicinity of the plants, *Heliotropium* ssp. and millet, but showed a significantly higher preference for ovipositing near egg pods laid by conspecific gregarious females than near the plants preferred without presence of gregarious egg pods. In contrast, gregarious females preferred to oviposit away from the plants in clean moist sand, but egg pods laid by gregarious females elicited significantly more oviposition.

Malual *et al.* (2001) suggested that the unsaturated ketones identified by Torto *et al.* (1999a,b) not only act as oviposition aggregating substances in *S. gregaria* but are also responsible for the transmission of gregarious behaviour from crowded parents to progeny, as discovered by McCaffery *et al.* (1998) and detailed in Section 16 of the present review. The review of Hassanali *et al.* (2005a) repeated this claim. However, a major problem in regard to the study of

Malual *et al.* (2001) is that they did not test directly the effect of these unsaturated ketones. Also, Simpson and Miller (2007), based on physicochemical difference between the material characterized by McCaffery *et al.* (1998) and the unsaturated ketones, refuted the claim by Malual *et al.* (2001). In addition, McCaffery *et al.* (1998) found the gregarizing agent in the foam of the egg pod, whereas Malual *et al.* (2001) found the unsaturated ketones mainly in that part of the egg pod that contained the eggs. Moreover, very recently, Miller *et al.* (2008) isolated activity to an aqueous fraction of the foam extracts and presented tentative structural features of the primary gregarizing agent responsible for transfer of gregarious behaviour from the gregarious parents to the progeny; the substance appears to be an alkylated L-dopa analogue. These issues are discussed in detail in Section 16.3.

In conclusion, more research is needed on the oviposition aggregating pheromone(s) in *S. gregaria*, as well as in other locusts. It also should be kept in mind that a pheromonal effect does not exclude possible effects of other stimuli, for example, visual stimulus as suggested by Norris (1963).

8.5 FECUNDITY AND FERTILITY

Fecundity is the number of eggs produced by a female during her lifetime, whereas fertility means the ability to reproduce or produce viable offspring (definitions from a dictionary of entomology by Gordh and Headrick, 2001). Obviously, the latter cannot be higher than the former and sometimes fertility is much less than fecundity. Eggs of locusts and grasshoppers are laid in synchronous batches, connoted egg pods (for details, see Uvarov, 1966, and references therein). The number of eggs in an egg pod is often termed as a clutch size. A female acridid usually lays several egg pods during her life. By the time of laying an egg pod, the oocytes, which give rise to the next egg pod, are already developing in the ovaries. In crowded *S. gregaria*, these oocytes are starting to take up vitellogenin after the eggs of the previous egg pod are ovulated, that is, chorionated eggs passed from the ovarioles to the oviduct (Tobe and Pratt, 1975). A detailed discussion of vitellogenesis is outside of the scope of the present review because no phase-related qualitative differences have been reported (though quantitative differences exist in the vitellin content of the eggs; see later). Concerning vitellogenesis, the reader may consult a recent study on crowded *S. gregaria* (Peel and Akam, 2007).

Fecundity is limited by the life span of a female and by the number of ovarioles in the ovaries of that female, as well as by some other factors (see later). The clutch size cannot be higher than the number of ovarioles. However, it seems that the life span plays the most important role in fecundity. It is especially so in the field, where abiotic factors (drought, flood, etc.) or biotic factors (pathogens, parasites, etc.) or both may drastically shorten the female's life, reducing fecundity and similar factors, affecting the eggs, may reduce fertility; for an example based on field data in *N. septemfasciata*, see Stortenbeker (1967, pp. 28–31). Human intervention may also exert some, often major, effect. Locust-related pest control may reduce fecundity and fertility, directly by various combinations of production of smaller clutch size, fewer egg pods and decreased fertility, or indirectly by shortening the female's life. Chemicals (for some recent examples, see Schmidt and Assembe-Tsoungui, 2002; Abbassi *et al.*, 2003a) and bioinsecticides may do both; a good demonstration of the effect is the fungal entomophagous pathogens, *Metarhizium anisopliae* var. *acridum* and *Beauveria bassiana* on locusts (see for example, Arthurs and Thomas, 2000; Blanford and Thomas, 2001; Gardner and Thomas, 2002; Klass *et al.*, 2007a,b). Since the subject is not related to locust phases (except in one instance, see later), it will not be detailed further.

The single known phase-related aspect of the effect of fungal entomopathogens is that infected locusts exhibit thermoregulatory behaviour, leading to a behavioural fever, and the fungal pathogens are more susceptible to high (38–40 °C) temperatures than the locust host (cf. Klass *et al.*, 2007a). Infested *S. gregaria* that had been able to develop such fever produced viable offspring (Elliot *et al.*, 2002). Offspring obtained from crowded *S. gregaria* that underwent such fever exhibited some shifts towards solitarious phase characteristics, namely, an increase in the ratio of green hatchlings (see also Section 7.2.1) and their behaviour (Elliot *et al.*, 2003; see Section 15). In addition, heat treatment, resembling behavioural fever, reduced flight performance of crowded *S. gregaria* males and reduced intensity of mating behaviour of the adults, which are again solitarious phase characteristics (Elliot *et al.*, 2005). However, as the authors themselves outline, they do not eliminate the possibility that the reduced flight performance and mating behaviour represent the cost of the fever and not related to phase (for further discussion, see Section 15).

The expression levels of heat shock proteins were found to be significantly upregulated in crowded hoppers of *L. migratoria* (Wang *et al.*, 2007). The relation of this finding to the claimed solitarizing effect of behavioural fever or heat treatment in *S. gregaria* (see earlier) remains to be explored (see also Sections 10.2 and 15 of the present review).

The older literature on fecundity and fertility of locusts is reviewed by Uvarov (1966, pp. 353–356, for certain aspects, see also pp. 320–321, 326–331, 362–363, p. 365); Uvarov (1977, pp. 297 and 311 for general considerations, as well as p. 459, pp. 464–466, p. 475, 481, pp. 506–507, 515–516 for data arranged according to different species); Dale and Tobe (1990, p. 394) and Pener (1991, pp. 22–24). In a latter review, Stauffer and Whitman (1997, pp. 257–262), dealing with the subject for the whole superfamily of Acridoidea, provide ample references. Not much recent work has been undertaken on fecundity and fertility of locusts, except for the studies on the relevant effects of locust control agents (see some examples earlier).

Laboratory data showed that the fecundity of isolated females is higher than that of crowded ones, especially in *L. migratoria* (Norris, 1950; Albrecht *et al.*, 1958), but also in *Nomadacris septemfasciata* (Albrecht, 1959; Norris, 1959) and *S. gregaria* (Papillon, 1960, 1970), though in the latter species Norris (1952) reported that clutch size did not differ between egg pods laid by crowded and isolated females. However, this result of Norris (1952) is exceptional; other authors (see later) found that clutch size is larger in isolated than in crowded locusts of this species.

The higher fecundity of isolated females is primarily due to the fact that the number of ovarioles (see Section 6) is higher in isolated than in crowded females of *L. m. migratorioides* (Albrecht *et al.*, 1958, 1959), *N. septemfasciata* (Albrecht, 1959) and *S. gregaria* (Papillon, 1960, 1970; Injeyan and Tobe, 1981a).

The number of ovarioles is determined already in the embryo (Albrecht *et al.*, 1958). Studying *L. m. migratorioides*, these authors found that changes in population density lead to changes in ovariole numbers of the offspring. The effect is cumulative over several generations. For example, Injeyan and Tobe (1981a) reported that the mean number of ovarioles in a crowded laboratory stock of *S. gregaria* females was 110 and it increased to 130 and then to 154, respectively, in the second and fourth generation of isolated locusts, but no further increase occurred in the next (fifth) generation of isolated rearing. This phenomenon is related to the transmission of the effect of population density from the parent generation to the progeny (see Section 16); for certain phase characteristics, such transmission takes place over several generations.

Because of the phase-related differences in the number of ovarioles in *L. migratoria*, *N. septemfasciata* and *S. gregaria*, the clutch size in solitarious females of these species is usually larger than in gregarious conspecifics. Hunter-Jones (1958) showed in the laboratory that two adults of *S. gregaria* kept together in a cage already exert an effect of crowding, and Papillon (1970) confirmed this conclusion. In Hunter-Jones' (1958) experiments, average clutch size was 85 and 67 in egg pods laid by a single female and by a pair (female and male), respectively. Albrecht *et al.* (1958) reported in laboratory studies on *L. m. migratorioides* that the mean number of ovarioles was 103 in hatchlings originating from isolated mothers, but only 83 in those from crowded mothers. These authors also studied the effects of population density of the grandparents and those of the nymphal and adult density. Solitarious eggs are smaller and lighter than gregarious eggs, but because of the larger clutch size, the average mass of an egg pod in *L. m. migratorioides* is about equal to that in crowded and isolated locusts (Albrecht *et al.*, 1958), and the average vitellin content of ovaries with mature oocytes does not differ significantly between crowded and isolated *S. gregaria* females (Injeyan and Tobe, 1981a, p. 100, Fig. 3). Therefore, the smaller size and the reduced vitellin content of the solitarious eggs are compensated by a larger clutch size.

Field data confirmed many of the conclusions obtained from laboratory research. Ashall and Ellis (1962) found that clutch size in non-swarming *S. gregaria* was 95–128, but in swarming conspecifics, it was less than 81. More recently, Abbassi *et al.* (2003b) reported field data from south Morocco; they found a larger clutch size in solitarious than in gregarious *S. gregaria*. Farrow and Longstaff (1986) found in an Australian race of *L. migratoria* that the average number of ovarioles was 104 and 59 in recession and plague population, respectively (see also Section 6 for estimation of these authors on relevant data for *S. gregaria* and *N. septemfasciata*). Latchininsky and Launois-Luong (1997, pp. 7–8, Table 1) present a list of Russian authors with data on the number of eggs per egg pod (clutch size) and conclude that the number of eggs per pod is less in gregarious than in solitarious *S. gregaria*.

Clutch size is also affected by the number of oocytes resorbed. Resorption may occur at any stage of oocyte development, up to, but not including, chorionated eggs (cf. Lūsis, 1963; Phipps, 1966). Non-functional oocytes were found relatively more frequently in crowded than in isolated females of *S. gregaria*; therefore, clutch size in crowded females was even less than that expected due to their smaller number of ovarioles (Injeyan and Tobe, 1981a). Also, Uvarov (1977, p. 515), citing Anderson's thesis, states that a higher proportion of oocytes are resorbed in crowded than in isolated *S. gregaria*.

Resorption rate much depends on the nutrition of the female. Clutch size is influenced by nutritional factors in several ways. Poor nutrition during nymphal life results in smaller adults with a high ratio of undeveloped ovarioles. In the adult stage, poor nutrition leads to oocyte resorption, and in the case of starvation, all oocytes may be resorbed (Stauffer and Whitman, 1997, p. 258) In addition, nutritional factors may affect the life span, therefore, exerting an effect on fecundity and may also affect the mass of the individual eggs, influencing fertility.

Jackson *et al.* (1978, p. 9, Table 8) presented data on fecundity and fertility of crowded *S. gregaria* fed on various natural food plants. Feeding on *P. typhoides* (Burm.) or *D. glaucum* Oecn. resulted in high fecundity and fertility, whereas feeding on *Sorghum* sp. and *Panicum turgidum* Forsk. led to absence of oviposition, that is, to zero fecundity and fertility.

Chen (1999, p. 44, Table 3), citing data of Chin *et al.* from 1957 (Chin *et al.* is not listed in the references given by Chen (1999)), presents data on the fecundity of *L. m. manilensis* fed on various plants. The largest clutch size was observed with locusts fed on *Echinochlora* sp., but fecundity was highest in locusts fed on *Sorghum vulgare*, because these laid more egg pods. Feeding adults on *Arachis hypogoea* or *Gossypium hirsutum* resulted in lack of oviposition, even when the hoppers that developed to these adults were fed on *S. vulgare* or *Triticum aestivum*, respectively. Unfortunately, Chen (1999) did not make a distinction between gregarious and solitarious populations.

Simpson *et al.* (2002) showed that in last-instar hoppers of *S. gregaria*, there are phase-dependent differences in nutritional regulation (see Section 17.2). These authors did not report the effect of the different nutritional regulatory strategies on the fecundity of subsequent adults, despite that such effect may be expected because nutritionally unbalanced diets affected the rate of survival (Raubenheimer and Simpson, 1999).

Other factors affect fecundity and fertility. Rainfall or soil moisture is important for species that need damp soil for oviposition, or for species that oviposit into dry soil, but in the egg stage have a diapause or quiescence period that is terminated by wetting the soil, that is, sufficient rainfall in the field. Photoperiod affects fecundity in certain species or geographic strains, and the effect of photoperiod may be density-dependent, as in the case of the 'Palavas' strain of *L. migratoria* (Verdier, 1967); in dense populations of this strain, maturation is inhibited under long days, whereas in isolated females it is not so inhibited. Moreover, in some instances, fecundity is affected by interactions between photoperiod, temperature and density, as in a Japanese strain of *L. migratoria* (Tanaka *et al.*, 1993).

Norris (1950) investigated density-dependent differences in the rate of oviposition and in the total number of egg pods laid by *L. m. migratorioides* during the female's lifetime in the laboratory. She found that the rate of oviposition of sexually mature locusts is higher in isolated than in crowded females. However, in Norris' (1950) experiments, crowded females lived longer and this compensated for their slower rate of maturation (see Section 8.1) and slower rate of oviposition, eventually bringing the average numbers of egg pods per female at both rearing densities to a similar number, approximately seven. Nevertheless, fecundity was lower in the crowded locusts because of the lower clutch size (see earlier).

In a similar study on *S. gregaria*, Norris (1952) found that the rate of oviposition was about one egg pod per week per mature female and was not affected by density. Also, the total number of egg pods laid by a female was similar in isolated and crowded locusts. In contrast, Papillon's (1970) laboratory studies showed a slightly higher rate of oviposition in isolated than in crowded *S. gregaria*.

Albrecht *et al.* (1958) reported that crowding in the laboratory reduced the number of egg pods in *L. m. migratorioides* and the average total number of eggs per female was about 300 and 1000 in crowded and isolated locusts, respectively. Fecundity was also found to be higher in isolated than in crowded *N. septemfasciata* (Albrecht, 1959).

Another factor that affects fecundity of different egg pods laid by the same female is age. Clutch size decreases with ageing in *S. gregaria* (Norris, 1952; Papillon, 1960), also in the field (Roffey and Popov, 1968) and in *L. migratoria* (Norris, 1950; Albrecht *et al.*, 1958); for reviews, see Uvarov (1966) and Stauffer and Whitman (1997). Recently, Maeno and Tanaka (2008c) confirmed the decrease of clutch size with ageing of crowded *S. gregaria* but found that clutch size slightly increases, or does not change, with age in isolated conspecific females (see also Section 16.3 for detailed comments on Maeno and Tanaka's (2008c) findings).

Fertility depends on fecundity minus the number of eggs that failed to hatch. This failure may be caused by intrinsic factors, or by extrinsic ones, for example, parasites or drought. Albrecht *et al.* (1958) observed that in a laboratory strain of *L. m. migratorioides*, failure of hatching occurred in 20% of the eggs laid by crowded locusts, but only in 7% of the eggs laid by a single pair of locusts.

The data accumulating from the literature, presented earlier, indicate that fecundity and fertility are lower in gregarious than in solitarious *S. gregaria*, *L. migratoria* and *N. septemfasciata*, mainly because of the higher number of eggs (and possibly egg pods) produced by the solitarious locusts. Although this conclusion was reconfirmed for *S. gregaria* by field data (Ashall and Ellis, 1962), some caution should be exercised in relation to laboratory results. It is practically impossible to maintain the same environmental conditions for isolated and crowded locusts. In the 1950s and 1960s, isolated locusts were usually kept in glass jars with a muslin or metal mesh lid (Hunter-Jones, 1961, p. 10, Fig. 6; Harvey, 1990, p. 23, Fig. 16). This practice led to an increased humidity in the jars, much above that experienced by crowded locust in considerably better aerated cages (Pener, unpublished observations). This problem was solved later, by using well-aerated small cages for isolated locusts (see for example, Roessingh *et al.*, 1993; Ochieng'-Odero *et al.*, 1994; Hoste *et al.*, 2002c). However, even these improved containers of isolated locusts still do not produce conditions equal to those experienced by crowded conspecifics. In the laboratory, crowded locusts may compete for oviposition sites, especially if the surface of the soil of the vessels for egg laying is limited. Also, the oviposition aggregating effect (see Section 8.4) may lead to too many egg pods laid into the same vessel, and over a certain limit, the increasing number of egg pods per vessel decreases the viability of the eggs (Chamberlain, 1980), presumably because of shortage and competition for oxygen and also because dead eggs may become sources of infection for healthy eggs. Moreover, isolated locusts in the laboratory have no competition for food, but such competition may occur in cages of crowded locusts. These, presumably better, conditions experienced by the isolated locusts may affect reproduction-related parameters such as life span of the female, number of egg pods laid by a female, ratio of resorbed oocytes, number of viable eggs, all influencing fecundity or fertility or both. Some of these differences originating from maintenance of isolated and crowded locusts in the laboratory may exist also in the field, though possibly in modified forms.

In spite of the higher fecundity of isolated (laboratory) or solitarious (field) *S. gregaria*, Cheke (1978), using a theoretical model, concluded that gregarious populations of this species have a faster rate of population increase in the field, because of their synchronous and faster sexual maturation (see Section 8.1). Solitarious *S. gregaria* have a longer period of maturation, and therefore, they are exposed for a longer period to predation and disease. It would be interesting to construct a model for *L. migratoria*, a locust species with a shorter maturation period in the solitarious than in the gregarious phase.

In a later article, Holt and Cheke (1996) modelled *S. gregaria* population dynamics, using the logistic equation, with switches in the values of r (intrinsic rate of increase) and *K* (the carrying capacity of the environment), to mimic phase changes. This model produced the bi-stable pattern of plagues (gregarious phase) and recessions (solitarious phase) as shown by observed records; the duration of plagues was dependent on the parameters' values. These authors defined r_s and r_g as the intrinsic rates of solitarious and gregarious locusts, respectively, keeping $r_s \leq r_g$. With increase of r_g, the duration of plagues became shorter and the variance in plague duration decreased.

More recently, Ibrahim (2001) modelled the population dynamics of *S. gregaria* adapted to metapopulations, by using a modified version of the model of Holt and Cheke (1996). Again, r_s and r_g, which depend on fecundity and fertility, were used (see Section 17.6).

Klass *et al.* (2007a) constructed a model for evaluating the effects of environmental temperature and thermal behaviour on fungal biocontrol (by *M. anisopliae* var. *acridum*, see earlier) of several locust species (*Locustana pardalina*, *N. septemfasciata*, *Chortoicetes terminifera*, *Dociostaurus maroccanus*, *Calliptamus italicus* and *Oedaleus senegalensis*) and a grasshopper pest (*Zonocerus variegatus*). The model used mortality rate data generated across a range of temperatures in the laboratory. It may be recalled that the efficiency of this fungal pathogen of locusts is reduced by behavioural fever or high temperatures, because the fungus is more sensitive to such thermal conditions than the locust host (see earlier). The predictions of the model were validated against field data and were found to be quite accurate. Klass *et al.* (2007a) did not investigate any phase-related aspect of the subject.

Despite species-dependent differences, the general trend in fecundity and fertility, especially the larger number of ovarioles in the solitarious than in the gregarious phase, is quite similar *in L. migratoria*, *S. gregaria* and *N. septemfasciata*. However, not all species of locusts have this general trend.

The Australian plague locust, *C. terminifera*, is a polyvoltine species that, according to Farrow (1977a, p. 458, Fig. 3), has about three generations per year, spring, summer and autumn generations in the field. Hunter *et al.* (2001) state that there are three or four generations per year. The rate of reproduction is much affected by the season and the rainfall. Eggs laid in the spring develop directly when soil moisture is available; if not, the eggs undergo a quiescent state. However, depending on the day length and temperature experienced by the parent generation (Wardhaugh, 1980a), the autumn layings (March–April in Australia) contain variable portions of diapause and non-diapause eggs in the same egg pod. The inception of the egg diapause depends also on the temperature and moisture experienced by the eggs after oviposition. This complex way of egg development, which may include quiescent states in both non-diapause and diapause eggs, is summarized in a diagram by Wardhaugh (1980b, p. 190, Fig. 1). Wardhaugh's (1980b) findings were based on egg pods laid by females, which had been obtained from a group collected in the field, but then each female was kept separately in the laboratory. Rain and humidity play a crucial role in the life cycle of *C. terminifera* (for a scheme used in a simulation model, see Wright, 1987, p. 426, Fig. 3), affecting egg development (see earlier), survival, oocyte resorption, migration and ability of migratory flights as mediated by rain-dependent presence of grass (Hunter, 1989).

Although there are many reports on plague development and plague history of *C. terminifera* (Magor, 1970; Farrow, 1977a; Symmons and Wright, 1982; Wright, 1987; and others; for a tangentially related review, see Hunter, 2004), comparative data on solitarious and gregarious locusts are confined mostly to the study of Farrow and Longstaff (1986), who distinguished between recession (presumably solitarious), outbreak and plague (presumably gregarious) populations. The mean number of ovarioles, about 72 (Farrow and Longstaff, 1986, p. 212, Table 1), was found to be similar in plague and recession populations, in spite of the somewhat smaller body size of the recession population (see Section 5.1). Clutch size rarely attained the number of ovarioles, presumably due to differential oocyte resorption, and clutch size was positively correlated with rainfall during the maturation period of the locusts, through the effect of the rain on the availability of green grass. In all three conditions of population density, clutch size was smaller in the autumn generation than that in the spring and summer generations (Farrow and Longstaff, 1986, p. 214, Table 2). In contrast to *S. gregaria*, clutch size was similar in the first, second and third egg pod laid by the same female, regardless of recession or plague populations (Farrow and Longstaff, 1986). The inter-oviposition period (the time elapsing between the laying of two consecutive egg pods) in migrating swarms is longer than in comparable solitarious females (Farrow, 1977a). Based on this and on some other evidence, Farrow and Longstaff (1986) concluded that developmental processes are slower in gregarious than in solitarious *C. terminifera*. However, the reproductive rate per generation (no. of female offspring:no. of female parents) is highly variable; Farrow (1979) obtained values between a maximum of 28.08 (his plot no. 3, autumn breeding, 1973) and a minimum of 0.33 (plot no. 1, summer breeding, 1972). Farrow (1979) also implied that a female may lay a maximum of three egg pods in the field.

In northern Australia, the spur-throated locust, *Austracris guttulosa* (Walker), formerly known as *Nomadacris guttulosa* (cf. Key and Rentz, 1994), has one generation per year (Hunter and Elder, 1999; Hunter *et al.*, 2001; and references in these articles). During the dry season (winter), the species exhibits a reproductive diapause, which is characterized by absence of vitellogenic growth of the oocytes in females and immature accessory glands in males (Elder, 1991, using the name *Nomadacris guttulosa*). Vitellogenic development of the oocytes commences when day length exceeds 13 h and the adults feed on green vegetation (Hunter, 1997). Temperature and humidity also play some role (Elder, 1991; Hunter and Elder, 1999). To complete vitellogenesis and oviposition, feeding on green vegetation is necessary, and for such vegetation to grow on cracking clay soil, over 40 mm of rainfall is required. In addition, to ensure feeding and survival of the resulting offspring, another rainfall, again over 40 mm, is needed. This second rain should follow the former rain within 6 weeks (Hunter and Elder, 1999).

Substantial outbreaks of *A. guttulosa* take several years to develop because the species is univoltine and requires regular rains. However, rainfall is variable in Australia; therefore, major plagues of this species are rare; they occurred in 1973–1975 and in 1996 (Hunter *et al.*, 2001).

Farrow (1977b), investigating reproduction-related effects of population density, compared three populations of *A. guttulosa*: swarms, high-density infestation and low-density infestation. He found that maturation, assessed by oocyte length, commenced earlier in high- and in low-density populations than in swarms. However, oocyte development, once it started, was quicker in swarms, so that ovulation eventually took place approximately at the same time in all three populations Average number of ovarioles was 156 in swarms, compared with 160 and 161 in high- and low-density infestations. However, the number of ovarioles was found to be affected by the population density of the parent and grandparent generations in *L. migratoria* and *S. gregaria* (see earlier and Section 6); therefore, variations in ovariole numbers in *A. guttulosa* may reflect density variations of previous generation(s). Average clutch size of locusts of low-density infestation was 136 as compared with 131 and 132.6 of locusts of high-intensity infestation and swarms, respectively (Farrow, 1977b, p. 35, Table 2). However, in contrast to *L. migratoria* and *S. gregaria*, eggs of *A. guttulosa* of low-intensity infestation were larger, 5.46 mm, than those of high-density infestation, 5.33 mm. These differences may reflect just differences in feeding of the two populations. Without making a distinction between population densities, Farrow (1977b) found that a female lays at least three egg pods and estimated that up to five egg pods may be laid per female.

A study by Elder (1997) showed that *A. guttulosa* does not exhibit clear phase polyphenism. Oviposition was observed to be dispersed, in contrast to oviposition aggregation found in *S. gregaria* and in some other locusts (see Section 8.4 and Uvarov, 1977, with references therein). The hoppers of *A. guttulosa* do not form dense bands. Swarms are made by overwintering adults in reproductive diapause, but the swarms are dispersed once sexual maturation has commenced in the field. It may also be mentioned that in another study, Elder (1996) found no density-dependent differences in the E/F and F/C morphometric ratios. It seems, therefore, that *A. guttulosa* represents a species that may be considered as a locust when forming swarms in the state of reproductive diapause, but always a grasshopper (albeit a grasshopper pest) in other stages of its life cycle.

Possibly, the case of *A. guttulosa* (see earlier) and that of the Brazilian *Rhammatocerus schistocercoides* (Lecoq and Pierozzi, 1996a; Lecoq *et al.*, 1999; see also Section 3.1 of the present review) indicate some intermediate stages in the evolution of full-scale locust phase polyphenism.

The Moroccan locust, *D. maroccanus* (Thunberg), is a univoltine species, with a long period of egg development lasting 8–10 months and including dormancy periods (Bodenheimer and Shulov, 1951; refer reviews by Uvarov, 1966, 1977; refer recent articles by Quesada-Moraga and Santiago-Álvarez,

2000, 2001b; Santiago-Alvarez *et al.*, 2003). The distribution of the species is Mediterranean, with extensions to the east (Uvarov, 1977, p. 461, Fig. 254), where it often constitutes a serious pest (see Section 1.2 for references). An improved map of distribution and an illustration of the life cycle as depending on the latitude are presented by Latchininsky and Launois-Luong, (1992, Fig. 4, p. 20, distribution, Fig. 20, p. 73, life cycle). Adults are found in the field in the spring and early summer.

Consulting Uvarov's (1966, p. 356; 1977, p. 475) brief descriptions of phase-dependent differences in fecundity for *D. maroccanus* does not reveal a clear picture. Some authors claimed that fecundity of solitarious *D. maroccanus* was lower than that of gregarious conspecifics, but data from other authors led Uvarov (1966, 1977) to consider the possibility that the fecundity of solitarious locusts exceeds that of the gregarious individuals. Latchininsky and Launois-Luong, (1992, p. 74) also noted that solitarious females appear to lay more frequently than gregarious females. However, the average clutch size was found to be higher in egg pods laid by females in a swarm, than in those laid by a less dense population. In contrast to *L. migratoria*, *S. gregaria* and *N. septemfasciata*, gregarious *D. maroccanus* are larger than conspecific solitarious locusts (see Section 5.1 and Uvarov, 1977, p. 471, Fig. 257), and because of this size difference, the gregarious females seem to have a higher number of ovarioles. Quesada-Moraga and Santiago-Álvarez (2001a) reported that the closer the female to the gregarious phase, the larger the number of ovarioles. However, the authors assessed the phase state only by the morphometric ratio E/F (connoted T/F according to their terminology), but did not base their conclusions on the actual density of the populations, neither that of the parent generation (see also Section 6). They found a considerable individual variations in the number of ovarioles; nevertheless, females with an average E/F ratio of 1.48 (a solitarious value) showed 31.3 ovarioles on average, whereas those with an average E/F ratio of 1.61 exhibited 36.5 ovarioles. Despite the high variation, the difference was significant. Quesada-Moraga and Santiago-Álvarez (2001a) obtained an average clutch size of 30–31 eggs in field-collected egg pods, without reporting the phase state of the parents. In conclusion, we still do not know all aspects of phase-related differences in the fecundity of *D. maroccanus* and relevant research is obviously needed.

The brown locust, *Locustana pardalina* (Walker), is an economically important pest in southern Africa. Its outbreak area covers the semi-desert Karoo region and southern Namibia and its invasion area extends to South Africa, Namibia, Botswana, up to Zimbabwe, western Mozambique and south-eastern Angola (map by Steedman, 1990, p. 108). The outbreaks and plagues of this locust are well documented (Lea, 1972; Price and Brown, 2000; Todd *et al.*, 2002; and references in these articles).

Under favourable weather conditions, the species may have two to three generations per year; up to four overlapping generations are thought to be possible (Uvarov, 1977, p. 316). However, in the case of drought, the eggs in the dry soil survive for long periods; although survival diminishes with time, if water becomes available, a few eggs hatch even after 3 years. Such egg survival may result in biannual or even triannual life cycle (Uvarov, 1977, p. 318). Based on ample references, Todd *et al.* (2002) present a recent summary of the life cycle of *L. pardalina*.

As in *C. terminifera* (see earlier), diapausing and non-diapausing eggs may be present in the same egg pod of *L. pardalina*. If non-diapause eggs absorb sufficient water, they hatch directly (Matthée, 1951). However, if non-diapause eggs experience shortage of water, or in the case of diapause eggs, embryonic development becomes complex, including quiescent and diapause states, as summarized in a diagram by Matthée (1951, p. 19, Fig. 4). This author reported that solitarious females lay 100% diapause eggs, whereas gregarious females lay only 42% diapause eggs. However, in another article, Matthée (1953) found 95%–100% diapause eggs laid by swarming, presumably gregarious, females. Therefore, the relation between phase and kind of eggs (diapause or non-diapause) is not clear.

As in *D. maroccanus* (see earlier), adults of gregarious *L. pardalina* are larger than conspecific solitarious adults (see also Section 5.1) and it seems that clutch size is larger in egg pods laid by gregarious females. Faure (1923) reported an average clutch size of 27.6 eggs (range, 2–68) in egg pods laid by females caged under isolation, whereas the average clutch size was 47.5 (range, 26–82) in egg pods laid by females from swarms. Uvarov, (1966, p. 357, Table 104) indicated that increased density leads to a larger clutch size in *L. pardalina*. Additionally, it seems that not only the clutch size but also the size of the individual eggs laid by solitarious females are smaller than those laid by gregarious females (Lea, 1962). The clutch size of 27.6 eggs from isolated females (see earlier) seems to be very small in light of Venter's (1966) findings, who reported that ovariole numbers of solitarious *L. pardalina* varied from 53 to 74. Venter (1966) also found that average ovariole numbers of five different solitarious populations varied between 60.4 and 66.4, perhaps as affected by the available forage of their former generations. The low clutch size reported by Faure (1923) may be explained either by drastic resorption of the oocytes or by a much higher variation in the number of ovarioles between different solitarious populations than that found by Venter (1966).

In a recent study, Arthurs and Thomas (2000), investigating the effect of the entomophagous fungus *M. anisopliae* var. *acridum* on the fecundity of *L. pardalina*, reported data on the fecundity of the controls (locusts not infected with the fungus). The control locusts were taken from a smaller band of newly fledged adults than the band of newly fledged adults from which locusts were taken immediately after field spraying. Both control and infected locusts were kept in field closures that provided both sunny areas and shade. Clutch size from younger control locusts was 34.4 ± 7.5 (mean $\pm$ s.e.) and it decreased with the age of the females to 28.5 ± 5.3. In contrast, the average number of egg pods per control female increased from 0.37 in young locusts (0–39 days after fledging) to 0.85 in older ones (50–61 days after fledging). Altogether, approximately two egg pods were laid per control female during the first 61 days after fledging, but the authors note that the experiment was stopped before the insects reached their full capacity of oviposition and cite various authors to claim that *L. pardalina* may produce four to nine egg pods in the field. Interestingly, treatment with the fungus led to earlier egg laying and a higher number of egg pods per female than those obtained from the controls, but because of a higher rate of mortality of the infected locusts, actual fecundity was similar in the two groups. The authors observed no significant difference in clutch size, neither in viability (fertility) between infected and control insects. Mortality was much affected by rearing conditions; all infected locusts kept in laboratory cages died within 10 days, but infected locusts maintained in the field enclosures survived much better and some of them were still alive on day 61 after fledging when the experiment was stopped.

Waloff and Pedgley (1986) as well as Hunter and Cosenzo (1990) reviewed the biogeography, biology and plague history of the South American locust, *Schistocerca cancellata* (Serville), formerly known as *S. paranensis* (see Section 1.2; and Harvey, 1981). The recession area of this locust is north-western Argentina (east of the Andes), south-eastern Bolivia and western Paraguay; its invasion area includes southern Bolivia, south-eastern Brazil, through Paraguay, Uruguay and Argentina to about 43°S (references and map by Waloff and Pedgley, 1986, pp. 2–3, Fig. 1). According to the summary of plague and outbreak history by Hunter and Cosenzo (1990), plagues were observed during 48 of the 58 years between 1897 and 1954. After this period, several outbreaks occurred, mostly in the Argentine provinces La Rioja and Catamarca that are considered to be major outbreak areas. Using available data and the rate of development of *S. cancellata*, Hunter and Cosenzo (1990) concluded that plagues developed when weather conditions were favourable for three generations per year and continued for as long as weather conditions permitted two or three generations per year. Presently, control measures made against outbreaks prevent development of plagues.

More recently, Sanchez *et al.* (1997) investigated the life cycle and fecundity of *S. cancellata* in the laboratory. The authors collected in the field older gregarious hoppers, maintained them in the laboratory at 30 °C and L:D 14:10 photoperiod up to oviposition of the adults and used the subsequent generation that hatched from the eggs for their study. Sanchez *et al.* (1997) stated that they investigated locusts in the gregarious phase, but the adults studied did not experience heavy crowding, as only 10 locusts (five males and five females) were kept in each of four separate cages. For *S. gregaria* and *L. migratoria*, 10 locusts in a cage constitute a crowd, but the density required for well discernible phase change is not known for *S. cancellata*. Therefore, it is possible that the authors investigated a transiens (dissociating) population. They found that fecundity was 160.7 ± 93.1 (mean $\pm$ s.d.) eggs per female and clutch size was 73 ± 18.9 eggs. From data on survivorship and fecundity, an age-specific life table model was constructed from which R_0, the net reproduction rate, meaning the average number of female offspring produced per female per generation, and other parameters were calculated. The value of R_0 obtained was high, 130.4 ± 9.7 (mean ± s.d.). Obviously, more research is needed to clarify phase-dependent differences in fecundity and fertility of this species.

The Central American locust, *Schistocerca piceifrons* (Walker), is distributed in tropical America. Its biology, biogeography and plague history was reviewed by Harvey (1983). According to him, there are two subspecies, *S. p. piceifrons* and *S. p. peruviana*. Solitarious populations are found in several permanent breeding areas, which may be considered as outbreak areas, especially the Yucatan Peninsula in Mexico, northern Honduras, and Guatemala in Central America and Peru in South America. The invasion area is larger; it covers Mexico, Belize, Honduras, El Salvador, Nicaragua, Costa Rica and Panama, as well as Peru, Ecuador, Columbia, Venezuela, Trinidad and western Guyana. For description and maps of the distribution, see Harvey (1983, pp. 178–180, Figs. 7 and 8). *S. p. piceifrons* seems to have two generations per year, and immature adults survive the dry season in a reproductive diapause.

Laboratory data on the fecundity of *S. p. piceifrons* (under the name of '*Schistocerca* sp.') were reported by Hunter-Jones (1967). In eight different groups of locusts, each with five males and five females in a 12-L cage, the number of egg pods laid per female was 2.0, 3.6, 4.8, 5.0, 5.8, 8.4, 8.5 and 8.8, yielding an average of 5.9 egg pods per female. Ten pairs of locusts, one male and one female in a 12-L cage, laid 8.4 ± 4.3 (mean ± s.d.) egg pods per female. A similar experiment resulted in 8.0 ± 5.8 egg pods per female. A parallel experiment with 10 individually isolated females that had been exposed to a mature male for 24 h laid 9.8 ± 6.4 egg pods per female. This value did not differ significantly from that obtained from pairs of locusts. Clutch size in crowded locusts was 70.8 ± 10.7 (mean ± s.d.); it was 70.2 ± 14.2 for pairs and 62.3 ± 13.8 for isolated females. These three averages were not significantly different. The rate of egg pod production was found to be independent of density.

Harvey (1983) cited the results of Hunter-Jones (1967). It is noteworthy that the difference in clutch size, albeit not significant, indicates a trend opposite to that found in *S. gregaria*, *L. migratoria* and *N. septemfasciata*, which have a larger clutch size in egg pods laid by isolated than by crowded females (see earlier).

9 Endocrinology

9.1 JUVENILE HORMONE

The role of juvenile hormone (JH) in locust phase polyphenism is subject to a major debate. Is it the primary intrinsic causal factor in locust phase polyphenism, inducing and maintaining the solitarious phase? Alternatively, is it just a minor factor that does not stand at the beginning of the process of phase transformation, but rather at the end of it? Is it a cause, or a result? JH does induce certain, though not all, solitarious phase characteristics; however, the importance of this action is controversial and JH even induces certain gregarious phase characteristics. The debate has been ongoing for about half a century and a brief description of its history is presented in the following paragraphs, leading to some necessary repetition of the review material already published by the senior author (Pener, 1983, 1990, 1991; Pener and Yerushalmi, 1998).

In the 1950s and early 1960s, Kennedy (1956, 1961, 1962) suggested and advocated a working hypothesis ''that the solitary phase is somewhat juvenile or neotenous compared with the gregarious phase which is more adult, as the result of a difference in the hormone balance during development. During the development of the *solitaria* the balance is shifted in favour of the effects of the juvenile hormone ...'' (Kennedy, 1956, p. 353). This working hypothesis of Kennedy was based on the extra nymphal moult exhibited by a portion of solitarious individuals of *S. gregaria* and *N. septemfasciata* (for references, see Section 5.2) and on the first relevant publications by P. Joly and L. Joly (1954) and L. Joly (1954), who had shown that implantation of the corpora allata (CA), a pair of endocrine glands that secrete the JH, induces solitarious green colouration and may lead to 'hypersolitary' E/F ratio in *L. migratoria*. Solitarious adults are relatively brachypterous as their E/F ratio is lower than in gregarious adults, and this fact served as an additional argument for Kennedy's working hypothesis (see Section 5.2 on the relation of the E/F ratio to phase and Section 7.2.1 for induction of green solitarious colour by CA implantation or JH or JH analogue (JHA) administration). It may be noted that before the chemical structure of JH was revealed, CA implantations had often been used to investigate effects of JH excess; therefore, the abbreviation 'CA/JH', later, means experiments employing CA implantation and (usually later and often by different authors) JH or JHA administration.

In spite of his own working hypothesis, Kennedy already reasoned that it is difficult to accept the suggestion of P. Joly and L. Joly (1954) that the CA constitute the ''véritable déterminant interne des phases'' (partial sentence cited by Kennedy (1956, p. 353) from the article of P. Joly and L. Joly (1954, p. 345)). Indeed, as already cited by Pener (1991), after additional studies, P. Joly (1962, p. 77) himself concluded that the ''problem of physiological determination of locust phases can not be explained on the basis only of differential activity of the corpora allata''.

However, the concept that higher CA activity and higher JH titres induce the solitarious phase in locusts gained acceptance (for some references, see Pener, 1991, p. 44), though some cautious authors expressed restrictions. Thus, for example, Rowell (1971, p. 182) firmly concluded that the ''simple apposition of the main endocrine glands is clearly an oversimplification even if coloration alone is considered and is certainly not a valid general statement on the endocrine basis of phase''.

The relevant history, findings and opinions in relation to CA/JH and locust phases were thoroughly analyzed by Pener (1983, 1990, 1991, pp. 37–45) who even tabulated (Pener, 1990, pp. 43–45, Table 1; Pener, 1991, pp. 41–43, Table 1) the effects of CA/JH on locust phase characteristics, distinguishing three alternatives, CA/JH: (1) promote solitarious features; (2) promote gregarious features and (3) do not exert a phase-related effect, or the effect is not clear. Pener's (1983, 1990, 1991) detailed analysis of the subject is not repeated here, but reference is made to the sections of the present review, which discuss the effects of CA/JH on certain phase characteristics: Section 7.2.1, green solitarious colour is induced by CA/JH; Section 7.2.2, combined effects of JH and DCIN on nymphal colouration; Section 7.3, induction of gregarious yellow colour with sexual maturation of adult males; Section 8.1, acceleration of sexual maturation of the adults (a solitarious characteristic for *L. migratoria*), induction of male pheromone production in *S. gregaria* (a gregarious characteristic) and maturation-retarding effect of young or allatectomized adults; Section 8.2, effect of CA/JH on male mating behaviour in different species, allatectomy abolishes or reduces the intensity of the mating behaviour in crowded *L. m. migratorioides*; such a reduction may be considered as a solitarious phase characteristic (cf. Pener, 1976a).

Pener eventually concluded that ''there is no clear evidence that endocrine factor(s) constitute(s) the physiological prima causa in locust phase transformation. Most of the information is related to the effect of the CA-JH that do not necessarily play a fundamental role'' (Pener, 1983, p. 390) and again ''Very probably, CA activity and JH titres stand not at the beginning, but somewhere in the middle, of a chain of events and physiological causal factors which are responsible for phase transformation. Thus, the hypothesis that the solitary phase is just a neotenous form induced by permanent or even specifically timed higher JH titres seems to be at best an oversimplification'' (Pener, 1991, p. 45).

It may be added that the green colouration is not a necessary characteristic of the solitarious phase, as demonstrated by the very existence of the solitarious 'brown' or homochrome colouration (Fig. 1A and 1B). Also, green-brown colour polyphenism and induction of the green colour by CA/JH is not specific to locusts (Pener, 1991); certain locust species do not exhibit colour polyphenism (*Dociostaurus maroccanus*, *Chortoicetes terminifera* and some others), whereas many species of non-locust grasshoppers do show green-brown colour polyphenism, with a definite green colour inducing role of the CA/JH (for some examples, see Section 7.2.1).

Other reviews dealing with locust phase polyphenism reflect the debate on the relevant role of the JH. Nijhout and Wheeler (1982) advocated a model of JH-induced gene switching mechanism, stating that ''phase differentiation depends largely on the presence or absence of JH at a critical period ...''

(Nijhout and Wheeler, 1982, p. 117). However, other authors were more cautious and outlined the difficulties in accepting the primary and sole role of JH in locust phase polyphenism. Hardie and Lees (1985, p. 458) stated that ''Although JH evidently has an important role to play in the development of pigmentation, the evidence does not necessarily indicate that this hormone has a causal role in the control of phase polymorphism''. Dale and Tobe (1990. p. 398) concluded that ''some phase characters have been shown to respond to manipulations of the JH titer but this was not possible for others, suggesting that differential JH titer cannot be the only factor of importance in the physiological control of locust phase polymorphism''. These reviews reliably cite in detail the (then) available data and references on which their conclusions are based.

The devastating plague of the desert locust in 1986–1989 (for references see Section 1.2) rekindled interest in locust research, including research on locust phase polyphenism, resulting in many new findings in relation to JH and locust phases.

Pener *et al.* (1992) showed that precisely timed precocene-induced chemical allatectomy of fourth-instar green isolated hoppers of *L. m. migratorioides* leads to disappearance or drastic reduction of the green colour, and the resulting fifth-instar nymphs exhibit solitarious homochrome colouration, without any component of the gregarious hopper colouration. The same treatment did not alter the gregarious colouration of crowded nymphs nor the homochrome colouration of isolated nymphs (for some additional details related to these experiments, see Section 7.2.1). In isolated hoppers, therefore, JH deficiency just shifts one solitarious characteristic (green colour) to another solitarious characteristic (homochrome colouration).

The behaviour of gregarious and solitarious locusts is vastly different, and it is well known that the first effect of change in density, especially from isolation to crowding, is change in behaviour (Ellis, 1959, 1963b and other publications by this author; more recently reconfirmed by Roessingh *et al.* (1993), as well as by Roessingh and Simpson (1994), and discussed in Section 11 of the present review). Not surprisingly therefore, the effect of JH on phase-related behaviour was studied by several authors.

Wiesel *et al.* (1996) investigated in the laboratory the effect of JH III (the authentic JH of locusts and apparently the whole order of Orthoptera, (cf. Trautmann *et al.*, 1974; Huibregtse-Minderhoud *et al.*, 1980; Bergot *et al.*, 1981; Loher *et al.*, 1983; Pener *et al.*, 1986) and fenoxycarb (a JHA), as well as two other JHAs, connoted KA 4580 and BASF 228743, on three phase-related behavioural patterns in *S. gregaria* and *L. migratoria*. The behavioural patterns studied were (1) aggregation (grouping) of hoppers; (2) marching of hoppers and (3) 'aggression', meaning reaction to confrontation with other individuals of hoppers or of adults. They found that JH III and JHAs dose-dependently and significantly reduced aggregation of crowded hoppers in both species, suggesting a solitarizing effect. In contrast, marching behaviour, a gregarious characteristic, was stimulated by administration of JHAs; the so intensified marching behaviour surpassed that of the crowded controls. Crowded control locusts showed mostly 'weak' responses in confrontation with fellow locusts, whereas isolated control locusts exhibited mostly 'strong' reaction. After treatment with JHAs, the test locusts showed an increase of strong reaction and a slight decrease of weak reactions; these responses may be considered as a solitarizing effect. Wiesel *et al.* (1996) found similar trends in both *S. gregaria* and *L. migratoria*, though the latter was slightly less responsive than the former. JH III was generally less effective than JHAs.

Applebaum *et al.* (1997) studied the effect of a JHA, methoprene, on the nymphal behaviour of *S. gregaria* and *L. m. migratorioides* by applying topically in acetone 50 μg of the substance to newly moulted fourth-instar crowded hoppers. They defined two variables to assess the effect: 'activity level', which is the duration of time spent in movement within a 2-min period, and 'associative index', which reflects the positional relations (grouping) of the hoppers. Details of the method and construction of a shifting scale of phase index from the foregoing variables are described by Heifetz *et al.* (1996) (see Section 11.1). Five hours after methoprene treatment, the gregarious nymphal behaviour was shifted towards solitarious behaviour in both species. However, the effect was temporary; 72 h after the treatment, the hoppers of both species exhibited gregarious behaviour. At this time, treated hoppers of *L. m. migratorioides* were even significantly more active than the controls, and the associative index was unaffected by the methoprene. The authors argued that Wiesel *et al.* (1996) (see earlier) did not test long-term effects of the JHAs, and this may explain the different conclusions of the two studies; also, the assay procedures were different. However, it cannot be ruled out that, despite relative persistence of methoprene, its activity was partially or completely lost 72 h after the treatment (see later).

Applebaum *et al.* (1997) also examined the effect of methoprene on nymphal colouration and the F/C ratio of the subsequent adults. The colouration of the crowded nymphs was shifted towards green or greenish tints in a portion of the hoppers of both species. The F/C ratio of the adults remained 'gregarious-like' in both species, reconfirming P. Joly's (1962) old conclusion that the F/C ratio is not affected by CA/JH. However, the gregarious-like F/C ratio implies appearance of morphologically normal adults after methoprene treatment in the fourth nymphal stadium. If so, such appearance may indicate that 50 μg of methoprene topically applied in acetone was a very small dose, especially for *S. gregaria*, and its activity might has been lost or reduced 72 h after the treatment. The green or greenish colour observed in a portion of the fifth-instar hoppers might have been induced in a short time window after the treatment. A sufficiently high dose of methoprene interferes with adult morphogenesis and results in metathetelic adultoids in the sixth stadium instead of normal adults. Such adultoids may even exhibit an extra moult from the sixth to the seventh (extra) stadium (cf. Pener *et al.*, 1997a).

In another experiment, reported in the same publication (Applebaum *et al.*, 1997), isolated newly moulted fourth-instar hoppers of *L. m. migratorioides* were chemically allatectomized by precocene. As expected, after the next moult, prothetelic adultiforms (precocious adults; cf. Pener *et al.*, 1978) were obtained. Although presumably lacking JH, the behaviour of the adultiforms remained solitarious and no induction of gregarious behaviour occurred. Applebaum *et al.* (1997, p. 387) eventually concluded that "the specific behavioral phase characteristics examined in our bioassay are not controlled by JH ..." and "it is not yet clear in all cases that JH is involved in the induction of other phase characteristics, and, if so, whether its effects are direct or indirect".

Ignell *et al.* (2001) investigated the effect of JH on behavioural patterns of adult *S. gregaria* in response to phenylacetonitrile (PAN), which is claimed to be the most potent adult aggregation pheromone component in this species (Mahamat *et al.*, 2000; but see Sections 8.1, 12.3, 12.4 and 12.5 for controversial opinions on the role of PAN).

Ignell *et al.* (2001) also studied the effect of JH on the responsiveness of olfactory interneurones in the antennal lobe to several substances, considered to be pheromones or pheromone components in adult *S. gregaria*. These effects are discussed in Section 12.3. The present section deals only with the behavioural effects of allatectomy and JH injection.

Ignell *et al.* (2001) found that adults injected with JH III on day 1, 2 or 3 and tested on day 8 after fledging exhibited little or no effect, except that females exposed to PAN showed a significantly lower aggregation index than female controls. However, Fig. 1B of Ignell *et al.* (2001) reveals that the aggregation index of JH-injected females did not differ much from that of allatectomized females and the aggregation index of the latter was also significantly lower than that of the controls. In contrast to females, allatectomized males exposed to PAN showed significantly much higher aggregation index than male controls at the age of 15 and 22 days after fledging. Also, allatectomized males exhibited a significantly higher level of aggregation than allatectomized females. Leg movements and antennal movements were also significantly higher on day 15, 22 and 29 in allatectomized males exposed to PAN than in male controls. These results may indicate induction of some gregarious behaviour of the allatectomized males, which would mean that JH has a solitarizing effect. However, the results are open to an alternative interpretation. Ignell *et al.* (2001) do not cite the study of Seidelmann *et al.* (2000) on the rival male repelling effect of PAN (see Sections 8.1 and 12.5), probably because their article that appeared in January 2001 was already with the reviewers when the study of Seidelmann *et al.* (2000, August) was published. PAN production seems to be under the control of JH (cf. Tawfik *et al.*, 2000). If so, no PAN is produced by allatectomized males. Therefore, control males of Ignell *et al.* (2001) were exposed to PAN provided by the experimental design, plus to PAN produced by themselves. Allatectomized males were presumably exposed only to the former; therefore, by not repelling each other, they exhibited a higher level of aggregation. Moreover, the difference found at 15 and 22 days after fledging between allatectomized and control males was not found at the age of 8 days; presumably because the controls were too young to emit PAN (cf. Tawfik *et al.*, 2000).

JH controls vitellogenin synthesis in the fat body of adult female insects and uptake of vitellogenin by developing oocytes (refer review by Wyatt and Davey, 1996). The higher fecundity of solitarious *L. migratoria*, *S. gregaria* and *N. septemfasciata* than that of the respective gregarious conspecifics (see Section 8.5) as well as certain effects of implanted CA were often regarded as evidence for promotion of solitarious phase characteristics by JH (refer reviews by Dale and Tobe, 1990; Pener, 1991, pp. 24–26). However, most of these findings ''may be related to the vitellogenetic-gonadotropic effects of the CA without any phase shift'' (Pener, 1991, p. 24).

Employing the old *Galleria* bioassay, L. Joly and P. Joly (1974) found higher haemolymph JH titres in isolated than in crowded fourth- and fifth-instar hoppers of *L. migratoria*. They also found that in isolated young adults, JH titres increased more rapidly with age than in crowded adults; nevertheless, eventual titre values were only slightly higher in the isolated adults (see figure in their paper, in which curves A and B seem to be mixed up; according to the text, A should show isolated locusts and B crowded ones).

Using the more reliable method of gas chromatography–mass spectrometry (GC–MS), Dale and Tobe (1986) found a higher haemolymph titre of JH, 4 days after fledging, in isolated than in crowded adult females of *L. migratoria*. In the same study, these authors also assessed JH biosynthetic activity of the CA *in vitro* and found that it was similar in crowded and isolated females of *L. migratoria* within the first 5–6 days after fledging. However, on day 8, the activity was much higher in isolated than in crowded females. Unfortunately, Dale and Tobe (1986) did not assess JH titres and JH biosynthetic activity of the CA in adult females older than 8 days after fledging. It is known that sexual maturation time of isolated adults of *L. migratoria* is shorter than that of crowded adults (see Section 8.1). Therefore, it is well possible that the authors stopped their observations before the crowded females reached maximum values, which might have been similar to the maximum values reached earlier by the isolated females.

Injeyan and Tobe (1981b) assessed by radiochemical assay, *in vitro*, JH biosynthetic activity of the CA in relation to vitellogenesis in *S. gregaria*. The activity was higher in penultimate and in last-instar isolated female hoppers than in crowded ones, especially on day 2 in the intermoult period. However, variations, as indicated by the standard errors of the means, were extremely high and on some days the differences might not have been significant. If the isolated hoppers tested were mostly green or greenish, a higher biosynthetic activity of the CA would not be surprising since JH induces green colouration (see Section 7.2.1 and Fig. 1A and 1B). Injeyan and Tobe (1981b, p. 205) stated that ''There was little difference in the overall synthetic abilities of the glands of gregarious and solitarious adult females and a great deal of overlap in the individual values''. Nevertheless, the overall mean JH synthetic activity was higher in isolated than in crowded females. Importantly, there were significant temporal differences; CA of isolated females exhibited relatively higher rates of JH biosynthesis early in the first gonotrophic period. The authors observed an earlier appearance of vitellogenin in the haemolymph in isolated than in crowded females, coinciding well with the earlier increase of JH biosynthesis in the former. Vitellogenin uptake by the oocytes also started 2 days earlier in isolated than in crowded females. However, despite the earlier onset of vitellogenesis and its uptake, oocyte maturation time (assessed by presence of chorionated eggs) was about 2 days longer in the isolated than in the crowded females. In crowded and isolated females, maximum vitellin content was found to be similar, about 120 mg per ovary. Injeyan and Tobe (1981b) did not draw a firm conclusion in regard to the causal role of JH in locust phase transformation.

Schneider *et al.* (1995) tested the effects of three JHAs, fenoxycarb, BASF 228743 and KA 4580, as well as JH III on oocyte maturation in adult females of *S. gregaria*. Treatment of crowded females with JHAs and JH III induced an earlier appearance of vitellogenin in the haemolymph and earlier onset of oocyte growth. JH III exerted the weakest effect. JHAs increased the number of mature eggs in the first gonotrophic cycle; in the case of fenoxycarb and BASF 228743, this number was quite similar to the number of eggs in isolated controls. However, all three JHAs induced oversized oocytes (see comment later on this effect).

Schneider *et al.* (1995) found that the first oviposition of crowded controls was observed on day 17 after fledging, whereas that of isolated controls occurred earlier, on day 12. JHAs and JH III administered to the crowded females induced earlier first oviposition, with a period similar or even shorter than that of the isolated controls. However, the authors mention that only in a few cases did first oviposition of the isolated control females occurred on day 11/12, and most isolated females retarded oviposition for some time, up to many days. But accelerated oocyte maturation and earlier egg laying is not a solitarious characteristic in *S. gregaria*. In this species, crowding accelerates maturation (Norris, 1952; Papillon, 1968). Also, in the experiments of Injeyan and Tobe (1981b, see earlier), oocyte maturation time was 2 days shorter in crowded than in isolated females. It is known that the time range of the first oviposition is very variable in isolated *S. gregaria* females (Norris, 1952). Were Schneider *et al.* (1995) assessing the average age of the first oviposition (technically difficult in crowded locusts, but see Norris, 1952), instead of the earliest age of the first oviposition, they probably would have found that the period between fledging and first oviposition is longer in isolated than in crowded *S. gregaria*.

Schneider *et al.* (1995) also reported that JHAs and JH III affected lipid metabolism in *S. gregaria* females, shifting the fresh weight of the fat body towards lower, solitarious, values and reducing the adipokinetic response observed after 60 min of flight. The authors interpreted the effects of JHAs and JH III on reproduction and lipid metabolism as solitarization of the crowded females. However, they found that all three JHAs resulted in oversized oocytes (Schneider *et al.*, 1995, p. 27, Fig. 3) and this is not a solitarious phase characteristic (see Section 8.5). Not only chorionated eggs but also the developing oocytes are larger in crowded than in isolated *S. gregaria* females (Injeyan and Tobe, 1981b) Therefore, in this instance, JHAs promoted a gregarious phase characteristic, and this finding strongly supports the claim (Pener, 1991; see earlier) that, in relation to locust phase polyphenism, the effect of JH is vitellogenetic-gonotrophic, rather than phase shifting.

Tawfik *et al.* (1997a) administered JH III to 3- to 5-day-old last-instar hoppers and, separately, 3- to 5-day-old young adults of crowded *S. gregaria* males, by three alternative routes, topical application in acetone, injection in olive oil and exposure to vapour ('fumigation'). They recorded the effect of the treatments on the timing and rate of PAN release. Topical application of a single dose of 50 μg of JH III to hoppers or young adults was ineffective; the onset of PAN release was similar to that of the controls. Topical application of 50 μg of JH III on each of the days, 3, 4 and 5 after the moult ($3 \times 50 = 150$ μg cumulative dose), to last-instar hoppers induced a large delay in the onset of PAN release, up to 30–35 days in the subsequent adult males, as compared with 10–15 days in the controls. Topical application of 50 μg of JH III to young adults, on each of the days, 3, 4 and 5 after fledging (again, $3 \times 50 = 150$ μg cumulative dose), also induced a slight delay, about 5 days, in the release of PAN. Injection of 3×50 μg of JH III (as earlier) into young adults induced a slight delay, similar to that induced by topical application of the same cumulative dose. Injection of 3×50 μg of JH III into last-instar hoppers resulted in a high proportion of mortality, or in morphogenetically disturbed locusts in the sixth stadium, and the authors did not present PAN release. Fumigation of last-instar hoppers, exposed to 400 μg of JH III vapour, induced maximum delay; the subsequent adult males started to release PAN as late as 35–40 days after fledging. The authors noted that in the case of topical application of 3×50 μg of JH III to the hoppers, as well as in the case of exposure of the hoppers to the vapour of 400 μg of JH III, the resulting adults showed a faded yellow colour instead of the bright yellow colour of the controls (see Section 7.3). The effects of these treatments on haemolymph absorbance ratio, 460/680 nm, were already discussed in Section 7.4. Tawfik *et al.* (1997a, p. 1177) concluded in the abstract of their article "that the effects of exogenous JH III on gregarious locusts represent a shift towards the solitarious phase".

After advocating the above conclusions that JH III causes delays in PAN release and promotes the solitarious phase, the findings reported in a later publication by Tawfik *et al.* (2000) came as a surprise. It may be noted that three (out of four) authors of the Tawfik *et al.* (1997a) and the Tawfik *et al.* (2000) articles were the same persons. Investigating *S. gregaria* adults, Tawfik *et al.* (2000) found consistently higher haemolymph JH titres in crowded than in isolated females. In the males, JH titres were higher in crowded than in isolated locusts on days 15 and 25. They also found that the onset of PAN release by the males coincided with the first high peak of haemolymph JH III titre. These findings imply a gregarizing effect of JH.

The apparent contradiction was selected by Zera (2007) as an example of problems with JH application. He concluded that JH-induced delay in the pheromone production reported in the earlier article of Tawfik *et al.* (1997a) appears to have resulted from a pharmacological effect of the JH due to inappropriately timed application. This conclusion is feasible, but an alternative explanation also seems to be available. It is known that JH maintains larval features; its name comes from this action. The early application of JH in the study of Tawfik *et al.* (1997a) might have been resulted in maintenance of a juvenile characteristic. Hoppers and young adults do not produce PAN (Hassanali *et al.*, 2005a, and references therein), and this absence of PAN production might have been continued longer in the JH-treated locusts. The faded yellow colour of the males (see earlier) could also be explained by the same reasoning. It is known that different target organs may respond differentially to JH (Pener and Lazarovici, 1979), and the system of PAN production and bright yellow colour induction might have been more sensitive to the maintenance of the juvenile characteristics by JH than other systems.

Dorn *et al.* (1997) reported on the effects of field spraying of the JHA, fenoxycarb, against hopper bands of *L. m. capito* in Madagascar. As expected, the spraying induced green colouration and metathetelic morphogenetic malformations (nymph-adult intermediates due to prevention of complete metamorphosis to adults; in mild cases, it is expressed just by deformed, crumpled wings after moult to the sixth stadium). Dorn *et al.* (1997) present data on the behavioural effect of the JHA for one hopper band out of six. In this band, normal marching is continued for 2 days after fenoxycarb spraying, but on day 3, the marching has stopped and the band broke up. Even 23 days after spraying, the locusts were still found in the vicinity where application had taken place; all were in the sixth stadium exhibiting different degrees of morphogenetic malformation. It is questionable whether this result is compatible with the laboratory results on the effect of JHAs on marching in *L. migratoria* and *S. gregaria* hoppers, as reported by Wiesel *et al.* (1996; see earlier), or as observed by Applebaum *et al.* (1997; see earlier). The behaviour as described by Dorn *et al.* (1997) seems to be a pharmacological or metathetelic effect or both, rather than an effect on phase. However, from the practical standpoint of control of locust bands, the effects of fenoxycarb spraying seem to be valuable.

Botens *et al.* (1997) reported phase-related JH III titre determinations in the haemolymph in three geographical races (subspecies) of *L. migratoria*. In one race, *L. m. capito*, from Madagascar, JH titres were investigated in field catches of fourth- and fifth-instar hoppers. The two other races, *L. m. migratorioides* and *Locusta migratoria gallica*, were laboratory-maintained colonies, in which JH titres were determined in isolated and crowded hoppers in all nymphal stadia (except of second-instar crowded hoppers in *L. m. gallica*). It is difficult to read this paper, because the figures, their captions and the reference in the text to the figures are all mixed up. After sorting out these blunders, the following results emerge. In *L. m. migratorioides*, haemolymph JH titres were higher in isolated than in crowded locusts in the first four hopper stadia, but no difference was found in fifth-instar hoppers. *L. m. gallica* yielded similar results (except that second-instar crowded hoppers were not tested); also, change from isolation to crowding, or vice versa, on day 0 of the third-instar, resulted in a change of the JH titre according to the new density conditions. In *L. m. capito*, haemolymph JH titre was higher in the solitarious than in the gregarious hoppers on day 3 of the intermoult period of the fourth stadium. In fifth-instar hoppers, JH titres were low in both phases, except on the last day before moulting to adult, when the titre suddenly increased in the solitarious locusts (according to the figure on page 200 which bears the caption connoted as Fig. 2, but its correct caption is found in the next page (201) as the caption of Fig. 3). No JH titres in adult locusts were reported by Botens *et al.* (1997).

Wedekind-Hirschberger *et al.* (1999) investigated haemolymph polypeptides in a laboratory strain and field catches of *S. gregaria*, revealing that some of the polypeptides are phase-specific. Their findings are discussed in Section 10.2.1. In the frame of this study, they applied topically a JHA, 150 μg of fenoxycarb per locust, to crowded males at day 0 after fledging and found that with sexual maturation ($\geq$15 days after fledging), 9 of 17 gregarious-specific polypeptides were repressed. They also found that two of three solitarious-specific polypeptides were expressed after JHA treatment of crowded males. The physiological roles of these polypeptides are unknown.

In a review on locust phases, Dorn *et al.* (2000) summarized the data on JH titres as assessed by L. Joly and P. Joly (1974) in isolated and crowded *L. migratoria* (see earlier) and the data published by Botens *et al.* (1997) in field-collected solitarious and gregarious hoppers of *L. m. capito*, as well as in isolated and crowded hoppers of the laboratory colonies of *L. m. migratorioides* and *L. m. gallica* (see earlier). Dorn *et al.* (2000) also referred to the results of Dale and Tobe (1986) in regard to JH biosynthetic activity of the CA in adults of *L. migratoria* (up to day 8; see earlier) and to the results concerning the activity of the CA in isolated and crowded fourth- and fifth-instar *S. gregaria*, as reported by Injeyan and Tobe (1981b) (see earlier). Based on all these data, Dorn *et al.* (2000) concluded that JH titres are higher in solitarious than in gregarious locusts and this difference is a legitimate phase characteristic. The authors do not say explicitly that JH is the primary physiological causal factor of locust phase polyphenism. Nevertheless, they say that ''In the endocrine control of phase polymorphism, a central role of JH cannot be doubted'' (Dorn *et al.*, 2000, p. 254). This sentence comes quite close to a conclusion that JH is the causative factor in locust phase polyphenism.

Dorn *et al.* (2000) also arranged data according to phase characteristics. They reported morphometrical parameters of crowded adult males of *S. gregaria*, which had been treated with a JHA in the last hopper stadium and concluded that ''Many parameters and indices of gregarious larvae adopted solitary-like values after the adult moult; others, however, ... behaved contrarily'' (Dorn *et al.*, 2000, p. 223). The F/C value, which is the best morphometrical ratio to reflect phase differences (see Section 5.2), was found to be slightly (probably not significantly) shifted in the adults towards gregarious direction after JHA administration in the last hopper stadium. Also, discriminant analysis of three groups of adults, crowded, isolated and crowded previously JHA-treated locusts, showed that these groups form separate classes that do not overlap. On the contrary, discriminant analysis showed that the JHA exerted an expected juvenilizing effect. Regarding phase-related behaviour, the authors mostly repeat the data and conclusions of Wiesel *et al.* (1996) (see earlier). Curiously, however, in their section on phase behaviour, Dorn *et al.* (2000) do not mention the study of Applebaum *et al.* (1997) (see earlier), although it is included in the list of references.

As already mentioned (see earlier), Tawfik *et al.* (2000), investigating JH titres in adults of *S. gregaria* within the age range of 10–30 days after fledging, found that haemolymph JH titres were significantly higher in crowded than in isolated males at the age of 15 and 25 days, though on day 30 the titre was higher in isolated males. In the females, consistently higher JH titres were found in crowded than in isolated locusts within the whole age range. These results constitute a strong deviation from the tendency based on the data as compiled by Dorn *et al.* (2000) (see earlier).

A recent review by Hartfelder and Emlen (2005) was strongly influenced by the claim of Dorn and coworkers (Schneider *et al.*, 1995; Wiesel *et al.*, 1996; Botens *et al.*, 1997; Wedekind-Hirschberger *et al.*, 1999; Dorn *et al.*, 2000; see earlier) that JH plays a crucial role in locust phase polyphenism and the authors even extended this claim. They state that ''for some time the role of JH has been controversial ...'' (Hartfelder and Emlen, 2005, p. 658), implying that it is no longer controversial. They return to Kennedy's (1956, 1961, 1962; see earlier) half century–old view, stating that ''Solitary animals normally resemble ''juvenile'' version of the gregarious form and ''juvenilization'' and ''solitarization'' may be overlapping concepts with regard to some traits ...'' (Hartfelder and Emlen, 2005, p. 659). However, they do not mention the results of the discriminant analysis by Dorn *et al.* (2000), which showed that three groups of adult locusts, crowded, isolated and crowded but JHA-treated in the last hopper stadium, form separate groups without overlapping (see earlier); that is. solitarization is not equal to juvenilization. They miscite the results of Tawfik *et al.* (2000) that females of *S. gregaria* ''raised under isolated conditions begin producing eggs sooner than females reared under crowded conditions ....'' (Hartfelder and Emlen, 2005, p. 660). Figure 2 of Tawfik *et al.* (2000) clearly shows that first oviposition occurred 2 days earlier in crowded than in isolated females of *S. gregaria*. Hartfelder and Emlen (2005) conclude that synthesis of PAN is stimulated by high levels of JH, and this results in high levels of pheromone production by gregarious, but not by solitarious, mature adult males of *S. gregaria*. However, they do not mention the logical outcome of this conclusion, that in this instance, JH promotes a gregarious phase characteristic. According to Tawfik *et al.* (2000, p. 1145, Fig. 1), JH titre is higher (though not significantly so) in isolated than in crowded males of *S. gregaria* at the age of 30 days; nevertheless, PAN release remains zero in isolated males even at this age. Although no relevant experiments were made, we doubt very much that even a massive dose of JH or JHA would induce PAN emission in isolated *S. gregaria* males. The case may be similar to the JH-induced yellowing in crowded or gregarious males and absence of such induction in isolated males, regardless of JH titres in the latter (see Section 7.3).

It should be stressed that Hartfelder and Emlen (2005) do not exclude effects of other hormonal factors such as ecdysteroids and DCIN (also termed [His^7]-corazonin) on locust phase polyphenism, but they grossly overestimate the causative role of JH. In contrast, based on publications by Pener (1983, 1990, 1991), Pener and Yerushalmi (1998) and Pener *et al.* (1992), Dingle (2002, p. 260) concluded that JH ''is not the primary physiological factor responsible for the solitary syndrome'' and Breuer *et al.* (2003, p. 6) stated that ''Juvenile Hormone is not the primary physiological factor inducing the solitarious phase''. The green colour–inducing effect of JH has never been doubted, but, as already outlined, although green colour is a solitarious characteristic, it is not a necessary trait of solitarious locusts, which under dry conditions exhibit non-green colouration; also, the green colour–promoting effect of JH is not restricted to locusts, it exists in many non-locust acridids (see Section 7.2.1).

9.2 ECDYSTEROIDS

A pair of ventral glands (VG) is located on each lateral side of the posterior part of the head in acridids (cf. Strich, 1954; P. Joly *et al.*, 1956; Staal, 1961). They are presumably homologous to the prothoracic glands of the Lepidoptera and other insects. Their secretion controls moulting in the hopper stadia (cf. Hirn *et al.*, 1979; Reichhart and Charlet, 1986). The VG secrete ecdysone or immediate precursors that are rapidly converted to ecdysone. In relation to the moulting process, ecdysone may be considered as a prohormone; it is converted in certain tissues to 20-hydroxyecdysone (20-OH-ecdysone), which is the actual moulting hormone. Ecdysone, 20-OH-ecdysone and closely related molecules are usually connoted as ecdysteroids. For an updated review on ecdysteroids, see Lafont *et al.* (2005).

The relation between ecdysteroids and locust phase polyphenism was discussed in the reviews by Pener (1983, 1990, 1991, pp. 45–47), Hardie and Lees (1985, p. 460) and Dale and Tobe (1990, p. 406). Brief descriptions of the older publications that are summarized in these reviews are presented in the following paragraphs.

Ellis and Carlisle (1961) and Carlisle and Ellis (1962, 1963) advocated some solitarizing effects of the VG in *L. m. migratorioides* and *S. gregaria*. Michel's (1972) claim that implantation of VG into crowded male adults of *S. gregaria* reduces flight performance may also be considered as a solitarizing effect. However, Staal (1961) did not find phase-related effects of implanted extra VG on hopper colouration and no appreciable effect on the E/F and F/C ratios in *L. m. migratorioides*. In some instances of implantation of extra VG into second-instar hoppers, Staal (1961) obtained 'giant adults' in the sixth stadium (the stadium of normal adults). These exhibited hypertrophic wings, leading to high, 'hypergregarious' E/F values, presumably reflecting disturbed morphogenesis, rather than an effect on phase.

Carlisle and Ellis (1959) reported that the VG degenerate in crowded adults of *L. m. migratorioides* and *S. gregaria* (a well-known phenomenon in insects after emergence of the adult; cf. Sedlak, 1985, p. 54; Nation, 2008, p. 134), but the glands persist in isolated adults. This claim was modified later in relation to *L. m. migratorioides*. Cassier and Fain-Maurel (1968, 1969) and Fain-Maurel and Cassier (1969) reported that in crowded adults, the VG degenerate under long photophase but persist under short photophase; in isolated green adults, the VG persist under high humidity, regardless of the photoperiod, but the glands degenerate in non-green isolated adults kept under dry conditions (see more recent findings later).

Independent laboratory studies demonstrated that ecdysteroid titres do not differ much in crowded and isolated hoppers of *L. migratoria* (L. Joly *et al.*, 1977a; Fuzeau-Braesch *et al.*, 1982) and *S. gregaria* (Wilson and Morgan, 1978). These findings led to the presumption that VG activity and ecdysteroids do not play important roles in locust phase polyphenism.

More recent publications confirmed that there are no major ecdysteroid-related phase-dependent differences in late-instar locust hoppers. Roussel (1993), using radioimmunoassay (RIA) to assess ecdysteroid biosynthesis by the VG *in vitro*, as well as haemolymph ecdysteroid titres, in crowded and isolated last-instar hoppers of *L. migratoria*, did not find marked phase-dependent differences, though the rate of ecdysteroid biosynthesis was somewhat higher, peaked earlier and the peak lasted longer in the crowded hoppers. In contrast, at the end of the stadium, haemolymph ecdysteroid titres were considerably higher in isolated than in crowded hoppers. Tawfik *et al.* (1996) studied phase-dependent ecdysteroid titres and characterized the ecdysteroids in penultimate- and last-instar hoppers of *S. gregaria*. They used two kinds of antisera in RIA for titre determinations, TLC and HPLC for characterization of the ecdysteroid components and hydrolysis to identify ecdysteroid components in ecdysteroid conjugates. The authors found a high proportion of 20-OH-ecdysone, some ecdysone, an ecdysteroid with TLC and HPLC retention time of makisteron A and highly polar metabolites not affected by hydrolysis. Tawfik *et al.* (1996) found just minor phase-dependent differences; in isolated locusts, the content of ecdysone and the unknown makisteron A-like ecdysteroid was higher in both sexes of penultimate-instar hoppers and in last-instar male hoppers (in last-instar female hoppers, the unknown ecdysteroid was not detected). In contrast, the content of the highly polar metabolites was higher in crowded than in isolated locusts in either sex and in both hopper stadia. Also, the moult-inducing peak lasted longer in crowded than in isolated hoppers. The studies of Roussel (1993) and Tawfik *et al.* (1996) reconfirmed the view (see earlier) that ecdysteroids have no major role in locust phase polyphenism, at least in the late hopper stadia. Nevertheless, in a review-like later publication, Tawfik and Sehnal (2003, p. 21) claimed that "Our data in larvae (Tawfik *et al.*, 1996) do suggest that the gregarious phase is associated with a lower titre of ecdysteroid than the solitary phase". Then, however, Tawfik and Sehnal (2003, p. 22) state that in the hoppers "the phases do not differ in the rate of ecdysteroid production or in ecdysteroid titre" (Tawfik *et al.*, 1996).

Roussel (1993) tested ecdysteroid biosynthetic activity of the persisting VG in green isolated adults of *L. migratoria*. He found that the glands do not produce measurable amounts of ecdysteroids. Tawfik *et al.* (1997b) reported that the VG are well preserved in both isolated and crowded adults of *S. gregaria*, at least for 1 week after fledging, then gradual but irregular regression of the glands takes place; the regression is quicker in males than in females. No phase-dependent differences in the persistence and rate of regression of the glands were detected. Testing the ecdysteroid biosynthetic activity of the adults' glands *in vitro*, the authors found very low amount of ecdysteroids in the medium and considered it as 'background noise'. Eventually, Tawfik *et al.* (1997b) concluded that the VG of adult *S. gregaria* do not produce physiologically significant amounts of ecdysteroids. Comparison of the findings of Cassier and Fain-Maurel (1968, 1969), Fain-Maurel and Cassier (1969) and Roussel (1993) (see earlier) in *L. migratoria* and the findings of Tawfik *et al.* (1997b) in *S. gregaria* may indicate some difference between the two species in regard to the persistence/degeneration of the VG in adults, but in both species, the adults' glands seem to be inactive.

Although the degenerating or persistent VG of adults are presumably inactive (see earlier), ecdysteroids are well present in adult locusts (for a review on alternative sites of ecdysteroid production in insects, especially in adults, see Delbecque *et al.*, 1990). In adult females of *L. migratoria*, ecdysteroids are produced by the ovaries (Lagueux *et al.*, 1977), especially by the follicle cell epithelium (Goltzené *et al.*, 1978). The adult ovaries produce ecdysteroids also in *S. gregaria* (Gande *et al.*, 1979; Dinan and Rees, 1981). Most ovarian ecdysteroids accumulate, mainly as conjugates, in the oocytes of the locusts (see later). In the subsequent eggs, relatively high amounts of ecdysteroids are found (Lagueux *et al.*, 1979; Scalia and Morgan, 1982), which seem to be responsible for the embryonic moults (cf. Mueller, 1963, study on *Melanoplus differentialis*; Lagueux *et al.*, 1979). The metabolism of the maternal conjugated ecdysteroids during embryonic development in *L. migratoria* was investigated by Sall *et al.*

(1983). For a review on ecdysteroids in embryonic development of insects, with much data on *L. migratoria* and *S. gregaria*, as well as more references on locusts than those mentioned here, see Hoffmann and Lagueux (1985) and Sbrenna (1991). All the preceding cited studies on adults were carried out in crowded locusts. Data on the source of ecdysteroids in adult male acridids are scarce. Gillott and Ismail (1995) found *in vitro* ecdysteroid synthesis by reproductive accessory glands, testis and abdominal cuticle in adult males of *Melanoplus sanguinipes*.

Ismail *et al.* (1979) and Fuzeau-Braesch *et al.* (1982) compared haemolymph ecdysteroid titres in crowded and isolated adults of *L. m. cinerascens*. Ismail *et al.* (1979) did not detect ecdysteroids in the haemolymph of the males and found no significant difference in the titres between isolated and crowded females. Crowded females treated with daily exposure to CO_2 exhibited higher ecdysteroid titres than either isolated or crowded females without CO_2 treatment. Regardless of density, in most females, the titre was low or under the level of detection (less than 10 pg ml^{-1} haemolymph), but in a few females, it was higher, over 100 pg ml^{-1}. Fuzeau-Braesch *et al.* (1982) reported low titres or absence of haemolymph ecdysteroids in both males and females within the age range of 0–18 days after fledging, except on day 12 and day 15 in crowded females when the titre reached 85 and 25 ng ml^{-1} haemolymph, respectively. These values (ng ml^{-1}) are almost thousand times higher than those reported (in pg ml^{-1}) by Ismail *et al.* (1979). It is noteworthy that the article by Fuzeau-Braesch *et al.* (1982) does not refer to that by Ismail *et al.* (1979), despite that two of the authors are authors on both publications.

More recent research (Roussel, 1993; Tawfik *et al.*, 1997b, 1999b), reviews by Dorn *et al.* (2000, pp. 215–218), Breuer *et al.* (2003, p. 5) and Tawfik and Sehnal (2003) and a publication by Hägele *et al.* (2004) reveal an updated picture. Roussel (1993), using RIA, assessed haemolymph ecdysteroid titres in isolated green adults of *L. migratoria*, without making distinction between males and females. On day 1 and day 5 after fledging, he did not detect ecdysteroids, but by day 10, the titre increased and an additional increase occurred by day 15. Roussel (1993) did not test crowded adults; nevertheless, he concluded that on days 10 and 15, isolated adults have higher haemolymph ecdysteroid titres than crowded adults, according to the data in the literature for the latter.

Tawfik *et al.* (1997b) studied haemolymph ecdysteroid titres in crowded and isolated adults of *S. gregaria*. They identified the ecdysteroids by RIA, with two different antibodies, both of them most sensitive to ecdysone, but one of them less sensitive to other relevant ecdysteroids. After hydrolysis, the authors found 20-OH-ecdysone, ecdysone, highly polar products and a compound resembling makisteron A, like in the hoppers (Tawfik *et al.*, 1996; see earlier). In the adult males, 20-OH-ecdysone constituted the most abundant ecdysteroid, 73% of the total in isolated and 94% in crowded locusts. In both isolated and crowded adult females, about 40%–50% ecdysone and about the same percentage of

20-OH-ecdysone were found, constituting together about 86%–90% of haemolymph ecdysteroids. Other ecdysteroids were in smaller amounts and did not exhibit consistent correlation, either with sex or with density.

Tawfik *et al.* (1997b) reported that haemolymph titres of ecdysteroids in the adults showed a brief peak in isolated and crowded males and in isolated females, on day 4 after fledging, but on day 6 in crowded females. This peak was considerably higher in isolated males than in other adults. After this peak, the titres dropped, but a major increase started on days 8–10, coinciding with the onset of vitellogenesis (not tested directly) in the females and with the preparatory stage or onset of adult pheromone production by the males, assessed by phenylacetonitrile (PAN) emission (see Sections 8.1, 9.1 and 12.1.3). The titres peaked on day 16 in crowded and isolated females and on days 16–18 in crowded and isolated males. These peaks reached in both sexes about 400 ng of 20-OH-ecdysone equivalent per millilitre haemolymph in the isolated adults, but only about 150 ng ml^{-1} in the crowded adults. The relatively high peak of the isolated males cannot be related to PAN emission, because isolated males do not produce PAN (see Sections 8.1, 9.1 and 12.1.3). This fact makes questionable the relation between the much lower peak in the crowded males and PAN production. This relation is further questionable because in the crowded males, ecdysteroid titre decreased to practically zero by day 40, but the crowded males still produced PAN, albeit in less quantity than that produced on days 16–24. Also, the somewhat decreasing ecdysteroid titre was still over 100 ng ml^{-1} in 40-day-old isolated males, which do not produce PAN. In the females of either condition of density, haemolymph ecdysteroid titres decreased by days 20 and 24, increased again by day 30, then declined by day 40. Crowded adults began mating on day 14 after fledging, and their first oviposition was observed on days 15–16. Isolated females laid their first egg pod on days 18–19 and their second egg pod on days 28–30. This timing of oviposition coincides reasonably well with the main and the subsequent peaks of haemolymph ecdysteroid titres. Transfer of ecdysteroids from the follicle cell epithelium to the oocytes is a well-known process (see earlier), but the function of the haemolymph ecdysteroids is unclear, and claims with regard to their effect on vitellogenesis in acridids are contradictory. J. Girardie and A. Girardie (1996) suggested promotion of vitellogenesis by 20-OH-ecdysone in *L. migratoria*, whereas Hatle *et al.* (2003) found that haemolymph ecdysteroids do not affect vitellogenesis in the lubber grasshopper, *Romalea microptera*. The physiological role of the phase-dependent difference in haemolymph ecdysteroid titres is unknown. Finally, the function of haemolymph ecdysteroids in male locusts is also unclear; it may affect the synthesis of some proteins in the reproductive accessory glands (cf. Gillott and Ismail, 1995) or may have a role in spermatogenesis (Dumser, 1980) or both; and see Delbecque *et al.* (1990) for various reports and assumptions in regard to adult male insects in general.

Employing RIA and HPLC, Tawfik *et al.* (1999b) compared, in *S. gregaria*, phase-dependent ecdysteroid content during oocyte development in the ovaries and throughout embryonic development of the subsequent eggs, including newly hatched hoppers. They also identified the free, polar conjugated and non-hydrolyzable polar metabolites and characterized the chemical nature of the ecdysteroid components present in the ovaries, then in the eggs of isolated and crowded females. The introduction and a part of the discussion of this article constitute a 'state-of-the-art' summary of relevant data known at that time on crowded locusts; Tawfik *et al.* (1999b) were the first authors to investigate the subject comparatively in crowded and isolated locusts.

Major phase-dependent differences were found by Tawfik *et al.* (1999b). Total ecdysteroid content of the ovaries with developing oocytes (first gonotrophic cycle) was about four times higher in crowded than in isolated females (see Fig. 1 by Tawfik *et al.*, 1999b). This trait is opposite to that found in the haemolymph of the adults; as outlined earlier, Tawfik *et al.* (1997b) reported considerably higher haemolymph ecdysteroid titres in isolated than in crowded adult females (as well as in crowded adult males). The difference found in the ovaries also existed during the embryonic development of the eggs; total ecdysteroid content was over six times higher in freshly laid eggs from crowded females than in eggs from isolated females. The difference was maintained in the hatchlings; ecdysteroid content was five times higher in hatchlings from eggs laid by crowded mothers than in those from eggs laid by isolated mothers.

Tawfik *et al.* (1999b) revealed that most ovarian ecdysteroids, over 80% of the total, were polar conjugates of ecdysone, 20-OH ecdysone and 2-deoxyecdysone. Some of the same ecdysteroids, up to 5%, existed also in a free state. The rest were non-hydrolizable highly polar metabolites. The summation of all these confirmed a four times higher accumulation of total ecdysteroids in crowded than that in isolated females' ovary. The only phase-related difference was found in the proportion of the conjugated ecdysteroids; both crowded and isolated females' ovary contained similar absolute amounts of 20-OH-ecdysone and 2-deoxyecdysone, but conjugated ecdysone was much higher in the ovary of crowded females.

Tawfik *et al.* (1999b) also reported that newly laid eggs contained 14 and 89 ng of 20-OH-ecdysone equivalents per egg in eggs from isolated mothers and in those from crowded mothers, respectively. Total ecdysteroid content of the eggs during the first 6 days of embryonic development did not change much and remained over six times higher in eggs from crowded mothers. Almost all ecdysteroids were maternal conjugates at this age range of the eggs, mostly of ecdysone plus a small amount of 20-OH-ecdysone and 2-deoxyecdysone. Small amounts of presumably newly produced 26-hydroxyecdysone (26-OH-ecdysone) appeared on day 2 in the eggs of crowded females. The amount of total ecdysteroids increased by about four to five times to maximum values on days 8–10, reaching 70 ng and nearly 400 ng of 20-OH-ecdysone equivalents per egg in eggs from isolated and in those from crowded mothers, respectively. Then the amount of total ecdysteroids decreased gradually until hatching. These events were quite parallel in eggs from crowded and in those from isolated females, although the ratio of the amount of total ecdysteroids in the eggs from crowded mothers to that in eggs from isolated mothers remained high. Developmental fluctuations and mutual ratios of free ecdysteroids, namely, ecdysone, 20-OH-ecdysone and 26-OH-ecdysone, were also similar. Massive synthesis of conjugates started on days 8–10. Phase-dependent differences in these conjugates were found in 26-OH-ecdysone and 2-deoxyecdysone. The conjugates of the former were present from day 2 onwards and constituted up to 6% of all conjugates in the eggs of crowded females, but in the eggs of isolated females, these conjugates were present only in small amounts in the latest stadia of the embryonic development and in the hatchlings. The accumulation of 2-deoxyecdysone conjugates in the middle of embryogenesis was about 50 times higher in eggs from crowded than in those from isolated mothers. All conjugates declined with hatching.

From the results of Tawfik *et al.* (1999b), the question arises whether the differences in the ecdysteroid content of the eggs from isolated and those from crowded mothers do or do not affect locust phase polyphenism; in other words, whether these differences constitute a causal factor for inducing/shifting other phase characteristics. Tawfik *et al.* (1999b) hinted that there may be some connection between the high ecdysteroid content of the eggs from crowded mothers and the effect of the rearing density of the parents on phase characteristics of the progeny. It may be recalled that the parental effect is mediated by a 'gregarizing factor', contributed by the crowded mother to her egg pod (cf. Islam *et al.*, 1994a,b; McCaffery *et al.*, 1998). This gregarizing factor transmits certain gregarious characteristics from crowded parents to the progeny (see Section 7.2.1 in relation to hatchlings' colour, Section 8.4 in relation to oviposition aggregating effect and Section 16 for a more detailed discussion).

Hägele *et al.* (2004) made additional studies on the effects of crowding and isolation on ecdysteroid content of eggs of *S. gregaria*. The authors aimed to investigate whether the maternal gregarizing effect upon the progeny and the gregarizing factor in the egg pod foam are associated with the different ecdysteroid content in eggs laid by crowded and by isolated mothers. Islam *et al.* (1994b) found that crowding of formerly isolation-reared females shortly before egg laying induces gregarious behaviour and gregarious hatchling colour in the progeny, similarly to the progeny of females kept continuously under crowding (opinions differ about the effect on hatchling colour; see Sections 7.2.1 and 16.3). Therefore, Hägele *et al.* (2004) tested whether transfer from isolation to crowding would lead to an increase of ecdysteroid content, similar to that found in eggs from crowded mothers, or would remain low, similar to that observed in eggs from isolated mothers. The results clearly showed that ecdysteroid content of the eggs from isolation-reared females, which were newly crowded for a while, remained the same as that found in eggs from continuously isolated females. This finding rules out the possibility that the induction of the gregarious phase characteristics mentioned earlier in the hatchlings by a brief crowding of the mother is associated with the amount of ecdysteroids in the eggs. Also, meanwhile, the chemical nature of the gregarizing factor affecting hatchling behaviour was partially clarified (Miller *et al.*, 2008; see also Section 16.3), and it has no similarity to ecdysteroids.

Concerning total ecdysteroid content and its changes along the development of the eggs from isolated and from crowded mothers, the results of Tawfik *et al.* (1999b) and those of Hägele *et al.* (2004) are compatible in most findings. Both studies found much higher ecdysteroid content in eggs from crowded females, with a peak on days 8–10 (Tawfik *et al.*, 1999b) or on day 10 (Hägele *et al.*, 2004). The absolute amount of ecdysteroids at this peak was found to be nearly 400 ng per egg and almost 250 ng per egg by Tawfik *et al.* (1999b) and Hägele *et al.* (2004), respectively. This difference may easily be related to different laboratory strains or different feeding and maintenance conditions of the locusts or both. Hägele *et al.* (2004) found a moderate brief peak on day 2 in the eggs from crowded mothers; Tawfik *et al.* (1999b) did not observe such a peak. The curve of the ecdysteroid content in the developing eggs from isolated mothers is somewhat different in the two studies. According to Tawfik *et al.* (1999b), it is declining by the end of the embryonic development, whereas Hägele *et al.* (2004) did not observe such a decline. However, careful comparison of the relevant curves reveals that the last point in the curve by Tawfik *et al.* (1999b) shows hatchlings, not eggs, and mainly this last point is responsible for the decline of the curve. If we assume that the curve by Hägele *et al.* (2004) does not include actual hatchlings, removal of the last point (hatchlings) from the curve shown by Tawfik *et al.* (1999b) makes the seemingly considerable difference between the curves from the two studies much less distinct.

Islam *et al.* (1994a,b) and McCaffery *et al.* (1998) tested transmission of gregarious behaviour and hatchling's colour in the progeny by the gregarizing factor produced by crowded females. They did not test possible effects on other phase characteristics. It cannot be ruled out that the differences in ecdysteroid content of the eggs from crowded and isolated mothers affect such other phase characteristics. For example, we do not know the causative factor for phase-related differences in the number of ovarioles (see Section 6), in the number of antennal and other sensilla (see Sections 5.3.1, 5.3.2 and 5.3.3) and so on.

Delayed effects of egg ecdysteroids may also be considered. As already mentioned, Staal (1961) implanted VG into second-instar hoppers of *L. migratoria* and obtained either complete adults in the fifth stadium (which is the last hopper stadium in normal development) or 'giant adults' (see earlier) in the sixth stadium (which is the stadium of the adult in normal development). Morphogenetic disturbances were also observed in crowded *L. migratoria* after feeding on a plant diet with a selectively modified sterol profile (Charlet *et al.*, 1988). The causative factor of the extra hopper instar in solitarious *S. gregaria* and *N. septemfasciata* is unknown.

The highest ecdysteroid peak was observed in the eggs of *S. gregaria* on days 8–10 by Tawfik *et al.* (1999b) and on day 10 in eggs from crowded mothers by

Hägele *et al.* (2004). At this age, about 60% of the development of the embryo is completed and organogenesis is well advanced. Therefore, some contribution of the embryo's own VG to the ecdysteroids in the eggs cannot be ruled out. This subject is clouded even for crowded locusts, and phase-related articles on ecdysteroids in locust eggs do not deal with it.

Gregg *et al.* (1987) found in crowded *Chortoicetes terminifera* that eggs laid by females reared under conditions that do not induce egg diapause contained three times more ecdysteroids than eggs laid by females reared under egg diapause–inducing conditions. Tawfik *et al.* (2002a) revealed in a Japanese strain of *L. migratoria* from Ibaraki that reproductively active crowded females kept under short days exhibited significantly higher ecdysteroid titres in the haemolymph and the ovaries than reproductively inactive crowded females kept under long days. In this strain, however, the reproduction-inhibiting effect of long days takes place only under crowding. Under isolation, reproductive activity occurs early under both long and short day length (cf. Tawfik *et al.*, 2002a, and references therein), but the females produce non-diapausing and diapausing eggs, respectively. In a subsequent article, Tawfik *et al.* (2002b) reported that non-diapausing eggs, produced by isolated females kept under long days, had three times higher ecdysteroid content than diapausing eggs produced by isolated females kept under short days. The fact that the effect of photoperiod may be density-dependent was already demonstrated by Verdier (1967) (see Section 8.5), who found that in dense population of the 'Palavas' strain of *L. migratoria*, sexual maturation is inhibited under long days, whereas in isolated females, it is not inhibited. The density-dependent effect of photoperiod seems to be similar in the 'Palavas' and the 'Ibaraki' strain of *L. migratoria*, at least in regard to delay or acceleration of reproductive activity. Although we do not know the effect(s) caused by the difference in ecdysteroid content of eggs from crowded and from isolated mothers, this difference may be associated, in an unknown way, with the photoperiod-controlled phase-dependent differences in the onset of reproductive activity and with the difference in the ecdysteroid content between diapausing and non-diapausing eggs.

9.3 NEUROPEPTIDES AND OTHER HORMONES

9.3.1 *General considerations*

Neurohormones are neuropeptides, usually just an amino acid sequence, although some have a peptide backbone with additional molecule(s), like the glycolysated prothoracicotropic hormone (PTTH) of some Lepidoptera (for chemical structure see Rybczynski, 2005). Biogenic or neurogenic amines may also act as neurohormones (Nijhout, 1994, p. 49), although in most instances, they are neuromodulators or neurotransmitters. There are many neurohormones, mostly, but not exclusively, produced by the median neurosecretory cells (NSC) in the pars intercerebralis (PI) of the brain, or by other brain NSC, or by the intrinsic cells of the corpora cardiaca (CC). In locusts, the CC quite clearly comprise two parts: the storage lobe and the glandular lobe. The former is composed of neurosecretory axons, coming from the NSC of the brain, as well as the terminals of most of these axons; the terminals store and release the neurohormones produced in the brain. The glandular lobe produces neuropeptides originating from its own intrinsic cells. These cells are, in fact, unusual neurones with short protuberances, probably stumped axons (see, for example, Smith, 1968, Plate 33, p. 102); in the old literature, the CC were named as oesophageal ganglia (see, for example, Albrecht, 1953, Fig. 108, p. 76, Fig. 117a, p. 86). Neurohormones are also produced by ganglia of the central nervous system, outside of the brain NSC-CC axis, as well as by peripheral ganglia. Specific endocrine organs such as the Inka cells in the epitracheal glands (cf. Zitnan and Adams, 2005) also produce peptide hormones.

There are many neurohormones and their number is growing as new ones are discovered (reviews by Gäde *et al.*, 1997; Nässel, 2002; Claeys *et al.*, 2005a, the latter with emphasis on neurohormone receptors). A general discussion of insect neurohormones is outside of the scope of the present review (much information is found in Vol. 3 of the book edited by Gilbert *et al.*, 2005).

A review by Veelaert *et al.* (1998) is available on neuropeptides in relation to the CC-CA complex of locusts, with fine and detailed illustrations of the relevant anatomy. A list of neuropeptides found in the median NSC and the CC on one hand and those observed only in the CC on the other hand was provided by Clynen *et al.* (2001).

Absence, or fragmentary knowledge, of some physiologically highly important neurohormones in relation to locusts are to be dealt first; then neurohormones for which phase-dependent effects or differences were reported will be discussed. To avoid repetitions, some reference will be made to Section 7.2.2 that also deals with the effect of a neuropeptide.

9.3.2 *The prothoracicotropic hormone and the ventral glands*

The prothoracicotropic hormone (PTTH) activates the prothoracic glands (ventral glands (VG) in acridids) to produce ecdysteroids for initiating the process of moulting (see Section 9.2). The PTTH of certain lepidopteran and dipteran species are slightly different, but their chemical structure and amino acid sequence are known (Rybczynski, 2005). In contrast, the structure of locust PTTH is unknown. A prothoracicotropic effect was demonstrated by Reichhart and Charlet (1986), who showed that brain-CC extracts made from penultimate- and last-instar hoppers, as well as from adult females, exert a dose-dependent ecdysiotropic effect, *in vitro*, on the VG of 1-day-old last-instar hoppers. No further advancement was made during two decades, except some findings which may imply that locust PTTH is structurally not closely related to the known lepidopteran PTTH. In an immunocytochemical study, using antibody against the PTTH of the saturniid moth *Antheraea pernyi* (Guéein-Méneville), no immunoreactivity was observed in the brain–suboesophageal complex of *L. migratoria* and *S. gregaria* (Závodská *et al.*, 2003). Although the authors used adult insects which do not moult, PTTH immunoreactivity was found in 10 other adult insect species belonging to the orders of Ephemeroptera, Odonata, Plecoptera, Heteroptera, Coleoptera, Hymenoptera and Diptera. Immunoreactivity was also found in the Archaeognatha, but this order of insects is primitive, apterygote, and exhibits moultings after sexual maturation by alternating gonotrophic cycles and moultings. However, in adult insects, the prothoracic (ventral) glands usually degenerate, and even if the glands persist, they do not produce ecdysteroids (see Section 9.2). Therefore, it is unclear whether the immunoreactivity observed in the ten non-acridid species was indeed to the PTTH or to a cross-reacting ecdysiotropic hormone that induces ecdysteroid production in adult insects (see Section 9.2). If so, a putative ecdysiotropic hormone in adult locusts may be sufficiently different from lepidopteran PTTH, or from putative locust PTTH, as not to produce immunoreactivity.

Recently, Vandersmissen *et al.* (2007) detected an 'autocrine factor' in the VG of last-instar hoppers of *L. migratoria* and *S. gregaria*. This factor originates from the VG themselves and enhances their ecdysteroid production. The report of Vandersmissen *et al.* (2007) rekindled interest in some older publications which claimed that the prothoracic glands of insects secrete other hormone(s), probably proteins (or peptides), in addition to ecdysteroids. Several articles indicated such function of the VG of *L. migratoria* (L. Joly *et al.*, 1969; Hoffmann and Weins, 1974; L. Joly and Schneider, 1976). It may be recalled that Staal (1961) (see also Staal and De Wilde, 1962) accelerated nymphal development by implanting extra VG into young second-instar hoppers of *L. migratoria* (see Section 9.2). By using this treatment, Staal (1961) obtained adults in the fifth stadium and also obtained some 'giant adults' in the sixth stadium (which is the adult stadium in normal development). These experiments were repeated by L. Joly *et al.* (1977b); yielding results similar to those of Staal (1961). Also, Charlet *et al.* (1988), feeding hoppers of *L. migratoria* on a plant diet with a selectively modified sterol profile, obtained adults in the fifth stadium, instead of the normal sixth stadium. The whole subject was recently discussed by De Loof (2008).

9.3.3 *Endocrine cascades in relation to ecdysis*

Ecdysis, also termed moult or moulting, is a complex process, integrating behavioural, biochemical and molecular events. It is initiated by a rise of ecdysteroid level that induces the pre-ecdysis state, in which certain motor programs are executed and gene expression is regulated to dissolve and reabsorb the inner layers of the old cuticle, as well as forming the new cuticle and tissues. In the ecdysis state, induced by decline of ecdysteroids, a cascade of peptide hormones induce typical sequential motor patterns, leading eventually to shedding of the outer layers of the old cuticle. Post-ecdysial processes include certain movements, expansion of the body, sclerotization (also termed tanning, hardening of the new cuticle) and melanization.

The whole subject of neuroendocrine regulation of these processes is reviewed by Truman (2005), Zitnan and Adams (2005) and Žitňan *et al.* (2007). Present knowledge is mostly based on ecdysis of the moths *Manduca sexta* (Johannsen) and *Bombyx mori* (L.), as well as that of the fruit fly *Drosophila melanogaster* Meigen. The most important peptides involved in the control of the moult are the pre-ecdysis-triggering hormone (PETH), ecdysis-triggering hormone (ETH), eclosion hormone (EH) and crustacean cardioactive peptide (CCAP). Additional neuropeptides, namely corazonin, FLRF amides, myoinhibitory peptides (MIHs) and ion transport peptides (ITPs) may also be involved. In the post-ecdysial state, bursicon induces sclerotization (tanning) of the new cuticle, resulting in the desired stiffness. PETH and ETH are produced by the so called Inka cells in the epitracheal glands; EH and CCAP are produced by NSC in the brain and ventral central nervous system. For more detailed information see Truman (2005), Zitnan and Adams (2005) and Žitňan *et al.* (2007).

With regard to locusts, six stages in the behavioural sequence of the last ecdysis (from last-instar hopper to adult) in crowded *S. gregaria* are described by Hughes (1980a). The first and the second stages constitute the pre-emergence behaviour. In the former, the hopper ceases to feed, reduces movements and selects a site for moulting. In the latter, the hopper hangs from a perch or similar object and slowly inflates its gut. Emergence starts with stage three; gut inflation processes more rapidly, the old cuticle splits, and the head, thorax and fore- and middle-legs emerge from the old cuticle. In stage four the hind legs emerge; the newly moulted soft and wet adult climbs up on its old cuticle, pulling the abdomen out and leaves the old cuticle (termed exuviae). The adult then selects a site for expansion of the new cuticle, where it settles in a vertical position, head upwards. The post-emergence behaviour includes stage five; the insect expands itself, the new cuticle stretches to adult size and shape. During the initial period of hardening, the wings expand and fold. Stage six is the end of the post-emergence behaviour; the adult does not change place as the new cuticle is hardening and darkening and the gut deflating.

The motor activity during the pre-emergence and emergence behaviour (stages 1+2 and 3+4, respectively, see earlier) is described in detail by Hughes (1980b). In the pre-emergence behaviour various synchronous or alternating contraction and relaxation of the skeletal muscles take place. Air swallowing starts and continues during emergence behaviour, the gut inflates and the abdominal muscles create waves of segmental contractions that move anteriorly along the abdomen, working the locust forward out of the old cuticle. Other motor programs extricate the legs and the mouthparts. The metathoracic ganglion controls the timing of the concurrently active motor patterns. The motor activity of the post-emergence behaviour (Hughes, 1980c) is characterized by dorsoventral abdominal contractions during which the gut deflates and the air sacs are filled. Expansion of the body and the wings, hardening of the new cuticle and deflation of the gut terminate the moulting. Hughes (1980d) also described in detail the role of the gut; its inflation contributes to emergence and expansion.

Bernays (1972a) described the muscles of newly hatched *S. gregaria* and the process of intermediate first moult (Bernays, 1972b) that occurs after the 'vermiform larva' reaches the surface of the ground and sheds the embryonic cuticle. She also reported the changes occurring in the cuticle of the first-instar nymphs before and after the intermediate moult (Bernays, 1972c). Truman *et al.* (1996) also observed the hatching and ecdysis of the vermiform larva, as well as the expansion of the first-instar hoppers of *L. migratoria*. These authors showed by immunocytochemical method that the behavioural patterns are associated with dramatic intracellular increase of cyclic GMP in identified neurones in the ventral central nervous system.

Although the behavioural sequence and underlying muscle motor activities of ecdysis are well known in locusts (see earlier), knowledge of the endocrine cascade and neuroendocrine regulation of the process is somewhat fragmentary.

PETH and ETH are secreted by the Inka cells and these peptides are characteristic to the Inka cells. Therefore, discovery of the sites where these peptides are located means discovery of the location of the Inka cells. Using immunohistochemistry with antiserum against PETH, Žitňan *et al.* (2003) identified Inka cells in representatives of all major orders of insects. In *L. migratoria* hoppers (presumably crowded, stadium not stated) they found simple oval Inka cells scattered over the surface of the tracheae. More recently, Clynen *et al.* (2006) examined the EST database of *L. migratoria* generated by Kang *et al.* (2004) and used BLAST search to explore neuropeptide precursors. Blasting the ETH precursor of *D. melanogaster* against the *L. migratoria* database, gave a hit in the gregarious whole body cDNA library. Eventually the authors constructed the amino acid sequence of the ETH precursor of *L. migratoria* and reported two ETHs cleaved from the precursor. These were designated as Lom-ETH-1, having the sequence of SDFFLKTAKSVPRamide, and Lom-ETH-2 having the sequence of SDLFLKSAKSVPRamide (Clynen *et al.*, 2006, Figs 4 and 5). Considering that ETH is secreted by the Inka cells of the epitracheal glands (see earlier), the authors dissected tracheae from last-instar hoppers of *L. migratoria*, just before ecdysis to adult and made a peptide extract of these tracheae. Employing complex physicochemical methods, they found the predicted masses of both ETH in the extract and confirmed their amino acid sequences.

Using immunocytochemistry and antiserum against eclosion hormone (EH) of *M. sexta*, Závodská *et al.* (2003) detected no EH-immunoreactivity in the brain-suboesophageal complex of adult (presumably crowded) *L. migratoria* and *S. gregaria*. In contrast, Zitnan and Adams (2005, p. 13) stated that immunohistochemical staining with antiserum against *M. sexta* EH, revealed EH-immunoreactivity in one pair of cells in the brain of *L. migratoria*. Unfortunately, Zitnan and Adams (2005) do not provide details, nor reference to the source of this information. The amino acid sequence of EH of locusts is not yet known, but a fragment of the EH of an acridid, *Romalea guttata* (Houttuyn) was revealed (see Zitnan *et al.*, 2007, Fig. 2).

The crustacean cardioactive peptide (CCAP) was originally isolated from a crab (Stangier *et al.*, 1987). Employing radioimmunoassay (Stangier *et al.*, 1988), an identical neuropeptide was found by Stangier *et al.* (1989) in the central nervous system (CNS) of *L. migratoria*. Using immunocytochemical methods and antiserum to CCAP, Dircksen *et al.* (1991) revealed several types of paired lateral neurones in the suboesophageal, thoracic and abdominal ganglia of the CNS of *L. migratoria* (review, including locusts, by Dircksen, 1998). In *M. sexta*, the release of CCAP from the CNS has been implicated as a crucial signal for the ecdysis motor program in the abdominal ganglia (cf. Zitnan and Adams, 2005, p. 18 and references therein). In *L. migratoria*, peripherally released CCAP acts as a neurohormone that stimulates oviduct contractions (Donini *et al.*, 2001) and as a neuromodulator it stimulates hindgut contractions (Donini *et al.*, 2002). Its possible function, if any, in locust ecdysis has not yet been clarified.

Detected by various methods, such as receptors to ETH, increase of the level in ecdysis, etc., several additional neuropeptides seem to be involved in ecdysis of *M. sexta* and/or *D. melanogaster* (cf. Kim *et al.*, 2006a,b). These neuropeptides are present in locusts, but their role, if any, in locust ecdysis is unclear; only some of them are mentioned here. Ion transport peptide, belonging to the crustacean hyperglycaemic hormones, was originally identified from the CC of *S. gregaria* (Audsley *et al.*, 1992a). In this species, it affects ion balance and fluid resorption in the hindgut (Audsley *et al.*, 1992b). Myoinhibiting peptide (also termed allatostatin B) was first isolated and identified from the brain-CC-CA-suboesophageal ganglion complex of *L. migratoria*; it suppresses the spontaneous contractions of the hindgut and the oviduct (Schoofs *et al.*, 1991). FLRF-amide-related peptides are also present in locusts (Lange *et al.*, 1994; review, including locusts, by Orchard and Lange, 1998) and even affect skeletal muscle contraction (Lange and Cheung, 1999).

It seems that no experiments have been conducted to detect locust polyphenism-related differences in the hitherto mentioned hormones and endocrine events that are, or may be, involved in ecdysis. Acceleration of the process of moulting would seem to be advantageous for gregarious locusts because during ecdysis they are vulnerable and exposed to cannibalism by fellow locusts. This subject, however, has not been investigated although cannibalism in gregarious marching bands of hoppers is known, even during intermoult periods (Bazazi *et al.*, 2008) (see Section 11.5).

9.3.4 *Corazonin*

Two additional neurohormones have a role in the post-emergence state of locust ecdysis; these are corazonin (dark-colour-inducing neurohormone (DCIN) in relation to locusts, see Section 7.2.2) and bursicon. DCIN is involved and bursicon may be involved in locust phase polyphenism.

Corazonin is a conserved neuropeptide that is found in insects and crustaceans. Presently, six isoforms of corazonin are known (Predel *et al.*, 2007, and references therein). The physiological functions of corazonin are widely different. In the crustacean *Precambarus clarkii*, it promotes tegumentary pigment migration (Porras *et al.*, 2003). It plays a presumably crucial role in the moulting process of *M. sexta*; Inka cells of this moth have receptors to corazonin, and it was reported that *in vitro* exposure of isolated Inka cells to corazonin induces secretion of PETH and ETH by these cells (Kim *et al.*, 2004). High level expression of corazonin receptor was also found in Inka cells of *B. mori* (Roller and Žitňan, unpublished, cited by Žitňan *et al.*, 2007, p. 89). Corazonin may be involved in the regulation of circadian clocks in moths (Qi-Miao *et al.*, 2003). Administration of corazonin reduces the spinning rate in *B. mori* (Y. Tanaka *et al.*, 2002b). Corazonin was discovered by Veenstra (1989) in *Periplaneta americana*; in this species, it acts as a cardioaccelerator. In acridids, it acts as a DCIN (see Section 7.2.2 and later). Insofar as it is known, corazonin-induced effects are specific in the sense that the effect induced in one group of insects is not induced in other groups. It also remains to be explored whether corazonin controls Inka cell secretion in insects other than moths.

The isoform that induces dark colour in locusts is [His^7]-corazonin, which is the DCIN (Tawfik *et al.*, 1999a). It is found in acridids, some phasmids and in certain Hymenoptera. In contrast to the specific effects of corazonin in different groups of insects, the isoforms do not show much effect-specificity. Thus, for example, [His^7]-corazonin (present in acridids) and [Arg^7]-corazonin (present in most insects; cf. Predel *et al.*, 2007) are equally efficacious at inducing dark colouration in Okinawa albino hoppers of *L. migratoria* (Yerushalmi *et al.*, 2002). Also, no difference was found in the effect of these two isoforms on the reduction of spinning rate of *B. mori* (Y. Tanaka *et al.*, 2003). Following the discovery of the isoform [Thr^4,His^7]-corazonin from the honey bee, *Apis mellifera* L., and its preprohormone (precursor) amino acid sequence (Roller *et al.*, 2006; Verleyen *et al.*, 2006), its dark colour–inducing activity was tested in Okinawa albino hoppers. Verleyen *et al.* (2006) stated that the honey bee corazonin is less active than [His^7]-corazonin of acridids. In contrast, Roller *et al.* (2006) found that the dark colour–inducing activity of the honey bee corazonin was as effective as [His^7]- or [Arg^7]-corazonin. Implantation of CC from 52 species, belonging to 10 orders of insects, induced darkening in Okinawa albino hoppers, but such implantations from eight species of Coleoptera did not induce darkening (Tanaka, 2000a). Also, Roller *et al.* (2003) found immunoreactivity to corazonin in eight species belonging to five orders of insects, but did not found such immunoreactivity in a coleopteran species.

As already discussed in detail, the dark-colour-promoting effect of DCIN (also termed [His^7]-corazonin) was discovered by induction of dark colouration, or dark gregarious patterns, in the Okinawa albino strain of

L. migratoria (Section 7.2.2). DCIN induces darkening also in normally coloured strains of *L. migratoria* and in all hitherto investigated other acridids that exhibit homochromy (Tanaka, 2000a, 2004b; Yerushalmi and Pener, 2001). DCIN shifts morphometric variables of locusts towards gregarious values (Section 5.2) and affects the number of antennal sensilla towards the same, gregarious, direction (Section 5.3.1).

Although DCIN is involved in locust phase polyphenism, it does not seem to be the primary gregarizing factor. Although it induces gregarious black patterns, it does not induce the bright yellow-orange background colour in late-instar *S. gregaria* hoppers. Moreover, in *L. migratoria*, it usually induces quite uniform dark colouration, similar to that shown by more or less dark homochrome solitarious hoppers (Section 7.2.2). Despite the effect of DCIN on morphometrics, it should be reminded that Okinawa albino locusts do exhibit density-dependent morphometric changes, albeit their morphometric ratios under crowding are close to those of conspecific normally coloured locusts under isolation and the range of their density-dependent morphometric changes is restricted (Section 5.2). This restriction, however, is not due to albinism, but due to a trait of the Okinawa strain of *L. migratoria* (Yerushalmi *et al.*, 2001; Grach *et al.*, 2004). Hoste *et al.* (2002a) investigated whether DCIN deficiency in Okinawa albino hoppers of *L. migratoria* results in solitarious behaviour. They found strain-dependent differences between the behaviour of crowded Okinawa albino hoppers and that of normally coloured hoppers of an African strain. Isolation of the hoppers resulted in a shift towards solitarious behaviour in both strains, even more markedly in the Okinawa albinos. These results demonstrated that absence of DCIN does not lead to solitarious behaviour in crowded Okinawa albino hoppers. In a subsequent study, Hoste *et al.* (2002b) injected three times 1 nmol of DCIN into fourth-instar isolated hoppers of *S. gregaria* and tested their behaviour in the next stadium. The injection of the hormone did not induce behavioural phase change; therefore, it is not involved in behavioural gregarization (see also Section 11.2). These findings support the conclusion that DCIN is not the primary gregarizing factor.

Several DCIN-related subjects need further research. Tanaka (1993, Table 1) found that implantation of CC from green solitarious hoppers, or from crowded hoppers of normally coloured *L. migratoria*, induced similar darkening in Okinawa albino hoppers. Tanaka and Pener (1994) injected brain extracts, or CC extracts, into Okinawa albino hoppers. These extracts were made from three different kinds of normally coloured *L. migratoria* donors, green solitarious, brown solitarious and crowded hoppers. The results showed that (1) there are no significant differences in the dark colour–inducing effect of brain extracts from the three different kinds of donors; (2) CC extracts are significantly more effective than brain extracts; and (3) CC extracts from green solitarious hoppers are significantly less effective than CC extracts from crowded hoppers. On the other hand, no difference was found between the effect of CC extracts from green solitarious and brown solitarious hoppers. Employing immunochemistry,

Baggerman *et al.* (2001, p. 154) showed that the brain-CC complex immunoreactivity is similar in hoppers of normally coloured crowded *L. migratoria* and in hoppers of crowded *S. gregaria*; they also found DCIN immunoreactivity in green isolated hoppers of *S. gregaria*. The question arises why green solitarious hoppers are not dark despite the presence of DCIN in their brain and CC. The logical answer could be that DCIN is not released, at least not in sufficient amounts, from the CC of solitarious hoppers, but retained to be released if there is a change in density from isolation to crowding and, in *L. migratoria*, also if the surroundings dry or darken (bush fire). Another possibility, rapid degradation of DCIN in solitarious locusts is improbable (see later). Receptors exist in isolated hoppers; they respond well to CC implantation or to injection of CC extract (see earlier). However, level of DCIN in the CC, its titre in the haemolymph, and the level of the receptors in the various forms (isolated green, isolated brown and crowded), as well as the rate of changing of these levels with changing surroundings (humidity, ground colour, density) have not been assessed. The need for DCIN titre demonstrations in crowded and isolated locusts was also stressed by Tanaka (2001), but no relevant advancement was made, except of the degradation profile of DCIN in crowded and isolated *S. gregaria* adults (see later).

Tanaka (2000c) investigated *in vivo* relationships between the green colour–inducing effect of JH (see Section 7.2.1) and the dark colour–inducing effect of DCIN (Section 7.2.2). He injected on different days of the intermoult period in the third stadium different doses of DCIN and JH into crowded Okinawa albino hoppers of *L. migratoria*. The results showed various colour shades in the subsequent fourth-instar hoppers, but the underlying mechanisms remained unknown. A single attempt to investigate the effect of DCIN on JH biosynthetic activity of the CA *in vitro* gave results that were difficult to interpret (Okuda and Tanaka, unpublished, cited by Tanaka, 2001, pp. 143–144).

The effect of DCIN injected in oil is long lasting, but the dark colouration eventually fades, at least in adults of *L. migratoria* (Grach *et al.*, 2004). Vandersmissen *et al.* (2006) injected 200 μg of [^{3}H]-labelled DCIN, dissolved in 35% ethanol, into each adult of crowded *S. gregaria*, to investigate the degradation profile of the peptide. *In vivo* tests showed a half-life of about 20 min (Vandersmissen *et al.*, 2006, p. 543 and Fig. 5), although the authors state in the abstract that the half-life was 30 min. They also studied *in vitro* proteolytic breakdown of [^{3}H]-labelled DCIN, by incubating the labelled peptide, dissolved in 35% ethanol, at 30°C, in 50 μl of cell free haemolymph of crowded and separately of isolated adults of *S. gregaria*. The authors found a phase-dependent difference in the rate of breakdown of the peptide. After 4 h of incubation, about 50% and 75% of intact DCIN was found in the cell free haemolymph of crowded and of isolated locusts, respectively. Also, a major *in vitro* degradation product, [Dopa^5,His^7]-corazonin, showed quantitative phase-dependent differences. After 4 h of incubation in haemolymph of isolated locusts, this modified peptide constituted less than 10%, whereas in haemolymph of crowded locusts, it constituted more than 20%. The total amount of other degradation products was also smaller in haemolymph of isolated locusts. These findings indicate that breakdown of the DCIN is slower in isolated than in crowded locusts.

Grach *et al.* (2003) substituted one amino acid each time, from residue 2 to residue 11 of DCIN, with D-phenylalanine, and tested the dark colour–inducing effect of these peptides, as compared with unaltered DCIN, in crowded Okinawa albino hoppers of *L. migratoria*. Substitution of any single amino acid reduced activity, but did not abolish it completely. Maximum inactivation was observed with substitution of Ser6, then with Gln4 and Trp9. However, residue 4 is Thr in the honey bee corazonin (see earlier) and this difference does not inactivate, or inactivate only moderately the effect of the peptide on darkening of Okinawa albinos (see earlier). In a similarly designed study, Y. Tanaka *et al.* (2003) also substituted one amino acid each time, from residue 1 (pGlu) to residue 11, with L-alanine. The results of Y. Tanaka *et al.* (2003) were different in many instances from those of Grach *et al.* (2003). The difference was especially large when Phe3 was substituted; Grach *et al.* (2003) found that this substitution moderately reduces the activity, whereas Y. Tanaka *et al.* (2003) found that this substitution results in maximum inactivity among all other substituted peptides. Grach *et al.* (2003) reported that substitution of Gln4 leads to extensive inactivation, whereas Y. Tanaka *et al.* (2003) found a moderate-medium effect on inactivity by this substitution. The levels of inactivation observed by substitution of other residues were also more or less different, although not as extremely different as in the instances of Phe3 and Gln4.

There are additional unclear issues in relation to DCIN and locusts. It is recalled that Tanaka (2007) found that DCIN-deficient CC, implanted from Okinawa albino donors into Okinawa albino recipients, shifted the morphometric ratios towards gregarious values in the recipients. Also, Baggerman *et al.* (2001) found that in Okinawa albinos of *L. migratoria*, which are DCIN-deficient, the typical DCIN-immunoreactive pars lateralis cells were not detected. This result was expected; unexpectedly, however, the CC of these albinos showed labelling of immunoreactive fibres, albeit much less than in the CC of normally coloured locusts. Mass spectrometric evidence clearly indicated that the immunoreactivity of the fibres in the albino CC is not due to the presence of DCIN. Tanaka's (2007) results and those of Baggerman *et al.* (2001) may be inter-related as outlined in Section 7.2.2. Perhaps some DCIN-altered substance is sufficiently close to DCIN to shift morphometric ratios towards gregarious values and to induce DCIN immunoreactivity, but has lost the dark colour–inducing activity of DCIN.

9.3.5 *Bursicon*

Tanaka (2001) indicated that DCIN is not responsible for the dark colouration of hatchlings and first-instar hoppers. He cited his unpublished results that green hatchlings of *S. gregaria* from isolated parents remained green after injection of DCIN. Bernays (1972c) suggested that induction of darkening and hardening of the cuticle of dark hatchlings of *S. gregaria* was due to a blood-borne factor, presumably bursicon. Padgham (1976a,b) studied the induction of the dark colour in dark hatchlings of *S. gregaria*. He found that a ligature around the neck or between the prothoracic and mesothoracic legs prevented darkening of parts of the body anteriorly to the ligature. By changing the site of the ligature along the body of the hatchlings, Padgham (1976a) concluded that a hormonal factor, released from neurosecretory axons of the metathoracic ganglion, is responsible for darkening of the hatchlings' cuticle. In additional experiments, ligatured non-darkening sections of the body were injected with blood from darkening hatchlings which resulted in darkening of the ligatured sections. The conclusion that this factor is bursicon was obtained by circumstantial evidence; darkening and hardening of the cuticle anteriorly to neck-ligated *S. gregaria* hatchlings was induced by injection of haemolymph from freshly emerged blowflies and *vice versa*, injection of haemolymph from darkening hatchlings into neck-ligated blowflies induced darkening of the cuticle of the blowflies posteriorly to the ligature (Padgham, 1976a). Very interestingly, green hatchlings or albino hatchlings of *S. gregaria* which exhibit hardening but no darkening, did not respond by darkening to injection of haemolymph from darkening hatchlings (Padgham, 1976b). Therefore, Padgham (1976b) concluded that the target organ, probably the epidermis, of green or albino hatchlings respond differentially to the putative bursicon, than that of the darkening hatchlings.

In an earlier article, employing ligature experiments, Vincent (1971) found that in the moult of fourth- and fifth-instar hoppers of crowded *L. m. migratorioides* bursicon is released ''from at least the terminal abdominal ganglion...'' (abstract, p. 625). In a subsequent article, Vincent (1972) found that bursicon is released from the abdominal ganglia in the moult from crowded last-instar hoppers to adults of *L. m. migratorioides*. Bursicon was identified by the available circumstantial evidence, based on the ligated fly bioassay, developed by Fraenkel and Hsiao (1965). Vincent's (1971, 1972) results and those of Padgham (1976a,b) seem to be contradictory in regard to the ganglia releasing the bursicon at the moult. However, this contradiction may not be very crucial. Beside the fact that different species were studied, Vincent (1972) found a much higher content of the putative bursicon in the metathoracic ganglion than in other ganglia of the ventral nerve cord, although the storage lobe of the CC also contained large amounts of bursicon. It is possible that bursicon is released from different ganglia in hatchlings and in late-instar moults.

Kostron *et al.* (1995) partially purified bursicon in *L. migratoria*; they found that it has a molecular mass of about 30 kD. Bursicon activity, based on the ligated fly bioassay, decreased after the moult to adults, but the thoracic ganglia contained more of this substance than the abdominal ganglia.

Recent advances in bursicon-related studies are substantial. However, these studies were carried out on cockroaches, crickets, flies and moths (see, for example, Honegger *et al.*, 2002; Dewey *et al.*, 2004; Luo *et al.*, 2005; Wang *et al.*, 2008; review by Zitnan and Adams, 2005, pp. 46–49). Insofar as locusts are concerned, knowledge is limited to the aforementioned studies. Locust phase polyphenism-related research on bursicon has not been carried out, except by Padgham (1976b), who found that, in *S. gregaria*, the putative bursicon acts differentially on dark hatchlings (sclerotization and darkening) produced mostly by gregarious parents, and on green hatchlings (sclerotization only), produced mostly by solitarious parents. The fact that a single egg pod may yield both dark and light hatchlings (see Section 7.2.1) makes this problem even more interesting.

9.3.6 *Allatotropins and allatostatins*

Although the CA/JH is more or less involved in locust phase polyphenism (Section 9.1), relatively little advancement was made on allatotropic or allatostatic effects in locusts (reviews on insects, including locusts, by Goodman and Granger, 2005, pp. 335–341, and by Raikhel *et al.*, 2005, pp. 465–466).

Allatotropic effects of the brain NSC in locusts were reported in many early publications (for references, see Tobe and Stay, 1985, pp. 380–382). Ferenz and Diehl (1983) found that CC and brain extracts, obtained from crowded females with maturing oocytes of *L. migratoria*, stimulated JH biosynthetic activity of the CA *in vitro*. Studying crowded adult females of *L. m. migratorioides*, Ulrich *et al.* (1985) concluded that a neurohormone, produced by the median NSC in the brain and released by the CC into the haemolymph, exerts an allatotropic effect. Gadot *et al.* (1987) verified the allatotropic effect; they made methanol extracts of brains, CC and suboesophageal ganglion of crowded adult vitellogenic females of *L. m. migratorioides*. The authors eluted the fractions, assayed the effect of these fractions on JH biosynthetic activity of the CA *in vitro* and reported partial purification and characterization of a putative allatotropin. Gadot *et al.* (1987) revealed that the substance is resistant to boiling and susceptible to trypsin and chemotrypsin inactivation. Lehmberg *et al.* (1992) found that CA, excised from male or female adults of *L. migratoria*, and preincubated for 24 h at 4°C, exhibit a reduced basal rate of JH biosynthesis *in vitro*. This condition allowed for reliable testing of allatotropic effects of conspecific brain-CC extracts. The authors found that extracts of 0.1–0.156 brain-CC equivalents exerted a strong allatotropic effect. However, with extracts of higher brain-CC equivalents, JH biosynthetic activity of the CA decreased. Lehmberg *et al.* (1992) reasoned that an allatostatic factor may be present in the extracts, which suppresses the activity of the CA with higher concentrations. A somewhat similar dose–response situation was described by Li *et al.* (2005) on the effect of brain extracts on *in vitro* biosynthetic activity of the CA of *Romalea microptera*, a non-locust acridid. Recently, it was found that proctolin (review in insects by Gäde *et al.*, 1997) increased JH release, *in vitro*, from CA of 20–25-day-old crowded adult females of *L. migratoria*. The increase was the greatest with CA that showed relatively low basal activity (Clark *et al.*, 2006).

Several allatostatins were isolated and identified from locusts (reviews on insects, including locusts: Stay, 2000; Stay and Tobe, 2007). There are three types of allatostatins in insects and crustaceans: (1) cockroach allatostatins (type A), with C-terminal sequence of Y/FXFGL/Iamide; (2) cricket allatostatins (type B), W(X_6)Wamide; and (3) lepidopteran allatostatins. Types A and B are found in locusts; type A is present in *S. gregaria* (Veelaert *et al.*, 1995, 1996; Vanden Broeck., *et al.*, 1996; Vitzthum *et al.*, 1996; Clynen *et al.*, 2003a) and in *L. migratoria* (Veelaert *et al.*, 1995). Type B was found in *L. migratoria* (Schoofs *et al.*, 1991). Lepidopteran allatostatins were not reported from acridids. The types of allatostatins are based on similarity in amino acid sequences of the C-terminal and not by their actual actions. Unfortunately, the 'allatostatins' reported from locusts (see earlier) do not inhibit or suppress JH biosynthetic activity of the CA of locusts, although some effect of a cockroach allatostatin on locusts was recently reported (see later).

The allatostatins found in locusts exert myoinhibitory effects (Schoofs *et al.*, 1991; Vanden Broeck *et al.*, 1996; Veelaert *et al.*, 1996). However, this situation does not mean absence of allatostatic peptides in locusts. Ferenz and Aden (1993) found, in crowded maturing adult females of *L. migratoria*, that a factor originating from the ovaries rapidly inhibits *in vitro* JH biosynthetic activity of the CA. Okuda and Tanaka (1997) reported that a factor in methanol extracts of brain-CC complexes, obtained from crowded adult females of *L. migratoria*, reversibly inhibits JH biosynthetic activity of CA *in vitro*, in both reproductively active (short-day) and inactive (long-day) females. This factor is heat resistant, but loses its activity after digestion with pronase, indicating a heat stabile peptide.

Recently, Clark *et al.* (2008) found some effects of an A-type allatostatin, *Diploptera punctata* allatostatin number 2 (Dippu-AST 2), on JH release *in vitro* from the CA of crowded 7–10-day-old adult virgin females of *L. migratoria*. This allatostatin significantly decreased JH release from CA exhibiting high basal rates, it significantly increased release of JH from glands having low basal rates, but did not affect CA with medium basal rates.

Unfortunately, none of these (quite fragmentary) findings on allatotropins and allatostatins in crowded locusts were investigated in isolated locusts; therefore, no information is available in relation to possible phase-dependent differences in these neurohormones and their effects.

9.3.7 *Other pars intercerebralis originating neuropeptides*

Ovary maturating parsin (OMP), a peptide with a sequence of 65 amino acids, was discovered by J. Girardie *et al.* (1991) in *L. m. migratorioides*. Two isoforms were described, differing only by a single amino acid at position 26. OMP is produced by the pars intercerebralis (PI) of the brain and found in the storage lobe of the CC. It is a gonadotropic neurohormone; daily injection of a specific polyclonal antiserum against Lom-OMP-inhibited vitellogenesis (J. Girardie *et al.*, 1992). The hormone is not needed for previtellogenetic development of the oocytes, nor for chorionation. J. Girardie *et al.* (1996) injected Lom-OMP into last-instar crowded female hoppers of *L. migratoria* and detected vitellogenin in the haemolymph which is not present in this stadium in normal development. They also injected Lom-OMP into crowded allatectomized last-instar female hoppers; this treatment also induced presence of vitellogenin in the haemolymph, demonstrating that the vitellogenesis promoting activity of the hormone is not mediated by CA activation. Lom-OMP did not induce vitellogenin production in last-instar male hoppers (J. Girardie *et al.*, 1996).

Employing several endocrine manipulations in last-instar female hoppers and adult females of *L migratoria*, J. Girardie and A. Girardie (1996) and J. Girardie *et al.* (1998a) concluded that Lom-OMP has two separate gonadotropic actions. Its C-terminal induced vitellogenin production in allatectomized last-instar female hoppers. However, no vitellogenin production was observed when last-instar hoppers were ovariectomized. In strong contrast, injection of Lom-OMP did not induce vitellogenin production in young allatectomized previtellogenetic adult females. However, when already vitellogenetic adult females were allatectomized, injection of Lom-OMP improved the maintenance of the circulating vitellogenin in the haemolymph, despite absence of the CA/JH. J. Girardie *et al.* (1998a) attributed these effects to an ecdysteroidogenic action of the C-terminal of the Lom-OMP on the ovaries and a protecting action of the N-terminal on vitellogenic mRNA in the fat body. However, the effect of injection of graded doses of Lom-OMP on haemolymph ecdysteroid levels did not show a clear dose–response relation (see Table 1 by J. Girardie *et al.*, 1998a).

Using electrophoresis, chromatography and immunological techniques, Ayali *et al.* (1996a) compared Lom-OMP content of the CC in three age groups of crowded and of isolated adult females, and separately, in adult males of *L. m. migratorioides*. These age groups were selected in accordance with the shorter period of sexual maturation in isolated than in crowded *L. m. migratorioides* (see Section 8.1). At the age of 5–9 days after adult emergence, isolated locusts are at the onset of sexual maturation, whereas crowded adults are completely immature. At this age and at the age of 12–16 days, no differences were found in Lom-OMP content of the CC between isolated and crowded adults. However, at the age of 22–28 days, when both isolated and crowded adults were fully mature, but not yet senescent, Lom-OMP content of the CC was considerable higher in crowded than in isolated locusts. No differences were found between males and females in any age group. The physiological interpretation of these findings is not clear. Higher OMP content in the CC of crowded locusts may indicate more storage-less release of the hormone. Laboratory data indicate that fecundity of isolated females of *L. migratoria* is higher than that of crowded females (see Section 8.5); perhaps the higher fecundity of isolated females is correlated with a higher turnover of OMP in the CC, leaving more of the hormone in the CC of the crowded females. However, the function of OMP in adult males is unknown, but OMP content of the male CC is not less than that of the female CC (see earlier).

J. Girardie *et al.* (1998b) also demonstrated the presence of OMP in crowded adults of *S. gregaria*. In contrast to *L. m. migratorioides*, in which two isoforms were detected (see earlier), in *S. gregaria* four isoforms, two long and two short ones, were found. In mature adults, only the two short isoforms were present. For amino acid sequence of Scg-OMP and for comparison of Scg-OMP to Lom-OMP, consult Fig. 4 and Fig. 6, respectively, of the article by J. Girardie *et al.* (1998b). Like Lom-OMP, Scg-OMP exhibits ecdysiotropic activity and it is claimed to induce earlier vitellogenesis, resulting in enhanced oocyte growth.

Neuroparsins (NPs) were isolated from the CC of *L. m. migratorioides* (Boureme *et al.*, 1987; J. Girardie *et al.*, 1987). Amino acid sequences of neuroparsin A (NPA) and of neuroparsin B (NPB) were revealed by J. Girardie *et al.* (1989). NPB was found to be shorter by five residues than NPA. Ayali *et al.* (1996a) demonstrated phase-dependent quantitative differences in NP content of the CC of crowded and isolated adult females and separately of males, in three age groups (see earlier) of *L. m. migratorioides*. NPA content of the CC was markedly lower in isolated than in crowded adults, whereas NPB content was more or less equal, or moderately lower in isolated locusts. Total NPs (NPA+NPB) content of the CC was lower in isolated locusts, especially in 23–31-day-old adults. J. Girardie *et al.* (1998b) discovered NPs also in *S. gregaria*; the peptides were found in the storage lobe of the CC. In later studies, T. Janssen *et al.* (2001) reported two NP precursors and Claeys *et al.* (2003) discovered another two; altogether four NP precursors (NPP1–NPP4) in *S. gregaria*. These peptides showed phase-dependent differences; in fact, they may be considered as molecular markers of locust phase polyphenism (De Loof *et al.*, 2006; Claeys *et al.*, 2006a; Badisco *et al.*, 2007).

NPs are peptides secreted by neurones which have their cell bodies in the PI of the brain and project to the storage lobes of the CC, from which they are released into the haemolymph and serve multiple physiological functions, including anti-juvenile, anti-diuretic, hyper-glycaemic, hyper-lipaemic and neuritogenic effects (Ayali *et al.*, 1996a; Clynen *et al.*, 2006; De Loof *et al.*, 2006; Badisco *et al.*, 2007). Claeys *et al.* (2003) reported four new members of the NP family in *S. gregaria*, termed *Scg*-NPP1–4 (*S. gregaria* NP precursors 1–4). These peptides share a number of structural features with vertebrate insulin-like growth factor binding proteins (IGFBP), which are key regulators of metabolism, growth, reproduction and ageing.

Claeys *et al.* (2005b) measured levels of *Scg*-NPP1 and *Scg*-NPP2 transcripts in the brain, fat body, gut, gonads and accessory glands of adult male and female, solitary- and crowd-reared *S. gregaria* and used quantitative real-time–polymerase chain reaction (RT–PCR) to follow changes in transcript level over time in fat body and brain. Assessments were made every two days from day 2 to day 24 after fledging and once in over 24-day-old adults. Claeys *et al.* (2006a) conducted a similarly detailed analysis for *Scg*-NPP3 and *Scg*-NPP4. Results implicated *Scg*-NPP1-4 in the regulation of reproductive physiology, with distinct phase-dependent differences being apparent, as well as tissue-, sex- and age-dependent differences.

In crowd-reared locusts, *Scg*-NPP1 and *Scg*-NPP2 mRNAs were found only in the brain, whereas in solitary-reared males and females both transcripts were found in the brain, as well as in the fat body, albeit transcripts in the fat body were much lower than in the brain (Claeys *et al.*, 2005b). An initial increase in both *Scg*-NPP1 and *Scg*-NPP2 mRNA was observed in the fat body of solitarious adults at the first appearance of immature oocytes in the ovariole terminals of females and of testis yellowing in males (days 6–8), after which *Scg*-NPP2 transcript levels fell in both sexes and *Scg*-NPP1 mRNA levels fluctuated. Whereas levels of *Scg*-NPP1 transcripts in the brain remained constant during adult development in gregarious males and females, in solitarious males and females there was a pronounced increase in *Scg*-NPP1 mRNA at the onset of sexual maturation (day 10 after adult ecdysis), which persisted until day 20 in males and at least day 24 in females. Levels of *Scg*-NPP2 transcript in the brain were consistently higher in solitary-reared than crowd-reared adults of both sexes. Gregarious adults of both sexes showed a peak on day 8. Solitarious males had highest transcript levels on days 10, 16 and 20, whereas female levels peaked on days 10–12. No transcripts of *Scg*-NPP1 and *Scg*-NPP2 were found in the gonads and accessory glands of either sex (Claeys *et al.*, 2005b).

In adults over 24 days after fledging, transcripts of *Scg*-NPP3 and *Scg*-NPP4 were found by Claeys *et al.* (2006a) in gregarious male and female brain and fat body, and in male testis and accessory glands, but not in female gonads. There were strong phase-related differences in tissue distribution of *Scg*-NPP3 but not *Scg*-NPP4 transcripts. Gregarious males had very much higher quantities of *Scg*-NPP3 mRNA in brain, fat body and accessory glands than did solitarious males. Gregarious females also had substantially higher levels in their fat body, but brain transcript levels of *Scg*-NPP3 in the females were more similar between the phases. Some transcripts of *Scg*-NPP4 were found in the gut of crowd-reared males and in isolation-reared females (see Fig. 1 by Claeys *et al.*, 2006a) Transcript levels in fat body varied over the reproductive cycle. In gregarious males, levels rose steeply from day 10 to day 14, then fell rapidly by day 16, remaining low until rising again on day 24. Solitarious males had generally lower levels of *Scg*-NPP3 mRNA, which were elevated during days 6–10. Fat body levels in gregarious females reached a maximum on day 6, whereas transcript levels were generally low and largely un-modulated in solitarious females. Amounts of *Scg*-NPP4 mRNA in fat body were higher in solitary-reared than crowd-reared males. Values reached a peak in solitarious males from day 6–10, falling by day 12 and increasing again to a lesser extent on days 14–16. Transcript levels were highest in the fat body of gregarious males from days 8–14. In contrast, values for *Scg*-NPP4 mRNA were low and unmodulated over time in the fat body of solitarious females, but elevated on days 6–10 in gregarious females.

Transcripts of *Scg*-NPP3 and *Scg*-NPP4 were monitored in brain, fat body and metathoracic ganglion in solitary-reared adults subjected to crowding for 6 days from the time of adult ecdysis (Claeys *et al.*, 2006a). The latter tissue was investigated in view of work demonstrating that mechanosensory stimulation of the hind legs triggers behavioural gregarization through neural pathways that project to the metathoracic ganglion (Simpson *et al.*, 2001; Rogers *et al.*, 2003; Section 14). Crowding solitarious adults resulted in mRNA values that were shifted towards gregarious levels in each of the tissue types. The effect was particularly strong for *Scg*-NPP4 in adult males. Whether the reverse transition occurs in gregarious adults that are isolated was not tested.

In conclusion, the NPs *Scg*-NPP1–4 would appear to be involved in phase-specific differences in reproductive physiology in *S. gregaria*, not only differing in transcript levels between the two extreme phases but also shifting in solitarious adults in response to crowding. It remains to be discovered exactly how they are implicated in the control of phase-related differences in reproduction, and this offers a fruitful avenue for future research.

Claeys *et al.* (2006b) also showed that in 5-day-old adults of *S. gregaria*, *Scg*-NPP1–4 transcript levels are affected by JH injection and by 20-OH-ecdysone injections. However, these experiments were carried out solely in crowd-reared locusts.

Insulin and related compounds are known not only from vertebrates but also from invertebrates where they are termed insulin-related peptides (IRPs), or insulin-like peptides (ILPs). IRPs in insects were reviewed by Claeys *et al.* (2002) and more recently by Wu and Brown (2006).

Hietter *et al.* (1990) isolated and sequenced a 5-kD peptide from the neurohaemal (storage) lobe of the CC of *L. m. migratorioides* adults. In an accompanying publication, Lagueux *et al.* (1990) reported that this 5-kD peptide is present in the brain, presumably in the PI NSC, and it is the C (connecting) peptide of an insulin preprohormone. Lagueux *et al.* (1990) also found that the domains of this preprohormone are similar to those of known IRPs: signal peptide, B chain, C peptide (the 5-kD peptide revealed by Hietter *et al.*, 1990) and A chain.

After cleavage of the signal peptide, the prohormone, consisting of B chain – C peptide – A chain, undergoes further processing; the A and B chains are linked by disulfide bonds and the C peptide is clipped out after translation. Hetru *et al.* (1991) reported the full characterization of the process and the amino acid sequences of the A and B chains in crowded *L. migratoria*. Kromer-Metzger and Lagueux (1994) cloned the IRP gene of this species and discovered at least two transcripts: LIRP T_1 (L stands for *Locusta*) and LIRP T_2 (see Fig. 4, p. 431, by Kromer-Metzger and Lagueux, 1994). They found both transcripts in the brain; LIRP T_1 was not detected in other organs, but LIRP T_2 was present in a wide variety of tissues. The authors concluded that the gene has two different promotors which could account for a differential regulation and expression of the IRP of *L. migratoria* in the NSC of the brain (a putative neurohormone?) and in other tissues and organs (a putative growth factor?). It may be added that the translated C peptide was found to exert electrophysiological activity, *in vitro*, by affecting membrane conduction of *L. migratoria* neurons, isolated from conspecific thoracic ganglia (Bermudez *et al.*, 1991). From the CC of adults of both *L. migratoria* and *S. gregaria*, Clynen *et al.* (2003b) isolated a translated decapeptide that is located between the signal peptide and the B chain. This decapeptide was found in the PI of the brain, in the CC and in the hypocerebral ganglion, but the authors suggest that the latter could be due to contamination. The peptide, termed as 'IRP copeptide', is able to reduce phosphorylase activity in locust fat body *in vitro*, an effect opposite to that of the adipokinetic hormone (AKH) (see Section 9.3.8).

Recently, Badisco *et al.* (2008) purified and characterized an IRP from *S. gregaria*. By employing immunohistochemical analysis, they found immunoreactivity in the median NSC of the PI, in the nervi corporis cardiaci 1 (NCC 1) and in the CC. The cDNA sequence of this preprohormone (Scg-IRP T_2) and the corresponding sequence of amino acids were revealed and compared with those of *L. migratoria* IRP (see Badisco *et al.*, 2008, p. 143, Fig.3). The IRP copeptide (see earlier), the B and A chains do not differ between the two species; one amino acid residue is different in the signal peptide, but the C peptide is less conserved.

Badisco *et al.* (2008) also studied phase-dependent, as well as gender-dependent, differences in the relative quantity of the Scg-IRP mRNA in brain and fat body of *S. gregaria* adults on day 4 and day 10 after fledging. They found (see Badisco *et al.*, 2008, p. 144, Fig. 4) neither phase-dependent, nor age-dependent differences in the brain of males, but in the brain of crowded females, the relative amount of mRNA decreased from day 4 to day 10. Major phase-dependent differences were found in the fat body of both sexes. About a threefold increase in the transcript level was found from day 4 to day 10 in crowded males, whereas in isolated males a decrease was observed during this period. In the fat body of crowded females, mRNA level greatly increased with age; on day 10 it was about 20 times higher than on day 4. This level also increased with age in the fat body of isolated females, but the increase was only about threefold, and the relative quantity of the transcript was much lower than that found in the fat body of the 10-day-old crowded females. Crowded adults of *S. gregaria* reach sexual maturation earlier than isolated conspecifics (see Section 8.1 of the present review) and the crowded females in the study of Badisco *et al.* (2008) were vitellogenic with well-developed ovaries at the age of 10 days after fledging. It should be reminded that the fat body produces vitellogenin which is eventually taken up by the oocytes. In crowded males the onset of mating behaviour was observed also by around day 10. The authors suggested, therefore, that there may be a link between Scg-IRP transcript level in the fat body and sexual maturation in gregarious *S. gregaria*. However, the available data for isolated locusts do not support such a link; as mentioned earlier, transcript level in the fat body of isolated males decreased from day 4 to day 10, and it was very low in isolated females even on day 10. Perhaps, due to some reason, sexual maturation was much delayed in the isolated locusts, but if so, the authors should have made tests later than day 10, preferably when isolated females reached vitellogenesis with well-developed ovaries and isolated males showed onset of mating behaviour. Also, it would be interesting to study the relation of IRP and maturation in *L. migratoria* that exhibits an opposite trend; the period between fledging and sexual maturation is much shorter in isolated adults (see Section 8.1).

Badisco *et al.* (2008) also found that recombinant *S. gregaria* neuroparsin (NP) 4 (Scg-NP4) was able to bind Scg-IRP. It may be reminded that in the fat body of crowded *S. gregaria* adults, Scg-NP3 and Scg-NP4 are upregulated during sexual maturation (Claeys *et al.*, 2003), similarly to Scg-IRP.

Most research on insulin and IRPs was made on vertebrates in relation to diabetes. Among insects, *D. melanogaster* is the best-studied species. However, findings based on *Drosophila* should be checked and not just extrapolated for locusts. Both recent reviews on IRPs in insects state that these peptides have a key role in metabolism, growth, reproduction and ageing (Claeys *et al.*, 2002; Wu and Brown, 2006). However, the exact physiological effects are somewhat blurred even in *Drosophila* and certainly not clear in locusts. Even the article by Badisco *et al.* (2008) that claims some relation between Scg-IRP and sexual maturation in crowded *S. gregaria* does not clarify the physiological processes underlying such a relation.

To some extent, this lack of exact physiological roles, as well as absence of knowledge of the mechanisms resulting in these putative roles, exists also in regard to some other PI originating neurohormones, like the NPs and to a lesser extent Lom-OMP. The present knowledge strongly indicates that more research should be made to clarify the exact primary or secondary roles of these neurohormones.

9.3.8 *Adipokinetic hormones and fuel for flight*

The effect of the first adipokinetic hormone (AKH), known today as AKH I, or Lom-AKH I, or Locmi-AKH I, was discovered independently in *L. migratoria* by Beenakkers (1969) and in *S. gregaria* by Mayer and Candy (1969). Both studies showed that injection of CC extracts substantially increases lipid, specifically diacylglycerol, levels in the haemolymph of adult locusts. Following this discovery, research on AKHs-related subjects greatly and rapidly advanced, yielding hundreds of research articles and ample reviews. Older reviews include those by Stone and Mordue (1980), Goldsworthy (1983, 1990), Beenakkers *et al.* (1985a,b), Orchard (1987), Wheeler (1989), Goldsworthy and Mordue (1989) and Gäde (1990, 1992). More recent reviews, among others, were presented by Candy *et al.* (1997), Gäde *et al.* (1997), Goldsworthy *et al.* (1997), Oudejans *et al.* (1999), Ryan and Van der Horst (2000), Van der Horst *et al.* (2001), Gäde and Auerswald (2003), Van der Horst (2003) and Van der Horst and Ryan (2005). Pener *et al.* (1997b) reviewed the subject of AKHs in relation to locust phase polyphenism. Very recently, Gäde and Marco (2009) reviewed again AKHs and related neuropeptides in insects and crustaceans, emphasizing the subject in Caelifera, that is, in grasshoppers and locusts, adding new findings and interesting conclusions.

A brief summary on AKHs, as presented here, is based on the aforementioned reviews. AKHs, together with the red pigment concentrating hormone (RPCH) in decapod crustaceans and hypertrehalosaemic/hyperglycaemic hormones in certain insects, constitute the structurally related AKH/RPCH family of neuropeptides (see later). The crustacean RPCH is responsible for concentrating pigment granules, leading to light colouration ('bleaching') of the animal. This neuropeptide was the first to be sequenced in invertebrates (Fernlund and Josefsson, 1972). In insects, neuropeptides of the AKH/RPCH family exert hyperlipaemic or hyperglycaemic response. AKHs are produced in the intrinsic cells of the glandular lobes of the CC in locusts. For the strategy of adipokinetic cells in the CC, see reviews by Van der Horst (2003) and Van der Horst and Ryan (2005).

AKHs have three functions in locusts. A major function is induction of hyperlipaemia, which is to produce lipids as the main fuel for flight. Another function is activation of inactive glycogen phosphorylase in the fat body. The third, possibly indirect, function is to promote lipid transport from the fat body to the flight muscles. Under natural conditions, AKHs are released in locusts from the CC by flight. For short flights and for the beginning of continuous (migratory) flights, locusts utilize carbohydrates, present in small amounts in the flight muscles and in the haemolymph, but then AKHs become effective and lipids constitute the major source of energy for flight. Lipids are stored in the fat body as triacylglycerols. AKHs bind to G-protein-coupled receptors in the cells of the fat body. Receptors were cloned from other insects, but not a single AKH receptor has yet been identified from an orthopteran species (Gäde and Marco, 2009).

The receptor induces two cascades of signal transduction. One is leading to the production of the sterospecific *sn*-1,2-diacylglycerols from triacylglycerols (for a proposed scheme, see Fig. 2 by Gäde and Auerswald, 2003). The other is resulting in the production of trehalose from fat body glycogen reserve, through the activation of glycogen phosphorylase (Fig. 1 by Gäde and Auerswald, 2003). The diacylglycerol produced by the fat body is strongly hydrophobic, but its efficient transfer through the haemolymph to the flight muscles is crucial and mediated by lipophorin conversion in the haemolymph. A high-density lipophorin (HDLp), with limited capacity, is present in the haemolymph, but it is insufficient to transport large amounts of diacylglycerol needed for flight fuel. An exchangeable apolipoprotein, apoLp-III, adheres to the HDLp and changes it to a low-density lipoprotein (LDLp), which has a much greater capacity of lipid transport (cf. Chino *et al.*, 1986 and references therein; review by Van der Horst and Ryan, 2005). Following transport of diacylglycerol to the flight muscles, apoLp-III separates from the LDLp and is available to join other HDLp, constituting an efficient shuttle of lipid transfer (see Fig. 5 by Van der Horst, 2003). Figure 1 by Van der Horst and Ryan (2005) summarizes all three functions of AKHs in locusts.

There are many peptides belonging to the AKH/RPCH family. Gäde *et al.* (1997) already listed 32 amino acid sequences then known; Gäde and Auerswald (2003) mention that there are almost 40, and meanwhile some more were discovered. In general terms, these are peptides with 8–10, exceptionally with 11, amino acid residues, all blocked by pGlu at the N-terminal, all, except one, are amidated at the C-terminal and have some additional common features (see Gäde *et al.*, 1997).

The amino acid sequence of an AKH, known today as Lom-AKH I or Locmi-AKH I, was first revealed by Stone *et al.* (1976). It was found to be a blocked decapeptide, pGlu-Leu-Asn-Phe-Thr-Pro-Asn-Trp-Gly-Thr amide. In *L. migratoria*, three AKHs are known: Lom-AKH I, II and III. In *S. gregaria*, there are two AKHs; AKH I is the same peptide as Lom-AKH I, but Scg-AKH II has a different amino acid sequence. However, in the Caelifera, grasshoppers and locusts, additional AKHs are known (see later and Gäde and Marco, 2009).

In the adipokinetic cells of the CC of *S. gregaria* two different mRNAs are translated to prepro-AKH I and prepro-AKH II. Further procession within these cells results in pro-AKH I and pro-AKH II. At this stage, three dimers are formed each by a disulfide bond; one is a homodimer of pro-AKH I, the other is a homodimer of pro-AKH II and the third is a heterodimer of pro-AKH I+pro-AKH II. In the final stage, AKHs are cleaved from the prohormone, leaving behind the AKH-precursor-related peptides (APRPs) (for details of these processes consult O'Shea and Rayne, 1992, and references therein). The situation is similar in *L. migratoria*, except that this species has three AKHs (Oudejans *et al.*, 1991; Bogerd *et al.*, 1995). There are few differences between APRP of AKH I and APRP of AKH II; the processes leading to these APRPs are similar to those in *S. gregaria*. However, APRP of AKH III substantially differ from the above two, and its amino acid sequence is much longer (see Fig. 2 by Van der Horst, 2003, for APRP sequences, termed in that figure as AKH-associated peptides and shortened as APPs). More recently, Baggerman *et al.* (2002) reported that the three dimers (see earlier) are further processed to AKH-joining peptide 1 (AKH-JP I) and 2 (AKH-JP II) before cleaving the actual AKHs. The physiological function, if any, of APRPs and AKH-JPs are unknown (Baggerman *et al.*, 2002, and references therein). In the CC of 12–16-day-old crowded adults of *L. migratoria*, Oudejans *et al.* (1999, and references therein) reported that the molar ratio, AHH I : AKH II : AKH III is 14:2:1.

There are many lipid- and AKH-effect-related phase-dependent differences. Ayali and Pener (1992) demonstrated that in adults of *L. m. migratorioides*, resting values of haemolymph lipid concentration are higher in crowded than in isolated males in age groups of 10–19 and 24–30 days after fledging. These authors also found that injection of CC extract, or injection of synthetic Lom-AKH I, resulted in a markedly higher increase of haemolymph lipid levels (hyperlipaemia) in crowded than in isolated males.

These findings were confirmed by Ayali *et al.* (1994) who reported that, similar to Lom-AKH I, injection of Lom-AKH II also induced markedly higher hyperlipaemia in crowded than in isolated adult males of *L. m. migratorioides* in the age groups of 12–16 and 27–29 days after fledging. Additional confirmation of higher adipokinetic response in crowded than in isolated adult males of *L. m. migratorioides* was provided by Ayali and Pener (1995).

Matthée (1945) found a higher total body lipid content in crowded than in isolated last-instar hoppers of *L. migratoria* and *Locustana pardalina*. Uvarov (1966, Table 25) summarized these and other data on fat content, as a percentage of wet weight, in locust hoppers. Cheu (1952) investigated total body lipid content in isolated and crowded males, and separately in females, of *L. m. migratorioides*, at different ages of adults. He found about two times higher body lipid content in crowded than in isolated males at the age of 1 day, 1 week and 2 weeks, but not at the age of 5 and 7 weeks after fledging. In females, body lipid content was again higher in crowded than in isolated locusts at the age of 1 day and 1 week after fledging. However, at the age of laying the first egg pod, isolated females showed more or less the same body lipid content as did crowded females and after laying the sixth egg pod, body lipid content was higher in isolated than in crowded females. Ayali and Pener (1995) also assessed total body lipid content in adult males of *L. m. migratorioides* in an age group of 14–16 days after fledging and found about three times more lipids, 0.280 and 0.091 g in crowded and isolated locusts, respectively. However, crowded males of this species are larger, and so heavier, than isolated ones (see Section 5.1) and after correction according to body weight, total lipid content of crowded males was about only twice as high than that of the isolated males at this age. This result is compatible to that found by Cheu (1952) for 2-week-old males.

The relationships between lipid- and AKH-related parameters and change of density were also investigated in 14–16-day-old adult males of *L. m. migratorioides* by Ayali and Pener (1995). Resting haemolymph lipid levels, hyperlipaemia induced by injection of Lom-AKH I and total body lipid content (see earlier) were assessed in isolation-reared hoppers that were newly aggregated and in crowd-reared hoppers that were newly separated, within 12 h after fledging. These parameters were assessed also in continuously isolated and continuously crowded males of the same age. Comparison of continuously isolated and continuously crowded males yielded a good correlation between total body lipid content and Lom-AKH I-induced hyperlipaemia. Hyperlipaemic response and total body lipid content of the newly separated and newly aggregated males did not differ significantly from the respective values obtained from continuously isolated males. Therefore, the period of 12–14 days elapsing between fledging and assessing the parameters was insufficient to alter hyperlipaemic response and total lipid content of the newly aggregated males. However, total body lipid content of the newly separated males was significantly higher than that of newly aggregated and continuously isolated males. The same period of 12–14 days was sufficient to decrease the hyperlipaemic response of the newly separated males to that shown by continuously isolated males and to lose (or not to gain) some of the total body fat content. In other words, change from crowding to isolation results in a more rapid adaptation of lipid- and AKH-related parameters than change from isolation to crowding. In contrast to the case of continuously isolated and continuously crowded males (see earlier), the data obtained for the newly separated and newly aggregated males yield no significant correlation between total body lipid content and hyperlipaemic response. Eventually, Ayali and Pener (1995) concluded that the differential adipokinetic response of crowded and isolated adult males of *L. m. migratorioides* is affected by both phase-dependent differences in total body lipid content and phase-dependent differences, not related to total body lipid content.

Ayali *et al.* (1996b) studied the effect of flight on adipokinetic responses of crowded and isolated adult males of *L. m. migratorioides* in two age groups, 12–16 days and 27–32 days after fledging. Locusts were flown in individual flight tunnels, for various durations up to 60 min and haemolymph lipid levels were assessed 90 min after the initiation of flight. As already mentioned, flight is the natural stimulus that induces release of AKHs from the CC, leading to hyperlipaemia (see earlier). In the younger age group, crowded locusts showed no adipokinetic response after 10 min of flight, but exhibited marked hyperlipaemic response already after 20 min of flight and this response did not increase further after 30, 45 and 60 min of flight. Isolated locusts of the same age group showed no response after 10 and 20 min of flight, showed a moderate response after 30 min and a maximum response after 45 and 60 min of flight. The maximum response of the isolated and the crowded locusts flying 45 or 60 min did not differ significantly, but it may be reminded that the isolated locusts reached this maximum after longer duration of flight. A different picture was obtained with the 27–32-day-old males. Adipokinetic response of the crowded males increased gradually from 20, through 30 and 45 min, to 60 min of flight. Maximum hyperlipaemia, obtained after 60 min of flight, was approximately 50% higher than the maximum exhibited by the younger males. Isolated males in the age group of older males showed no hyperlipaemic response after 20 and 30 min of flight, a very moderate response after 45 min and a moderate response after 60 min of flight. The latter was only about 50% of the response of the crowded males of the same age group.

The age of the older group of adult males, used by Ayali *et al.* (1996b) to study phase-dependent differences in the effects of flight on hyperlipaemia, was 27–32 days after fledging. This age is roughly comparable to the age of

Cheu's (1952) 5-week-old locusts, in which total body lipid content was found to be more or less similar in isolated and crowded males. Despite this similarity, Ayali *et al.* (1996b) found marked differences in flight-induced hyperlipaemia between isolated and crowded males in the older age group (see earlier). These findings further emphasize the conclusion (see earlier) that there are phase-dependent differences in adipokinetic response which are not related to total body lipid content. In other words, the correlation between total body lipid content and adipokinetic response seems to be less in flight-induced than in AKH (or CC extract) injection-induced hyperlipaemia. Nevertheless, phase-dependent differences in flight-induced hyperlipaemia may also be affected by different flight performance between isolated and crowded locusts in laboratory experiments. Investigating neural correlates to flight-related density-dependent phase characteristics in *S. gregaria*, Fuchs *et al.* (2003) found that in 7-day-old adult females wind stimuli resulted in a significantly higher probability of flight initiation in crowded than in isolated females. In 30-day-old females, no flight was induced in isolated locusts at any wind stimuli tested (see also Section 11.6).

Studying adult females of *S. gregaria*, Schneider and Dorn (1994, p. 887) stated that resting haemolymph lipid levels of isolated and crowded locusts are 'roughly comparable' during ageing of the adults from day 3 to day 28 after fledging. However, inspection of Fig. 4 of Schneider and Dorn (1994) indicates a somewhat higher resting lipid concentration in crowded than in isolated females. The comparability claimed by the authors is based mainly on the values of older, 24- and 28-day-old, females. If these are disregarded, resting haemolymph lipid levels are higher in crowded than in isolated females in 8 of 9 tests carried out with females younger than 24 days, and in one instance they are similar. Ogoyi *et al.* (1996), exploring the effect of AKH on hyperlipaemia, found no differences in resting haemolymph lipid levels of crowded and isolated males of *S. gregaria* at the age of 3–4 weeks after maturation. This finding is compatible with the results of Schneider and Dorn (1994) who found no phase-dependent differences in resting haemolymph lipid levels of 24- and 28-day-old females of the same species, although the locusts of Ogoyi *et al.* (1996) were of different sex and presumably older than those of Schneider and Dorn (1994).

Schneider and Dorn (1994) also investigated the resting fatty acid composition of the triacylglycerols in the fat body of crowded and isolated adult females of *S. gregaria*, every few days, along a period of up to 28 days after fledging. They found in both isolated and crowded locusts that more than 95.5% of the examined fatty acids have chain lengths of 12–18 carbon atoms. No qualitative phase-dependent differences were revealed, but a significantly higher proportion of linoleic and linolenic acids were found in isolated than in crowded females, whereas on most days a significantly higher proportion of oleic acid was detected in crowded females.

Schneider and Dorn (1994), as well as Schneider *et al.* (1995), tested adipokinetic response after 60 min of flight in crowded and isolated adult females of *S. gregaria*, again every few days, along the period of up to 28 days (Schneider and Dorn, 1994), or only up to 22 days (Schneider *et al.*, 1995) after fledging. Crowded females exhibited a considerably higher adipokinetic response than isolated females up to the age of 20 days. However, after this age, phase-dependent differences in flight-induced adipokinetic response decreased, and in one instance, at the age of 24 days (see Fig. 4 of Schneider and Dorn, 1994), isolated females showed a higher adipokinetic response than crowded ones. Schneider and Dorn (1994) found a higher fat body weight, expressed as a percentage of locusts' wet weight, in crowded than in isolated females of *S. gregaria*, up to about 20 days post-fledging. However, after 20 days, the phase-dependent difference decreased. Considering the similarity of this decrease in fat body weight, the decrease of adipokinetic response, and even haemolymph lipid resting levels (see earlier) in females older than 20 days, it may be inferred that there is some positive correlation between fat body weight and adipokinetic response. It may be recalled that a positive correlation was found between total body lipid content and AKH-injection-induced adipokinetic response in *L. m. migratorioides* (see earlier). In contrast, however, in *L. m. migratorioides* flight-induced hyperlipaemic response appears to be less related to fat content (see earlier and Ayali *et al.*, 1996b) than in *S. gregaria*.

Ogoyi *et al.* (1996) also assessed AKH I-induced adipokinetic response in crowded and isolated adult males of *S. gregaria* at the age of 3–4 weeks after maturation. They found no phase-dependent difference in resting haemolymph lipid levels (see earlier). Injection of 2 pmol of AKH I per locust resulted in a somewhat higher adipokinetic response in crowded than in isolated males. Injection of 5 or 10 pmol of AKH I resulted in the absence of phase-dependent adipokinetic response up to 60 min after the injection and in a slightly higher adipokinetic response in isolated than in crowded males 90, 120 and perhaps 150 min after the injection (Ogoyi *et al.*, 1996, Fig. 1). In other experiments (Ogoyi *et al.*, 1996, Figs. 2 and 3), adipokinetic responses were assessed 90 min after the injection of graded doses of AKH. Up to the dose of 2 pmol per insect, adipokinetic responses of crowded males were slightly higher than those of the isolated males, but with higher doses isolated males exhibited a higher hyperlipaemia than crowded ones.

The results of Ogoyi *et al.* (1996) were obtained with *S. gregaria* males, at the age of 3–4 weeks after maturation, following injection of AKH I, whereas those of Schneider and Dorn (1994) were obtained with 24–28-day-old conspecific females following flight. Despite these differences, the results seem to be quite similar. It may be recalled that in 24-day-old adult females of *S. gregaria*, Schneider and Dorn (1994) found a higher adipokinetic response in isolated than in crowded locusts (see earlier).

In later experiments, Dorn *et al.* (2000) tested adipokinetic response, induced by 60 min of flight, in crowded and isolated males and females of *S. gregaria*, at the age of 14 days after fledging. In the same study, the authors also tested hyperlipaemic effects of graded doses of Lom-AKH I (identical to Scg-AKH I) on crowded and isolated females. Their results are presented in Fig. 16 of Dorn *et al.* (2000). Resting haemolymph diacylglycerol levels did not differ between crowded and isolated males, but were significantly lower in isolated than in crowded females. Adipokinetic response was phase dependent, being considerably higher in crowded than in isolated adults of either sex after 60 min of flight. These results are contrasting to those of Ayali *et al.* (1996b), obtained in 12–16-day-old adult males of *L. m. migratorioides*, which showed no difference in adipokinetic response after 60 min of flight (though marked differences were recorded after 20 or 30 min of flight; see earlier). Dorn *et al.* (2000) found that injection of 1 pmol of AKH I did not induce hyperlipaemia in 14-day-old adults of *S. gregaria*; injection of 10 and 100 pmol resulted in a dose-dependent adipokinetic response in crowded females, whereas isolated females showed a maximum response already with injection of 10 pmol. The maxima obtained for crowded and isolated females injected with AKH I were comparable to the maxima observed, after 60 min of flight, by crowded and isolated females, respectively.

Undoubtedly, flight or injection of AKH induce strong adipokinetic response in both *L. m. migratorioides* and *S. gregaria*. This response is higher in crowded than in isolated adults, at least up to the age of 32 days after fledging in *L. m. migratorioides* (no older locusts were tested) and up to the age of 20 days in *S. gregaria*. Nevertheless, as discussed earlier, many details are different in these two species. There may be several reasons for these differences. Ayali *et al.* (1996b) assessed adipokinetic response in males of *L. m. migratorioides*, 90 min after initiation of flight, regardless of flight duration, whereas Schneider and Dorn (1994) did this immediately after cessation of the flight in females, or after cessation of a 60-min flight in both sexes (Dorn *et al.*, 2000). Also, 14 days after fledging, when flight tests were carried out by Dorn *et al.* (2000), crowded *S. gregaria* females have vitellogenetic oocytes and were more or less sexually mature (see maturation time in Schneider *et al.*, 1995), whereas 12–16 days after fledging, when flight tests were carried out by Ayali *et al.* (1996b), crowded adults of *L. m. migratorioides* are at an early state of sexual maturation (see also Section 8.1). Finally, two different species were tested, probably with species-related differences in phase-dependent migratory strategies (see later).

Fatty acid compositions of diacylglycerols in the haemolymph before and after 60 min of flight in crowded and isolated adult females of *S. gregaria* were assessed by Schneider and Dorn (1994) every few days from 3 to 28 days after fledging. Compared with resting levels, the percentage of polyunsaturated fatty acids (linoleic and linolenic acids) was generally lower after flight and the percentage of oleic acid increased in both crowded and isolated females. Phase-dependent differences were also related to these fatty acids; in isolated females, the relative proportion of polyunsaturated fatty acids was higher than in crowded females, both before and after flight.

Ogoyi *et al.* (1998) assessed phase-dependent triacylglycerol lipase activity in the fat body of *S. gregaria*. This lipase is necessary for hydrolysis of triacylglycerols to diacylglycerols that are released from the fat body (see earlier). Unfortunately, the authors do not state the age and sex of the locusts used; they refer to the rearing conditions as described by Ogoyi *et al.* (1996), but do not mention sex and age. Assuming that these were the same as in the study of Ogoyi *et al.* (1996), males were used 3–4 weeks after maturation. This is, however, uncertain and it is possible that the authors used mature locusts without paying attention to sex and age.

Ogoyi *et al.* (1998) found that resting lipase levels do not differ substantially between isolated and crowded locusts. Injection of 10 pmol of AKH activated fat body lipase in both isolated and crowded locusts with maximum response occurring 30 min after injection. Up to this 30 min, AKH-induced lipase activity was higher in crowded than in isolated locusts. After 30 min lipase activity declined and no more difference was seen between crowded and isolated adults. Injection of 2 pmol of AKH also increased lipase activity in isolated and crowded locusts 30 min after injection, but the activity in crowded locusts was only moderately higher than that in isolated ones. Kinetic data suggested that the lipase had a higher catalytic ability, but a lower affinity to the substrate, in crowded than in isolated locusts.

Ayali *et al.* (1996c) studied phase-dependent differences in AKH content of the CC in adult males of *L. m. migratorioides* in two age groups: 12–19 and 25–30 days after fledging. In the age group of 12–19 days, less Lom-AKH I and Lom-AKH II were found in the CC of crowded than of isolated males. In the 25–30-day-old group, AKH I and AKH II content of the CC did not differ significantly between isolated and crowded males. AKH I and AKH II content of the glands increased from 12–19 to 25–30 days in crowded males, but no such increase occurred in isolated males. Consequently, there were no significant differences in AKH I and AKH II content of the CC between the age of 12–19 and 25–30 days in isolated males, neither between these isolated males and 25–30-day-old crowded males. The ratio of AKH I : AKH II was about 6.5 in the CC of crowded males in the younger age group, but it decreased to about 4.5 in the older crowded males. The latter ratio did not differ significantly from the ratio of about 5.0 in the CC of the younger and the older isolated males. The absolute values (in pmol) of AKH I and AKH II content of the CC, as found by Ayali *et al.* (1996c), were somewhat lower than relevant data in the literature on crowded males, but the trend of increasing AKH content of the CC with ageing and the AKH I and AKH II ratio in crowded locusts were compatible (see, for example, Siegert and Mordue, 1986, for males and Oudejans *et al.*, 1993, for females).

In the same study, Ayali *et al.* (1996c) presented a dose–response curve of adipokinetic response of 11–15-day-old crowded adult males of *L. m. migratorioides* to injection of CC extracts made by the same method as assessing AKH content of the CC. The ED_{50} was found to be about 0.01 CC equivalents and maximum response was elicited by 0.05 CC equivalents.

Ayali *et al.* (1996a) found a consistently higher level of AKH precursor-like peptides (APRPs = AKH I and II APRP together) in the CC of isolated than in crowded males of *L. m. migratorioides* ageing 5–9, 12–16 and 22–31 days and in females ageing 6–9 and 26–30 days after fledging. Higher APRP content indicates a higher AKH content in the CC. The findings of Ayali *et al.* (1996c; see earlier) for 12–19-day-old adult males of *L. m. migratorioides* support this conclusion, but the results obtained for 25–30-day-old males do not. Unfortunately, we cannot offer a satisfactory explanation for the inconsistency in relation to the older males; except that probably AKH content of the CC is variable (see later).

The ergoplasmatic (intracisternal) granules in the AKH cells of the CC of *L. m. migratorioides* represent accumulating storage of adipokinetic prohormones. Diederen *et al.* (1999) found that size is larger and number of these granules is more frequent in isolated than in crowded males of *L. m. migratorioides* 43 days after fledging. These results seem to be compatible with the findings of Ayali *et al.* (1996a). However, it should be kept in mind that storage of the prohormone, which contains AKH+APRP before their cleavage, may not reflect the amounts of AKH and APRP after their cleavage.

Different results were reported by Dorn *et al.* (2000, pp. 212–214, including Fig. 4). These authors investigated AKH I content of the CC, every 2 days, from day 2 to day 10 after fledging, in males, and separately in females, of isolated and crowded adults of *L. m. migratorioides*. In contrast to Siegert and Mordue (1986) and Oudejans *et al.* (1993), Dorn *et al.* (2000) reported that at the age of 10 days, AKH I content of the CC was less than at the age of 2 days in crowded males, at the age of 4 days in isolated males and at the age of 2 and 4 days, in both isolated and crowded females. Also, on day 6, 8 and 10, AKH I content of the CC was higher in crowded than in isolated males and females, contradicting the results of Ayali *et al.* (1996c) in regard to AKH I, albeit in older, 12–19-day-old males. The findings of Dorn *et al.* (2000) also contradict the circumstantial evidence arising from the results of Ayali *et al.* (1996a). The latter study reported higher APRP content of the CC in isolated males and females than in crowded males and females even at the age 5–9 days, similar to the age of the locusts (see earlier) used by Dorn *et al.* (2000).

Employing HPLC separation and mass spectrometry, Clynen *et al.* (2002) assessed AKH and APRP levels in the CC of crowded and isolated males, and separately of females, of *S. gregaria*. Comparison of the findings of Clynen *et al.* (2002, p.112 and Table 1) in *S. gregaria* to those observed in *L. m. migratorioides* by Ayali *et al.* (1996a,c) reveals similarities and dissimilarities. The ratio of AKH I : AKH II was higher in crowded than in isolated males of *S. gregaria* at all ages tested, 2, 8, 16 and 25 days after fledging, and it was similarly higher in 2-day-old females. These results of Clynen *et al.* (2002) for *S. gregaria* males are compatible with the results obtained by Ayali *et al.*

(1996c) for *L. m. migratorioides* males. However, in females the results were different. Clynen *et al.* (2002) found that in 8- and 25-day-old females, AKH I : AKH II ratio was similar in crowded and isolated females, whereas Ayali *et al.* (1996c) found that this ratio was higher in isolated than in crowded females at the age of 6–9 and 26–30 days (see earlier). In 16-day-old males of *S. gregaria*, AKH I and AKH II, as well as APRPs content of the CC was higher in isolated than in crowded locusts. The latter findings are consistent with those of Ayali *et al.* (1996a,c) in *L. m. migratorioides*. However, Clynen *et al.* (2002) reported higher APRPs content of the CC in crowded than in isolated males of *S. gregaria* at the age of 8 and 25 days, contrasting with the results obtained for *L. m. migratorioides* males by Ayali *et al.* (1996a). Also, AKH I and AKH II content of the CC were higher in crowded than in isolated males, but similar in crowded and isolated females of *S. gregaria* on day 25. Ayali *et al.* (1996a) found such similarity in 25–30-day-old males (females were not tested).

Comparison of the results reported by Clynen *et al.* (2002) in *S. gregaria* to those found by Dorn *et al.* (2000) in *L. m. migratorioides* regarding AKH I also reveals similarities and dissimilarities. Dorn *et al.* (2000) found higher AKH I content of the CC on day 2 in isolated than in crowded females, whereas Clynen *et al.* (2002) obtained this result in males, but not in females of the same age. Both Clynen *et al.* (2002) and Dorn *et al.* (2000) found higher AKH I content in crowded than in isolated males on day 8. However, the results in females at this age were different; in *S. gregaria* AKH I content of the CC was similar in crowded and isolated females, whereas in *L. m. migratorioides* crowded females exhibited a higher AKH I content of the CC than did isolated females.

Clynen *et al.* (2002) concluded that in *S. gregaria*, the relative amounts of AKHs and APRPs between isolated and crowded adult locusts vary with age and depend on the gender. Obviously, the inconsistency between the results in *S. gregaria* and those in *L. m. migratorioides* may be related to the fact that two different species were studied. However, there are inconsistent findings also for *L. m. migratorioides*. Different methods of extraction and assessment of AKH from the CC may lead to different results (see Table 1 by Ayali *et al.*, 1996c). Also, there may be differences between different laboratory strains of the same species. Finally, AKH and APRP content of the CC depends on biosynthesis, storage and release of the hormones and their precursor peptides and the balance between these factors may differ according to possible, yet unknown, environmental influences. In conclusion, AKH and APRP content of the CC may not be a very consistent phase characteristic. Dorn *et al.* (2000, p. 214) also concluded that "the amount of stored AKH in the CC is not, or not strongly, phase-related".

AKH is known to activate inactive glycogen phosphorylase in the fat body (cf. Gäde, 1981; Van Marrewijk *et al.*, 1983), and this effect is also induced by flight (Van Marrewijk *et al.*, 1980). Ayali *et al.* (1994) found a significantly higher total (active+inactive) phosphorylase activity in isolated than in crowded adult males of *L. m. migratorioides* in an age group of 11–18 days after fledging. These authors observed no phase-dependent differences in the proportion of fat body glycogen phosphorylase activity15–20 min after injection of 10 pmol of AKH I.

Ayali *et al.* (1996b) investigated resting haemolymph carbohydrate levels, effect of AKH-injection and effect of 30-min flight on these levels in crowded and isolated males of *L. m. migratorioides* in two age groups, 12–16 and 27–30 days after fledging. They found no significant differences in resting carbohydrate levels, approximately 20–22 mg ml^{-1}, between crowded and isolated males ageing 12–16 days, nor between younger and older crowded males. However, this level was significantly higher, approximately 26 mg ml^{-1}, in 27–30-day-old isolated males than in the younger or older crowded males and in the younger isolated males.

In the same study, Ayali *et al.* (1996b) injected 1 or 20 pmol of AKH I, or 10 pmol of AKH II into the 12–16-day-old males, whereas the 27–30-day old males were injected with saline and graded doses of AKH I, from 0.5 to 10 pmol. Carbohydrate content of the haemolymph was assessed 90 min after injection. Decreases of up to 5 mg carbohydrate per millilitre haemolymph were observed with the higher doses, 2, 5 and 10 pmol of AKH I in the older locusts and with 10 pmol of AKH II in the younger ones. However, in strong contrast, injection of 20 pmol of AKH I into the younger males resulted in a decrease of only 1–2 mg ml^{-1} of haemolymph carbohydrate content.

The effect of 30-min flight on haemolymph carbohydrate levels was tested by Ayali *et al.* (1996b) only in the younger, 12–16-day-old, males. These tests were made 90 min after onset of flight. Flight-induced decrease of carbohydrate levels in the haemolymph was practically the same in crowded and isolated males, −44.6% and −43.3% relative to resting levels, respectively. This decrease was similar to those found by other authors for crowded locusts (cf. Jutsum and Goldsworthy, 1976; Van der Horst *et al.*, 1980).

Dorn *et al.* (2000) found that the amount of carbohydrates, mostly glycogen, in the fat body of crowded and isolated males and females of adult *S. gregaria* at the age of 14 days after fledging is phase dependent. Isolated males and females contained more than three times as much carbohydrates than the respective genders of crowded locusts. The causative factors for this difference may be related to restricted movements of the less-active isolated locusts, which are usually confined to small cages/containers, or to a possibly higher competition for food in crowded cultures, or both (for phase-related nutritional effects see Section 17.2.1).

Dorn *et al.* (2000) recorded no phase- or gender-dependent differences in resting haemolymph trehalose levels, which were approximately 20 μg ml^{-1}. This value and the absence of phase-dependent differences are compatible with the results of Ayali *et al.* (1996b) obtained for resting haemolymph carbohydrate levels in 12–16-day-old adult males of *L. m. migratorioides* (see earlier). Dorn *et al.* (2000) reported a considerable decrease of haemolymph trehalose concentration after 1 h of flight in *S. gregaria*, while carbohydrate content of the fat body decreased between approximtely 15% in isolated males and more than 50% in crowded females. The decrease of haemolymph trehalose concentration after 1 h of flight in *S. gregaria* (Dorn *et al.*, 2000) was somewhat higher than the decrease of haemolymph carbohydrate concentration after 30 min of flight in *L. m. migratorioides* (Ayali *et al.*, 1996b).

As already outlined, transport of diacylglycerol through the haemolymph, from the fat body to the flight muscles, is mediated by LDLp, which is produced by reversible joining of apoLp-III to HDLp (see earlier). On removal of the fatty acids in the flight muscles, LDLp and apoLp-III dissociate and the free apoLp-III is available to join again HDLp, creating an efficient shuttle for diacylglycerol transfer.

Chino *et al.* (1992) investigated phase-dependent differences in the production of LDLp. These authors found that field-collected solitarious adult males of *L. migratoria* were unable to produce LDLp, even following injection of 10 pmol of AKH-I, despite having apoLp-III in the haemolymph. Chino *et al.* (1992) found very small amounts of triacylglycerols in the fat body of the solitarious males, only about one twentieth, or even less, than in crowded males from a laboratory colony. Therefore, Chino *et al.* (1992, p. 101, Summary) concluded that in solitarious locusts ''the stores of fuel in the fat body are insufficient to maintain prolonged flight''.

However, migration of solitarious locusts does take place in the field (cf. Farrow, 1990; see also later) and for migration solitarious locusts must have fat reserves. The solitarious and crowded locusts compared by Chino *et al.* (1992) probably originated from two different geographical strains (subspecies), with different life cycles, of *L. migratoria*. Solitarious adult males were collected in the field, near Sapporo in Hokkaido, the northernmost main island of Japan. The Hokkaido strain of *L. migratoria* is univoltine, with an embryonic diapause (Tanaka, 1994) and a long pre-oviposition period under long days (Hakamori and Tanaka, 1992). The age and history of the solitarious males used by Chino *et al.* (1992) are uncertain, and it is possible that these locusts were collected after solitarious migration, resulting in depletion of their lipid reserves. In contrast, the crowded males of Chino *et al.* (1992) were taken from laboratory colonies. Although not stated, probably these laboratory colonies belonged to the polyvoltine strain of *L. m. migratorioides* (sensu Farrow and Colless, 1980), which has no embryonic or adult reproductive diapause. Other studies demonstrate that isolated adult males of *L. m. migratorioides* have substantial lipid reserves, even if these reserves are less than those of crowded males, at least in the first parts of the adult's life (see earlier). Unfortunately, no data seem to exist on the fat content of solitarious young (before migration) adults of *L. m. migratorioides* under field conditions; data compiled by Uvarov (1966) does not distinguish between laboratory-maintained isolated locusts and field-captured solitarious locusts.

Ogoyi *et al.* (1995) investigated phase-dependent differences in lipophorins of adult males of *S. gregaria*, at the age of 3–4 weeks after fledging. They found that after 90 min following injection of 10 pmol of AKH I, there was formation of LDLp, from HDLp and apoLp-III, in both isolated and crowded locusts. In isolated locusts, lipid content of HDLp and LDLp was estimated to be 48.4% and 57.1%, respectively. In crowded locusts, lipid content of 51.8% and 59.8% was estimated in HDLp and LDLp, respectively. The densities of lipophorins were found to be lower in crowded locusts. Diacylglycerol content in LDLp was considerably higher in crowded than in isolated males. Ogoyi *et al.* (1995) concluded that these differences may reflect adaptations associated with high metabolic requirements for migration by gregarious locusts.

In a subsequent article, Ogoyi *et al.* (1996) investigated again haemolymph lipophorin profiles in isolated and crowded adult males of *S. gregaria*, at the same age and by the same experimental procedures as in their former article (Ogoyi *et al.*, 1995).They found that in response to injection of 10 pmol of AKH I, the LDLp formed in crowded locusts exhibited a larger molecular size than the LDLp formed in isolated locusts.

In contrast, Dorn *et al.* (2000), citing results of Sickold (1997) Ph.D. thesis, reported that mature adults of *S. gregaria* form lipophorin of identical densities and structures in both sexes and under both crowded and isolated conditions. HDLp of resting locusts did not differ between males and females, nor between isolated and crowded locusts, and injection of AKH I produced the same LDLp, regardless of sex and phase.

Orchard *et al.* (1993) reviewed the multifunctional role of octopamine in locust flight; according to these authors, octopamine acts as neuromodulator and neurotransmitter; it induces release of AKHs from the CC and, acting as a neurohormone, it directly induces an adipokinetic response in the fat body in the initial period of flight before the action of AKHs takes place. Goosey and Candy (1980) found that octopamine level in the haemolymph of sexually mature, 20–32-day-old, adults of *S. gregaria* (under the name *S. americana gregaria*) increases within the first 10 min of flight.

Fuzeau-Braesch and David (1978) reported higher octopamine content in whole heads of isolated than of crowded *L. migratoria*. Presumably, their study was carried out on *L. m. cinerascens* from Sardinia because they refer to another study (Fuzeau-Braesch, 1977) in which *L. m. cinerascens* from Sardinia was employed. In another article, Fuzeau-Braesch *et al.* (1979) assessed octopamine content of heads during the last hopper stadium and in adults at three different ages, again in *L. m. cinerascens* from Sardinia. They found higher octopamine content in isolated than in crowded females, both in hoppers and in adults, except at ecdysis, when octopamine levels were similarly low in crowded and isolated females. In males, octopamine levels were higher in isolated than in crowded last-instar hoppers; regardless of phase, the levels were low and similar in adult males on day 1 and day 5 after fledging and again higher in isolated than in crowded sexually mature males. These results and those of Fuzeau-Braesch and David (1978) were cited by Fuzeau-Braesch and Nicolas (1981, p. 294, Table 5). Benichou-Redouane and Fuzeau-Braesch (1982) investigated octopamine content in several components of the nervous system in isolated and crowded adults of *L. m. cinerascens*, 15–20 days after fledging, and again reported higher octopamine content in isolated than in crowded locusts.

Morton and Evans (1983) studied octopamine distribution in isolated and crowded *S. gregaria* (under the name *S. americana gregaria*). They found neither phase-related, nor sex-related differences in octopamine content of several components of the nervous system and of several muscles, as well as of whole heads. Morton and Evans (1983) strongly criticized the works of Fuzeau-Braesch and coworkers (see aforementioned references).

Dorn *et al.* (2000), citing Ress' (1997) Ph.D. thesis, reported that resting haemolymph octopamine content in a laboratory strain of crowded *S. gregaria* and in conspecific field catches of solitarious locusts, did not reveal significant phase-related differences. The authors added that octopamine may also act as a neurotransmitter or neuromodulator, but this aspect was not investigated by them.

Rogers *et al.* (2004) studied phase-dependent differences and changes in neurotransmitters and neuromodulators, including octopamine, in last-instar hoppers and adults of *S. gregaria* (see Section 10.3). They investigated three samples, brain, optic lobes and thoracic ganglion chain, in last-instar hoppers and the same samples in adults, except that pro- and metathoracic ganglia only were made into separate samples. Octopamine content of the central nervous system (CNS) differed by less than 10% between long-term crowded and third generation of isolated last-instar hoppers. Similarly, no significant differences were found in the CNS between long-term crowded and second generation of isolated adults. Crowding of last-instar hoppers in the third generation of isolation, resulted in a decrease of octopamine levels in the CNS 4 h after initiation of crowding. However, 24 h after crowding, a 13-fold increase in octopamine content of the optic lobes and a sevenfold increase in that of the thoracic ganglia were observed (see Section 10.3). Rogers *et al.* (2004) stated that their results and those of Fuzeau-Braesch and coworkers (see aforementioned references) are incompatible, except if it is assumed that handling procedures inadvertently led to behavioural gregarization of the isolated locusts of Fuzeau-Braesch and coworkers. This means, however, that the timing between putative behavioural gregarization and assessment of octopamine content was well over 4 h, which is improbable. Nevertheless, it should be kept in mind that different species, *S. gregaria* and *L. migratoria cinerascens*, are concerned.

Unfortunately, no comparison has been made between isolated and crowded locusts to explore possible flight-related differences in octopamine function(s), release, content and its possible changes in the haemolymph, or in the nervous system, although a related study was reported in *Melanoplus sanguinipes* by Min *et al.* (2004; see also later).

Veelaert *et al.* (1997) found that crustacean cardioactive peptide (CCAP) stimulates, in a dose-dependent way, *in vitro* release of AKH from the glandular lobes of the CC of crowded adults of both *S. gregaria* and *L. migratoria* at the age of 12–14 days after fledging. The authors do not exclude other factors with similar effect, but claim that CCAP is the most efficient factor. Unfortunately, no phase-related studies of this effect have been carried out.

Migratory flight of locusts requires appropriate fuel reserves. The Australian plague locust, *Chortoicetes terminifera*, constitutes a well-demonstrated case. Hunter (1989) showed that in Australian arid areas, occasional adequate rainfall induces both hatching of the eggs of this species and germination of ephemeral grasses. If rainfall is sufficient to keep the grass green during hopper development, the subsequent adults swarm and migrate. However, if rainfall is insufficient to keep the grass green during late hopper development, the resulting adults have no sufficient lipid reserves for migratory flight (see also Section 5.1 in relation to the effect of such conditions on size of the adults). More recently, Clissold *et al.* (2006) reinforced the conclusion that rainfall during hopper development affects adult migration. They found that the effect of rainfall is mediated through the availability of the annual grass, *Dactyloctenium radulans*. This grass produces green shoots rapidly after rainfall. However, if rainfall is then insufficient and this grass dies before the hoppers complete development, the adult locusts do not have resources for migration and outbreak.

In both *L. m. migratorioides* and *S. gregaria*, as well as in some other locust species, gregarious adults exhibit reiterative movements during their active life. Usually, displacement of gregarious field population takes place in the so-called rolling swarm motion (Uvarov, 1977, pp. 206–207; Farrow, 1990, pp. 276–277; and references in these reviews; see also Section 11.5). In such motion, most locusts are feeding on the ground, but those at the rear of the swarm fly forward and land at the front edge of the swarm. Farrow (1990, p. 277) estimated some data for a 1-km-long swarm on the ground from front to rear edge. According to him, the locusts would spend more than 90% of the day on the ground and would undertake approximately 10 flights of 4 min each. Under such conditions, swarm displacement on the ground would add up to approximately 10 km per day. For such short flights locusts may utilize carbohydrates, without adipokinesis, and the carbohydrates may be replenished during a period, estimated to be approximately 60 min, spent on the ground between flights.

Although this rolling motion may be the most frequent movements of locust swarms, it is certainly not the sole kind of swarm displacement. When a swarm reaches a barrier, such as wide open water courses, or areas where food is absent, long-distance migration takes place. It is known that locust fly on the wind (Uvarov, 1977; Rainey, 1989; Farrow, 1990); therefore, long-distance migration may occur also as a result of high-velocity atmospheric currents. Such wind currents bring the locusts to areas of low atmospheric pressure, where rainfall probably occurs, providing sufficient conditions (wet soil) for egg laying. In *S. gregaria*, daily swarm displacements of up to 130 km were observed (Rainey, 1989, p. 63) and displacements of 50–100 km per day are not exceptional (Uvarov, 1977, pp. 210–211; Rainey, 1989, pp. 63–64). In October 1988, swarms of *S. gregaria* crossed the Atlantic, from West Africa to the Caribbean and the northern coastal areas of South America, in a continuous flight of several thousands kilometres (Kevan, 1989; Ritchie and Pedgley, 1989; Richardson and Nemeth, 1991; Rosenberg and Burt, 1999; Lorenz, 2009) unless they landed on floating drowned insects on the ocean and cannibalized them. This theory was created by Waloff (1960) who cited ship records on floating dead locust patches in the Atlantic. She speculated that living locusts may stand and feed on the floating dead locusts. However, Lorenz (2009, p. 6) concluded that the ''possibility that locusts could stop over and feed on a floating mass of dead locusts that had fallen into the sea due to exhaustion has not been substantiated by any direct evidence''. Rainey (1989, p.168) mentions that *L. m. migratorioides* may cross eastward the Mozambique Channel, implying a continuous flight over 400 km. Obviously, for long flights locusts need and use fat as fuel.

Some molecular biologists, biochemists and even physiologists, unfamiliar with locust field biology, do not realize that solitarious locusts also migrate, or at least may migrate, although their migration is individual and usually occurs at night (Uvarov, 1977, pp. 74–75, 363–367 and amply references therein). In Farrow's (1990) extensive study on flight and migration in acridoid insects, he concluded that ''population displacement can be the result of both solitarious individuals (particularly during recession) and gregarious aggregations during plagues'' and ''swarm movements are generally less extensive than those undertaken by solitary individuals of migratory species'' (Farrow, 1990, p. 300 and 304, respectively).

However, there are differences in the migratory strategies of solitarious *L. m. migratorioides* and *S. gregaria*. In the former, ''possibly only one night of prereproductive flight occurs *between* fledging and breeding areas, but it is supplemented by several nights of less extensive movement by maturing and ovipositing individuals *within* the breeding areas'' (Farrow, 1990, p. 240). When habitats are short lived, emigration of maturing adults with large fat bodies to complementary breeding areas is the norm (Farrow, 1975).

The migratory strategy of solitarious *S. gregaria* seems to be more complex and less well understood. There are migrating and non-migrating solitarious adults (Waloff, 1962) and reproduction without migration may occur when weather conditions are locally suitable (Farrow, 1990, p. 241). However, migratory displacement of solitarious *S. gregaria* may be similar to that of conspecific gregarious swarms (Waloff, 1966), possibly when rain-bearing weather systems are in developing stages. Rainfall in the areas of distribution of *S. gregaria* is usually more sporadic, and the habitats are living even shorter than in the distribution area of *L. m. migratorioides*. Consequently, solitarious individuals of *S. gregaria* may take reiterative migrations, even during the reproductive period, to change the breeding area when the soil dries out at one location. In these instances, solitarious individuals should migrate, presumably with atmospheric currents that bring them, like the gregarious swarms, to low pressure areas where rainfall and wet soil may be available, However, if the first breeding area does not dry up, no migration of solitarious *S. gregaria* may take place.

Laboratory data in relation to adipokinetic response of these two species may fit their migratory strategies in the field. Considering the relatively more limited lipid reserves in immature, maturing, or freshly matured, isolated than in crowded adults of *L. m. migratorioides* (Cheu, 1952; Ayali and Pener, 1995; see earlier), these reserves may be focused on a single long-distance migration from fledging to breeding area. If so, induction of a full adipokinetic response for short, trivial flights would be disadvantageous. Indeed, 12–16-day-old isolated adults reach full adipokinetic response as late as 40 min after initiation of flight, whereas crowded adults reach that already after 20 min of flight (Ayali *et al.*, 1996b, Fig. 2A; see also earlier). In older isolated adults, well over the physiological age of long-distance migration of solitarious locusts in the field, even longer initial flights do not induce an adipokinetic response as high as that observed in crowded locusts (Ayali *et al.*, 1996b, Fig. 2B; see earlier), or in the younger isolated locusts, despite that at this older age, isolated locusts have not less fat reserves than crowded ones (cf. Cheu, 1952 and see earlier) The strategy of gregarious *L. m. migratorioides* may be characterized by an initially higher lipid content as fuel for possible reiterative migrations. It seems to be feasible that in younger adults, flight-induced AKH release results in an earlier adipokinetic response in gregarious than in solitarious locusts, and in older adults, it results in a more intense adipokinetic response in the former than in the latter.

Phase-dependent differences in adipokinetic response and lipid-related parameters exist also in *S. gregaria* (cf. Schneider and Dorn, 1994; Dorn *et al.*, 2000; see also Ogoyi *et al.*, 1995, 1996 [though doubted by Dorn *et al.*, 2000, see earlier]), but these differences seem to be less extreme, at least in older adults, than in *L. m. migratorioides*. Possibly, this situation reflects the more opportunistic migratory strategy of solitarious adults of *S. gregaria*. On one hand, they should be ready for considerably long-range displacements, like conspecific gregarious adults, but on the other hand, if food and wet soil are available at their fledging locality, they may not migrate at all.

Ziegler *et al.* (1988) made an interesting discovery. Sexually mature, flightless adult males of *Barytettix psolus* exhibited very feeble adipokinesis in response to injection with extracts of their own CC, or to injection of Lom-AKH I, despite having fat body with sufficient amounts of lipids for a strong response. However, injection of 0.1 pair equivalents of CC of *B. psolus* into *Melanoplus differentialis* and *Schistocerca americana* adults induced a considerable adipokinetic response in the former and a high response in the latter. These authors also showed that the haemolymph of *B. psolus* contains apoLp-III, but this apoprotein does not associate with HDLp, which is also present in the haemolymph, and after injection of CC extract or AKH, no LDLp is formed. However, injection of ^{35}S-labelled HDLp of *B. psolus* and AKH into *S. americana* led to the formation of labelled LDLp in the recipients. Injection of ^{35}S-labelled apoLp-III and AKH into *S. americana* also resulted in ^{35}S-labelled LDLp. Therefore, the feeble adipokinetic response of *B. psolus* to injection of extracts of their own CC is not caused by the absence of the necessary components of adipokinesis, namely, adequate fat body, AKH and its release from the CC, HDLp and apoLp-III. Extracts of the CC of *B. psolus* induced glycogen phosphorylase activity in its own fat body, but haemolymph carbohydrate level did not change.

It turned out that the defective adipokinetic response of *B. psolus* is not unique. In a recent article, a mixture of a review and of original studies, Gäde and Marco (2009, see their Table 3 and references therein) showed that grasshoppers, especially flightless species, do not exhibit, or exhibit a feeble, or a very moderate adipokinetic response to injection of extracts of their own CC, or to injection of Lom-AKH I. The only exceptions are two species of the genus *Melanoplus* that do exhibit a considerable adipokinetic response, which is nevertheless less than that shown by most locust species (see later). However, *M.sanguinipes* and *M. differentialis* may be considered as migratory grasshoppers, or even as less typical locusts (see Section 3.1). *M. sanguinipes* utilizes lipids for flight by the usual adipokinetic process (cf. Kent *et al.*, 1997; Taub-Montemayor *et al.*, 1997, 2002). In *M. sanguinipes*, there are good-flier and bad-flier adults. The former, making at least 60 min of continuous tethered flight, are considered as migrants and have a higher adipokinetic response; the latter, making only a short tethered flight, are considered as non-migrants and have a less intense adipokinetic response (Min *et al.*, 2004 and references therein).

Gäde and Marco (2009) speculate that the defective adipokinetic response of the grasshoppers could be related to the fat body; perhaps triacylglycerol lipase is not activated, or the AKH receptors may be downregulated, or have different affinity (but it may be remembered that glycogen phosphorylase is activated, so it is probably not the receptor, but the secondary messenger for lipase activation is affected). Obviously, additional experimental studies are needed to discover the cause(s) of such defective adipokinetic response.

Gäde and Marco (2009, Table 2) also list the sequences of all known AKHs of Caelifera, according to superfamilies, and in Acridoidea also according to families and subfamilies. Finally, they present the primary structure of AKH/RPCH family peptides in Caelifera, in three groups of peptides, with a putative 'ancestral' peptide in each group (Gäde and Marco, 2009, Fig. 2). According to their assumptions, these 'ancestral' peptides gave rise, mostly (but not exclusively) by single point mutation to other caeliferan AKH sequences. These three ancestral peptides are:

(1) pGlu-Leu-Asn-Phe-Thr-Pro-Asn-Trp-Gly-Ser-NH_2, a decapeptide of *Phymateus morbillosus* (Phymo-AKH);
(2) pGlu-Leu-Asn-Phe-Ser-Thr-Gly-Trp-NH_2, an octapeptide of AKH II of *S. gregaria* (Schgr-AKH-II); and
(3) pGlu-Leu-Asn-Phe-Thr-Pro-Asn-Trp-NH_2, an octapeptide of AKH of *Pyrrhocoris apterus*, a heteropteran bug, but this octapeptide may be considered as a truncated form of Phymo-AKH.

9.3.9 *Additional neuropeptides and some remarks*

There are many neuropeptides in locusts (for some reviews see Section 9.3.1). However, no phase-related studies have been carried out on most of these neuropeptides, possibly because maintenance of isolated locust colonies is time- and laboratory-space consuming, and therefore, expensive. Even when phase-dependent differences were discovered in regard to some neuropeptides, knowledge is often fragmentary. Also, new locust neuropeptides may be discovered in the future. Although the subject attained interest and considerable advancement has been made in the last decade or two, relevant research is still in its infancy. There are some candidates that may yield important findings. The putative neurohormone involved in yellowing of adult males of *S. gregaria*, as suggested by Sas *et al.* (2007; see also Section 7.3), is such a candidate to reveal phase-dependent differences, if this neurohormone does indeed exist. Another interesting subject is gonadotropic neuropeptides (review by De Loof *et al.*, 2001), especially because reproduction-related differences exist between locust phases (see Section 8). Also, the recent review by Verlinden *et al.* (2009) outlines that the neurohormones, FMRFamide and sNPF, are involved in locust reproduction, but possible phase relations of these neurohormones have not been explored. Although some phase-dependent differences have been discovered in relation to Lom-OMP, locust NPs and Scg-IRP (see Section 9.3.7), these cover only very limited aspects of this subject. Isaac *et al.* (2009) reviewed data on neuropeptidases and metabolic inactivation of neuropeptides in insects, including locusts. The review by Verlinden *et al.* (2009) also mentions angiotensin-converting enzyme (ACE) in *L. migratoria*, although not emphasizing its neuropeptidase activity, but its presence in reproductive tissues, mainly in testes. However, Isaac *et al.* (2009) state that strong immunoreactivity to ACE was found within the NSC of locusts. This statement is based mainly on the study by Isaac *et al.* (1998), reporting such immunoreactivity in certain NSC of the brain, in the nervi corpori cardiaci, the storage part of the CC, the nervi corpori allati and in the suboesophageal ganglion of *L. migratoria*. The exact roles of the various neuropeptidases in locusts are not well known, and no information is available on possible phase-related aspects of this subject. In conclusion, more relevant research is needed on locust neurohormones and it may result in fruitful findings.

10 Biochemistry and molecular biology

In recent years, there has been an increasing effort to seek biochemical and molecular markers of locust phase polyphenism. In the words of one of the main research teams involved in this search, ‘‘For physiologists, the most intriguing question is whether there is such a thing as a primordial molecular driver of phase transition’’ (De Loof *et al.*, 2006). Although much valuable data have been collected, no such driver has yet been found – nor, it might be argued, would a single master controller be expected, given the complex make-up of phase transition, with its multiple elements that change over different time scales and do not seem all to share the same mechanisms. The search for the biochemical and molecular mechanisms of phase in locusts is further complicated when it is considered that the factors involved in the initiation of phase transition need not be the same as those controlling its subsequent maintenance and that differences exhibited between extreme phase insects may be the result, rather than the cause, of phase change.

10.1 LARGE-SCALE GENE EXPRESSION STUDIES

Uncovering the molecular genetics of phase change remains as the 'final frontier' in locust research. Such an understanding is necessary to explain the mechanisms and evolution of phase polyphenism and might provide the opportunity for future development of gene-targeted control techniques. Progress has been limited by the fact that the genome has yet to be sequenced for any locust species – a problem which is exacerbated by the huge size of the genomes of grasshoppers and locusts. Acridids possess the largest of all known insect genomes, spanning 6–17 picograms mass of haploid DNA content. *L. migratoria* has one of the smallest acridid genomes (6 pg), whereas the genome of *S. gregaria* is 9 pg. These values compare with 0.18 pg for *D. melanogaster*, 0.53 pg for *B. mori*, 0.2 pg for *Apis mellifera*, and 3.5 pg for the human genome (Animal Genome Size Database; www.genomesize.com).

Despite the absence of a sequenced locust genome, Kang *et al.* (2004) have recently provided an important opportunity for exploring the molecular genetics of phase transition. These authors used whole carcasses and samples of head, hind legs and gut to derive cDNA libraries from which they produced an expressed sequence tag (EST) library and database for *L. migratoria*. They generated 76,012 ESTs comprising 12,161 unigene clusters. Kang and collaborators have since used their EST library to develop a public transcriptomic database (http://www.locustdb.genomics.cn.org) (Ma *et al.*, 2006) and a 60-mer oligonuclotide microarray, representing 10,000 unigenes, for additional gene expression analyses (Kang *et al.*, unpublished). A complementary project is underway to develop an EST library from the CNS of *S. gregaria*, based on head, thoracic and abdominal ganglia (De Loof *et al.*, 2006; Verlinden *et al.*, 2009).

Kang *et al.* (2004) used the EST library to compare gene expression patterns in 2–3 generation solitary-reared and crowd-reared fifth-instar *L. migratoria* nymphs, followed by validation using quantitative polymerase chain reaction (Q-PCR). They identified 532 genes that were differentially expressed between the two phases, most of which were downregulated in gregarious relative to solitarious insects.

The challenge is to discover which of the 532 genes that were differentially expressed between the phases are involved in particular phase traits and which genes are the causes rather than the consequences of phase transition. Since approximately 70% of the 532 genes have no homologue in any sequenced insect genome, the task of narrowing the list down to those which initiate and/or maintain phase change, and then attributing a function to these genes, is daunting. So far as they could be attributed a biological function, genes that were differentially expressed between the phases served a wide range of roles. In the midgut, genes involved in morphogenesis, cell growth, nucleic acid metabolism and cell communication were downregulated in gregarious nymphs, whereas genes for carbohydrate metabolism were upregulated. In the hindlegs, genes involved in morphogenesis, muscle development, cell growth, metabolism, cell motility and muscle contraction were all downregulated in gregarious nymphs; whereas in the head, genes involved in responding to external stimuli were upregulated in gregarious nymphs, and metabolic genes and those involved in protein and nucleic acid biosynthesis were downregulated relative to solitary-reared insects (Kang *et al.*, 2004). Zhou *et al.* (2006) further reported that transcripts of various putative chemoreceptive proteins were expressed at very different levels in the heads of the two phases.

In recent years various methods have become available that can extend the analysis of large-scale gene expression patterns to organisms for which the genome has not been sequenced, and these offer promise for studies on locust phase transition until full genomic data become available (Feder and Mitchell-Olds, 2003; Fitzpatrick *et al.*, 2005; Robinson *et al.*, 2005; Vasemägi and Primmer, 2005). Several studies have demonstrated the use of heterologous microarray hybridizations using both cDNA and oligonucleotide arrays as tools for functional gene analyses in related species for which an array does not exist (Enard *et al.*, 2002; Gu and Gu, 2003; Ji, W. *et al.*, 2004b; Renn *et al.*, 2004; Nowrousian *et al.*, 2005; Kassahn *et al.*, 2007). Hence, the arrays for *L. migratoria* might be used to explore gene expression patterns in other oedipodine locust species, such as *Chortoicetes terminifera*. A complementary approach to this would be to conduct genome-wide cDNA amplified fragment polymorphism gene expression assays (cDNA-AFLP) (Bachem *et al.*, 1996; Breyne *et al.*, 2003) to screen for candidate genes that are differentially expressed between the solitarious and gregarious phases. This protocol has been widely applied in non-model organisms [plants (Bensch and Akesson, 2005), vertebrates (Rubinstein *et al.*, 2000) and insects (Reineke *et al.*, 2003)] and can detect differences in expression patterns in more than 10,000 genes (Donson *et al.*, 2002; Breyne *et al.*, 2003). Having identified differentially expressed candidate genes and confirmed their expression patterns using quantitative real-time polymerase chain reaction (qRT–PCR), the effects of these genes could then be assessed using RNA interference (RNAi) gene knockout (Fire *et al.*, 1998) in conjunction with appropriate assays of locust phase state. Fortunately, although RNAi appears to be ineffective in some insect lineages (Roignant *et al.*, 2003), systemic RNAi competence was recently demonstrated in grasshoppers by Dong and Friedrich (2005), who suppressed the expression of an eye colour pigment gene for 10–14 days following the injection of corresponding dsRNA into the haemolymph of early-instar nymphs. Another promising development is the recent profiling of the small RNA transcriptome in the two phases of *L. migratoria* by Wei *et al.* (2009). These authors found marked differences in small RNA expression levels between the phases (embryos, hoppers and adults were pooled) and suggested that these might mediate gene expression differences between the phases.

As mentioned earlier, a project is underway to complete an EST library for *S. gregaria* (De Loof *et al.*, 2006; Verlinden *et al.*, 2009). Prior to that, Rahman *et al.* (2003c) used differential display reverse transcriptase polymerase chain reaction (DDRT–PCR) in combination with semi-quantitative RT–PCR to seek differences in gene expression in the brains of young adult solitary- and crowd-reared *S. gregaria*. PCR amplification with 26 primers resulted in eight bands that were differentially expressed between the phases, of which two were analysed further, one that was particularly expressed in solitary-reared locusts and the other in crowd-reared insects. These two bands were partially sequenced and compared with existing genomic databases. The 'solitary-specific' gene had no homologue, whereas the 'gregarious-specific' gene shared 80% sequence homology with *D. melanogaster* SPARC protein (secreted protein, acidic and rich in cysteine). This glycoprotein is associated with the extracellular matrix and binds Ca^{++}. It is known to affect mobility and morphology in nematode worms, but any role in phase change remains to be determined.

10.2 PEPTIDES AND PROTEINS

Recent developments in mass spectrometry, new genomic databases such as the *L. migratoria* EST library and rapid advances in bioinformatical techniques have provided new opportunities to seek peptides and proteins associated with phase polyphenism. So far most of the work involved describing differences with rearing density, rather than exploring the functional relevance of different compounds to phase polyphenism, but progress is being made. Neuropeptides are discussed in Section 9.3 and a recent review of the literature is provided by Verlinden *et al.* (2009).

10.2.1 *Phase-related haemolymph proteins and peptides*

Colgan (1987) sought isozyme markers of phase in the haemolymph of young nymphs of *L. migratoria*. Nymphs from eggs laid by crowd-reared parents were reared either alone or in crowds and sampled for electrophoresis at days 1, 2, 4, 6 and 14 after hatching. Of the 23 enzymes surveyed, differences were found between solitary- and crowd-reared nymphs in levels of two families of glycolytic enzyme. Two glycerol-3-phosphate dehydrogenase isoenzymes were found at higher levels in 1-day-old solitary-reared nymphs, although the significant difference between solitary- and crowd-reared nymphs had disappeared by day 4. Solitary rearing also induced a different haemolymph aldolase isozyme profile over days 1–6, which had disappeared by day 14. Oral dosing of crowd-reared first-instar nymphs with JH 1 did not result in a shift towards the solitary-reared isoenzyme condition, nor did injection of second-instar nymphs with JH 1. The relevance of these results to phase polyphenism, if any, remains obscure.

Three recent studies have used screening techniques to compare polypeptide profiles in the haemolymph of solitarious and gregarious locusts. Wedekind-Hirschberger *et al.* (1999) used 2D gel electrophoresis to generate polypeptide maps for mature male *S. gregaria*. A total of 238 silver stained spots (equating to that number of polypeptides) were recorded, ranging in size from 14 to 205 kD and 3–6 pH. Of these, 20 peptides were differentially displayed in solitarious and gregarious insects, of which three were found in the blood of isolated-reared laboratory insects and wild-caught solitarious males from Mauritania but were not found in crowd-reared laboratory animals or field-collected gregarious locusts. The remaining 17 polypeptides were present only in crowd-reared laboratory and field-caught males. When solitarious locusts from Mauritania were reared under crowded conditions in the laboratory for two generations, 14 of the 17 'gregarious' polypeptides had appeared, while one of the 3 'solitarious' polypeptides was no longer expressed. When crowd-reared laboratory males were treated on the day of adult ecdysis with a single dose of JH analogue, fenoxycarb, nine of the 17 polypeptides were not present 15 days later, whereas two of the solitarious polypeptides had appeared in the haemolymph (see also Section 9.1). Until the identity and functions of the various compounds is known, it is not possible to interpret what these changes in polypeptide profile mean.

Clynen *et al.* (2002) used HPLC and matrix-assisted laser desorption ionization time of flight mass spectrometry (MALDI-TOF MS) to compare peptide profiles in the haemolymph of solitary- and crowd-reared male and female adult *S. gregaria* (2 days of age after adult ecdysis). One peak, comprised of several masses, was detected only in solitary-reared adults of both sexes. There were quantitative differences between the phases in other peaks, which were found in higher concentrations in gregarious haemolymph.

Rahman *et al.* (2002a), who also used HPLC and MALDI-TOF MS to explore haemolymph peptide profiles in adult *S. gregaria*, reported somewhat different phase-related patterns in haemolymph peptide profiles to those documented by Clynen *et al.* (2002). No mention is made, for example, of Clynen *et al.*'s 'solitarious' peak. However, Rahman *et al.* (2002a) did find a peptide of molecular mass 6.08 kD that occurred in substantially higher concentrations in the blood of gregarious insects of both sexes (up to 0.1 mM). This was later confirmed by Rahman *et al.* (2008b) to be the same compound as the 6075-kD peptide originally reported by Clynen *et al.* (2002) to be in higher concentrations in the haemolymph of gregarious than solitarious insects. Haemolymph concentrations fell with four successive generations of solitary rearing (Rahman *et al.*, 2002a). This 54 amino acid peptide was not recognizable in the *Drosophila* genome database and its function in locusts remains unknown, even though some possible effects on phase characteristics were tested. Hence, injection of the compound on days 0, 2 and 4 of the fourth stadium in solitary-reared nymphs had no effect on their coloration, morphometry or behavioural phase state as fifth-instar nymphs. The compound was not a protease inhibitor [it did not demonstrate inhibition of chymotrypsin or angiotensin converting enzyme (ACE) activity (see Verlinden *et al.* (2009) for a brief review of the literature on ACE in locusts), nor did it have any antibiotic activity against various bacteria and fungi]. The peptide was reported to be found in the freshly laid eggs of crowd-reared females at higher concentrations than in solitary-reared females (Rahman *et al.*, 2003b). When injected into solitary-reared females after laying their first egg batch, the peptide appeared in somewhat elevated concentrations in the next egg batch, indicating incorporation from the haemolymph into developing eggs and prompting the suggestion that it may play a role in determining the phase state of hatchlings (Rahman *et al.*, 2003b). Such an effect has yet to be measured, however.

Rahman *et al.* (2008b) measured levels of the 6.08-kD peptide in the haemolymph of fourth and fifth-instar and adult *S. gregaria* and studied whether it had a role in synthesis of yellow colour protein, or in the production of the adult male-produced volatile, phenylacetonitrile (PAN) (see Section 12). In nymphs, titres of the peptide were highest approaching the moult, whereas in both solitary and crowd-reared adults, there was a peak at around day 15 after fledging, with another peak at around day 9 in males but not females. Levels were less variable and somewhat lower in solitarious than in gregarious adults, but differences appear marginal (no statistical comparison was made between the phases). Injection of the peptide into adults at 48-h intervals from day 6 until day 12 resulted in an increase in yellow protein mRNA (YP mRNA), which was apparent by day 15 but not earlier, and declined soon afterwards without visible yellowing of the cuticle. We have already outlined that the 6.08-kD peptide claimed by Rahman *et al.* (2008b) to induce yellow protein transcription is not identical to the putative brain-CC peptidergic hormonal factor that was shown by Sas *et al.* (2007) to induce yellow protein mRNA transcription. These presumably controversial issues were fully discussed in Section 7.3, including the fact that Rahman *et al.* (2008b) do not cite the article by Sas *et al.* (2007), despite that both articles are from the same laboratory and authored partially by the same persons. There was no effect of peptide injection on the amount of PAN released by crowd-reared adult males. In short, the role of the 6.08-kD peptide remains to be discovered, but whether it plays a part in phase change seems unlikely.

In a further study, Rahman *et al.* (2008a) investigated the site of synthesis of the 6.08-kD peptide using immunocytochemical techniques, followed by confirmation of the presence of peptide using HPLC and mass spectrometry. No immunoreactivity was found in fat body, haemocytes, midgut, or Malpighian tubules, but strong activity was discovered in mature follicle cells of the ovary and the seminal vesicle tubules of the male accessory gland complex. A pair of neurosecretory cells (NSC) in the sub-oesophageal ganglion connected to the brain through nervi corporis allati-II (NCA-II) also stained positive, as did fibres in nervi corporis cardiaci (NCC-I), which run from the brain to the storage lobes of the CC. No activity for the peptide was found in haemolymph or other tissue samples from *L. migratoria*.

Another HPLC peak which differed in size in the haemolymph of solitary- and crowd-reared adults of *S. gregaria* was identified by Rahman *et al.* (2002a) as a serine protease inhibitor, SGPI 2, which had previously been isolated from ovaries and the CC of *S. gregaria* by Hamdaoui *et al.* (1998) and by Clynen *et al.* (2002), respectively. In contrast to the 6.08-kD peptide, concentrations increased in the haemolymph with successive generations of solitary rearing (Rahman *et al.*, 2002a). This peptide is one of the so-called 'pacifastin' family.

10.2.2 *Pacifastins*

The pacifastins are a family of serine protease inhibitors found in the blood and central nervous system (CNS) of arthropods. The name pacifastin derives from the original isolation of these peptides from the crayfish, *Pacifastacus leniusculus* (Liang *et al.*, 1997). The inhibitory subunit of members of the pacifastin family characteristically comprises six cysteine residues, which form three disulphide bridges to produce a globular folding pattern. The biological functions of pacifastins are incompletely understood, but they are known to play a role in the innate immune system, inhibiting the prophenoloxidase (PO) activating system (see Simonet *et al.*, 2003 for a review). Phenoloxidase catalyses the oxidation of phenolic compounds to melanin, resulting in wound healing, melanization and encapsulation of invading organisms (Siva-Jothy *et al.*, 2005).

To date, 8 pacifastin-like precursors encoding 22 different peptides have been identified in locusts (5 in *S. gregaria* and 3 in *L. migratoria*) (Hamdaoui *et al.*, 1998; Vanden Broeck *et al.*, 1998; Simonet *et al.*, 2002a–c, 2003, 2004a, 2005; see Verlinden *et al.*, 2009). When the *L. migratoria* EST database was searched, *L. migratoria* pacifastin-like precursor (LMPP) mRNAs were found to be widespread, being identified in midgut, hindleg and head tissues (Kang *et al.*, 2004; Clynen *et al.*, 2006).

Kang *et al.* (2004) found from their EST library that the unigene cluster coding for LMPP 2 was expressed at higher levels in solitarious than in gregarious *L. migratoria* nymphs. Consistent with this, as discussed earlier, Rahman *et al.* (2002a) reported higher levels of one of the pacifastins (SGPI 2) in the haemolymph of solitary-reared than crowd-reared *S. gregaria* adults, with concentrations rising across four progressive generations of solitary-rearing. A peptide of the same mass as SGPI 2 was reported at elevated levels in the CC of solitarious but not gregarious adult *S. gregaria* (Clynen *et al.*, 2002), perhaps suggesting that they may play some role in neurosecretory processes (Simonet *et al.*, 2003).

Given that pacifastins are known to inhibit the PO-activating system, at least in *L. migratoria* (Brehélin *et al.*, 1991; Boigegrain *et al.*, 1992), the implication is that elevated levels of pacifastins would correspond with reduced immunocompetence in solitarious relative to gregarious locusts. Wilson *et al.* (2002) reported just such a difference, with gregarious adult *S. gregaria* demonstrating higher survival than solitarious adults after topical application of spores of the fungal pathogen, *M. anisopliae* var. *acridum*. However, no difference was found between the phases in phenoloxidase activity or encapsulation response, although gregarious locusts showed higher antibacterial activity and had marginally higher haemocyte counts. Both phases showed similar behavioural fever responses. Franssens *et al.* (2008) explored the effect of SGPIs on PO activation in haemolymph of 10-day old crowd-reared adult *S. gregaria*. Neither SGPI 1 nor SGPI 2 inhibited the induction of PO activity in response to challenge by the immune elicitor laminarin (a glucan from brown algae); however, increased levels of transcripts of two pacifastin-like peptide precursors (SGPP 1 and 2), which encode SGPI 1-3, were found in the fat body of locusts that were injected 20 h previously with laminarin. No other tissue (brain, foregut, hindgut or gonads) exhibited this increase in transcript levels, according to unreported data. Following reports that *L. migratoria* serine protein inhibitor LMPI 1inhibits fungal trypsins (Kellenberger *et al.*, 2003), Franssens *et al.* (2008) suggested that perhaps pacifastin-related protease inhibitors in locusts serve a different immune function to the PO system, for example, preventing fungal penetration.

Simonet *et al.* (2004a,b, 2005) quantified levels of transcripts of several pacifastin-like peptide precursors (SGPP 1-3, two isoforms of SGPP 4, 4a and 4t, and SGPP 5) in various tissues of mature adult *S. gregaria*. Among these precursors, SGPP 1 encodes the SGPI 2 peptide that was measured in the haemolymph by Rahman *et al.* (2002a). Marked phase- and tissue-dependent differences were found, by Simonet *et al.*, but contrary to expectations based on haemolymph SGPI 2 levels, *higher* levels of mRNA were found in the tissues of gregarious than solitarious adults, with levels of SGPP 1, SGPP 3 and SGPP 4 being especially high in the fat body of gregarious males, and SGPP 2 reaching its highest level in the brain of gregarious males. SGPP 5 mRNA was only measured in gregarious insects (Simonet *et al.*, 2004a) and was highest in foregut of both sexes, with males but not females possessing relatively high levels in fat body.

Having higher levels of SGPP mRNA in tissues of gregarious than solitarious insects is somewhat surprising if it is presumed that levels of peptide precursor mRNA are positively related to levels of product peptides. Simonet *et al.* (2004b) comment briefly on this apparent anomaly, suggesting that the explanation may lie in the fact that the solitarious locusts used by Rahman *et al.*

(2002a) were only up to four generations solitary-reared, yet by the time of Simonet *et al.*'s (2004b, 2005) experiments, the culture had reached greater than 20 generations of solitary rearing. This would require that pacifastin peptide levels initially rise above gregarious levels with early generations of rearing in solitary culture, but then fall back to below gregarious levels with more extended solitary rearing. Whether this happens has not been tested. As an aside, it is noteworthy that a feature of the solitary-rearing system used in the Leuven laboratory (Hoste *et al.*, 2002b,c) is to maintain animals in continuous culture, rather than maintaining a steady turnover of insects through reintroductions from the crowded culture as happens in some other solitary-rearing locust facilities (e.g. Roessingh *et al.*, 1993). Given that there is inevitably some pre- and peri-reproductive mortality during each generation, there is a risk that genetic selection, drift or founder effects may, over time, account for an increasing portion of the differences that are attributed to phenotypic plasticity.

Breugelmans *et al.* (2008) conducted a further study of SGPP 2 and SGPP 4 transcript levels in brain and fat body of male and female crowd- and solitary-reared adults throughout the first 24 days of adult life. These authors found, as did Simonet *et al.* (2004b, 2005), that tissues of gregarious insects had higher levels of transcripts than did solitarious insects. In solitarious insects levels of both transcripts were consistently low in both fat body and brain throughout the 24 days after the final moult. Patterns across time in gregarious adults were similar for both SGPP 2 and SGPP 4 but differed between the sexes. In brain tissue of males, both transcripts were initially at relatively high levels after the moult, before falling over days 4–6, increasing again on days 8–10, then dropping to remain at intermediate levels until day 18, after which they fell to very low levels until day 24. In female brains, there were two periods of elevated transcript levels, from days 8–12 and 18–22, with low levels being found at other times. In fat body of males, transcript levels were somewhat elevated immediately after the moult, fell to very low levels on day 6, rose sharply to day 10, fell again on day 12, then rose to a peak on days 14–16 before falling progressively until day 24. Levels in female fat body were lower than in males, rising from a very low level after the moult to a modest peak on day 10 falling back to low levels on day 14, then rising to their highest levels from days 16–20 and falling again. Breugelmans *et al.* (2008) conducted a further experiment in which they placed solitarious adults together in groups after moulting to adults. Brain and fat body were collected on day 8 and compared with gregarious and fully solitarious insects at the same age. The brains of regrouped males had SGPP transcript levels equivalent to fully solitarious males, and hence substantially lower than in gregarious males. In females, however, brain levels in regrouped adults were elevated to values approaching (but not reaching) those of the gregarious insects. Regrouping caused a partial increase in transcript levels in the fat body of both males and females, relative to values in solitarious and gregarious insects.

10.2.3 *Heat shock proteins*

Heat shock proteins (Hsps) are a family of molecules produced by organisms in response to various environmental stressors such as temperature shock, nutritional stress and immune challenges. Hsps serve to protect ('chaperone') other proteins, helping maintain their folding state and competence, hence protecting vital cellular functions. Crowded rearing induces upregulation of *Hsp* gene expression in *D. melanogaster* (Sørensen and Loeschke, 2001) and a higher number of copies of all *Hsp* gene families are found in gregarious than in solitarious fifth-instar nymphs of *L. migratoria* (Kang *et al.*, 2004).

Wang *et al.* (2007) compared the expression profiles of *Hsp* genes in the two phases of *L. migratoria* throughout embryogenesis and nymphal development. The full-length cDNAs of *Hsp20.5, 20.6, 20.7, 40, 70* and *90* were cloned and sequenced. Results from real-time PCR showed that the six *Hsps* were expressed in all developmental stages except for the early-stage embryo. Expression patterns during development were complex, but several general conclusions can be drawn. First, levels of expression tended to be higher in solitarious than gregarious mid-stage embryos, but were similar in late stage embryos. Second, except for *Hsp 40*, there was a general trend for *Hsp* expression to increase throughout nymphal development, rising especially rapidly in the fourth and fifth stadia. Third, expression levels were typically higher in gregarious than solitarious nymphs during the first, fourth and fifth stadia, but the opposite was the case in second and third-instar nymphs. Higher expression levels in gregarious nymphs make sense in view of crowding being a potential source of several environmental stressors, including increased competition for resources and elevated exposure to disease (Wilson *et al.*, 2002).

Wang *et al.* (2007) measured *Hsp* expression in head, thorax and hind leg of fifth-instar nymphs (see also Section 8.5). Thorax and leg showed the most marked elevation of gene expression in gregarious nymphs, a finding which the authors related to the fact that hind leg mechanoreceptors, projecting to thoracic ganglia, are important sources of input causing behavioural gregarization (Simpson *et al.*, 2001; Rogers *et al.*, 2003; see Section 14). The causal link, if any, is not clear, but the association is certainly interesting. To observe changes in *Hsp* gene expression during the early stages of phase change, Wang *et al.* (2007) subjected solitary-reared fourth-instar nymphs to a 32-h period of crowding, and gregarious nymphs to 32-h isolation. Isolation of gregarious nymphs resulted in pronounced reduction in *Hsp* expression towards solitarious values. In contrast, crowding solitarious nymphs resulted in elevated *Hsp* expression in three cases (*Hsp 20.5, 20.6* and *70*), a reduction in *Hsp 90*, and no significant change in *Hsp 20.7* and *40*. Such hysteresis effects, whereby changes during gregarization and solitarization differ in their time-course, have been reported for other phase characteristics, notably behaviour in *S. gregaria* (Roessingh and Simpson, 1994), but in that case solitarization proceeded more slowly than gregarization, rather than vice versa (see Section 10.4). A case of more rapid change from crowding to isolation than from isolation to crowding was reported by Ayali and Pener (1995) concerning changes of lipid- and AKH-related parameters with changes of density (see Section 9.3.8).

As mentioned earlier, it seems to make sense that *Hsp* expression would be elevated as locust nymphs become crowded, simply as a response to increased levels of competition and stress. However, a more interesting speculation is that *Hsp* genes may actually control aspects of phase change. In particular, Hsp 90 has properties which might be relevant to the evolution and/or proximate control of phase polyphenism [see studies on *D. melanogaster* (Rutherford and Lindquist, 1998) and *Arabidopsis thaliana* (Queitsch *et al.*, 2002) and the recent review of the role of Hsps in the control of phenotypic plasticity in insects by Williams *et al.* (2009)]. Most of the cellular targets of Hsp 90 are signal transducers involved in the regulation of cell-cycle and development. Hsp 90 keeps these protein 'switches' ready to flip into one or other stable conformation upon receipt of the appropriate environmental or cellular signals. Target molecules such as individual members of the large families of steroid hormone receptors and various protein kinases differ markedly in their response to Hsp 90. The net result is that Hsp 90 could, in theory at least, have the capacity to orchestrate biochemical elements involved in the induction and maintenance of phase polyphenism.

It has been proposed that Hsp 90, and potentially other Hsps, may serve as 'capacitors' for the evolution of complex developmental phenotypes (Rutherford and Lindquist, 1998; Queitsch *et al.*, 2002; Williams *et al.*, 2009) – perhaps including locust phase. Because of the buffering effects of Hsps, variant developmental responses that arise by mutation will remain unexpressed and will accumulate in the genome over evolutionary time. However, when the buffering effects of Hsps are compromised by a particular environmental stressor (perhaps associated with crowding in the case of ancestral locusts), these hidden developmental variants reach phenotypic expression. Most of these variants will be maladaptive, but some may be advantageous and will be naturally selected. Selection could then result in the continued expression of these traits even in the absence of the stressors that originally compromised the Hsps, or else result in the expression of an alternative developmental pathway only under certain environmental conditions – such as during crowding.

10.3 NEUROTRANSMITTERS AND NEUROMODULATORS

The behavioural components of phase change in locusts are highly labile. In nymphs and adults of *S. gregaria*, gregarious behavioural tendencies are expressed within a matter of hours when solitarious insects experience crowding (reviewed by Simpson *et al.*, 1999; see Section 11). Such rapid changes are unlikely to result from gene expression differences, but rather would be expected to be mediated by neuromodulation of existing neural circuitry

(Simpson *et al.*, 1999; Rogers *et al.*, 2004). Rogers *et al.* (2004) compared relative amounts of various neurotransmitters and neuromodulators in the CNSs of solitarious and gregarious *S. gregaria* (final-instar nymphs) as they underwent behavioural phase transition. Locusts were taken at one of nine stages during the cycle from behavioural gregariousness to solitariousness and back again (long-term crowded, 24 h isolated, one stadium isolated, one generation isolated, two generations isolated, three generations isolated, 4 h crowded, 24 h crowded and one stadium crowded). Thirteen substances (aspartate, glutamine, glycine, GABA, arginine, taurine, acetylcholine, tyramine, citrulline, dopamine, serotonin, octopamine and *N*-acetyldopamine) were measured using HPLC in the optic lobes, remaining brain tissue, and thoracic ganglia. Eleven substances differed between long-term solitarious and gregarious locusts, including excitatory (glutamate, acetylcholine) and inhibitory (GABA) neurotransmitters as well as the neuromodulators/neurohormones, dopamine and serotonin. The amounts of glutamate, glycine and aspartate in solitarious nymphs were at least double those in gregarious nymphs, while GABA, taurine, serotonin and dopamine were elevated by 20–35%. Acetylcholine, tyramine and citrulline occurred at lower levels in solitarious than gregarious nymphs, with citrulline being the most extremely reduced (a 90% decrease). *N*-acetyldopamine did not differ significantly between long-term crowd-reared and 3-generation solitary-reared nymphs, although tended to be at lower (17%) concentrations in solitarious insects. Levels of octopamine did not differ significantly between the phases. A qualitatively similar difference in the profiles of the 13 substances was found between long-term crowd-reared and 2-generation solitary-reared adults as in final-instar nymphs.

In nymphs, 12 of the 13 substances were found to differ in at least one of the nine measured stages of isolation or crowding. However, only serotonin underwent a large increase (ninefold) during the 4-h period during which behavioural gregarization is established in solitarious nymphs. Tellingly, this increase occurred in the thoracic ganglia but not the brain. The thoracic ganglia receive mechanosensory inputs from leg receptors that stimulate behavioural phase change (Simpson *et al.*, 2001; Rogers *et al.*, 2003). The involvement of this transient rise in serotonin in causing behavioural gregarization is discussed in detail in Section 14. Serotonin was also substantially elevated during the initial stages of the reverse transition from long-term gregariousness to solitariousness: 24-h isolated gregarious nymphs had eightfold higher amounts of serotonin, but only in the optic lobes of the brain. These regional peaks in serotonin with the early stages of behavioural gregarization and solitarization were transitory. Dopamine also showed a pronounced peak in 24-h isolated gregarious nymphs, and a transitory depression in the optic lobes and thoracic ganglia during the very early stages (4 h) of crowding of solitarious nymphs.

Earlier studies reported elevated levels of octopamine in solitary-reared relative to crowd-reared *L. migratoria* (Fuzeau-Braesch and David, 1978; Fuzeau-Braesch and Nicolas, 1981). Rogers *et al.* (2004) did not find this difference between long-term crowd-reared and solitary-reared nymphs or adults of *S. gregaria*, but there was a decrease to undetectable levels in brain, optic lobes and thoracic ganglia during the first 4 h of crowding, followed by a large increase, especially in the optic lobes and thoracic ganglia, during the period from 24-h to 1 stadium of crowding. Inadvertent gregarization of solitary-reared insects through the mechanical effects of handling, or stress induced by handling or husbandry procedures, may well explain the results reported by Fuzeau-Braesch and David (1978) and Fuzeau-Braesch and Nicolas (1981). Another study in which octopamine was measured in relation to phase state was that of Morton and Evans (1983), who found no difference between crowd- and solitary-reared *S. gregaria* (denoted by them as *S. americana gregaria*). For additional discussion concerning octopamine and locust phases also in relation to adipokinetic response and flight see Section 9.3.8.

In conclusion, measurement of the detailed time-course of change in the CNS with crowding and isolation suggests that, of 13 neuroactive substances, serotonin possesses characteristics that make it the most promising candidate for an enabling/causal agent in the initial stages of behavioural phase transition (Rogers *et al.*, 2004). Antibody staining to serotonin is found in a small number of neurones in crowd-reared locusts, but these have processes which branch extensively throughout the CNS (Burrows, 1996), offering abundant possibilities for neuromodulation of neural circuitry controlling key aspects of behavioural phase change.

Anstey *et al.* (2009) subsequently explored the role of serotonin in the mediation of behavioural gregarization in *S. gregaria* in a series of detailed pharmacological manipulations. Their data clearly indicate that elevation of serotonin in the thoracic ganglia is both necessary and sufficient for initiation of behavioural phase change (see full discussion in Section 14).

Simpson *et al.* (1999) and Anstey *et al.* (2009) drew the analogy between behavioural phase change and the mechanisms of short- and long-term memory. Both processes involve an initial, easily reversible phase, which with continued reinforcement becomes translated into a change that becomes increasingly resistant to reversal (Roessingh and Simpson, 1994; Simpson *et al.*, 1999). Serotonin is involved in the establishment of short-term memory through its modulatory effects on synaptic transmission and leads through second messenger cascades [protein kinase A (PKA), mitogen-activated protein kinase (MAPK), and transcription factors such as cyclic AMP response element binding proteins (CREB)] to synthesis of proteins that produce structural synaptic changes that result in long-term memory (Kandel and Pittenger, 1999).

One further study that warrants mention is that of Lenz *et al.* (2001), who used high resolution proton nuclear magnetic resonance spectroscopy (^{1}H NMR) to compare the composition of the haemolymph of crowd- and solitary-reared nymphs of *S. gregaria*. More than 20 compounds were identified and compared, including amino acids, organic acids, lipids, ethanol and the polyamine putrescine. The most interesting difference reported was in putrescine, which was found at twice the concentration in solitary-reared than in crowd-reared nymphs. This polyamine is known to act as a growth promoter for insects cells (Mitsuhashi, 1998) and can stimulate mitogenic activity in adult neuroblasts, mimicking the effect of JH (Cayre *et al.*, 1997). Any role in phase transition has yet to be discovered.

10.4 cGMP/PKG

It has emerged in recent years that the cyclic guanosine monophosphate/protein kinase G pathway (cGMP/PKG) is involved in the control of natural variants of foraging and locomotory behaviour in *Drosophila* (the 'rover' and 'sitter' phenotypes) (Osborne *et al.*, 1997; Riedl *et al.*, 2005), density-dependent aggregation in nematode worms (Coates and de Bono, 2002; Cheung *et al.*, 2004), and developmental behavioural plasticity (i.e. division of labour) in honey bees (Ben-Shahar, 2005) and ants (Ingram *et al.*, 2005). Perhaps the cGMP/PKG pathway may represent a common mechanism controlling other examples of density-dependent and social behaviour, including locust phase? However, while it is tempting to look for common molecular mechanisms across species, drawing an association between elements of such a basic and richly multi-functional signalling pathway, and phenomena as diverse and complex as phase change in locusts, division of labour in social insects, foraging behaviour in flies, and aggregative feeding behaviour in nematode worms may turn out to be overly simplistic.

The cGMP/PKG pathway is activated by soluble guanylate cyclases (sGC), which in insects (but not nematodes) are principally stimulated by the gaseous messenger, nitric oxide (NO). NO is produced by the enzyme nitric oxide synthase (NOS) and is synthesized by many neurones in the CNS of locusts, including population of neurones in the optic lobes, antennal lobes, and the mechanosensory processing neuropiles in the thoracic ganglia (Ott *et al.*, 2001). The precursor of NO is arginine, with citrulline being produced in stoichiometric amounts during NO synthesis. Rogers *et al.* (2004) reported a large (90%) and persisting fall in the amount of citrulline in brain, optic lobes and thoracic ganglia of gregarious *S. gregaria* nymphs during the first 24 h of isolation. Upon crowding of 3-generation solitary-reared nymphs, there was an increase in brain levels during the first 4 h, but not in optic lobes or thoracic ganglia. When crowd-reared insects were isolated, a decline in arginine level was evident from nymphs isolated for 24 h to those reared in isolation for an entire stadium, but levels rose with increasing periods of isolation thereafter, reaching highest concentrations after 2–3 generations of solitary rearing. When 3-generation solitarious insects were crowded, arginine levels gradually declined over 1 stadium. There is thus some suggestion of reciprocal changes in arginine and citrulline levels, which suggests that NO synthesis occurs in a phase-dependent manner (Rogers *et al.*, 2004). A causal link with phase transition has yet to be established.

11 Behaviour

Gregarious and solitarious locusts differ substantially in behaviour, particularly in their activity levels and responses to other locusts (reviewed by Uvarov, 1966; Pener, 1991; Pener and Yerushalmi, 1998; Simpson *et al.*, 1999). There are also phase-related differences in feeding behaviour and nutrition, which are considered in Sections 7.2.3 and 17. Behavioural phase state changes rapidly when locusts experience a variation in local population density. Behaviourally gregarized locusts are much more active than solitarious locusts and are attracted rather than repelled by others, setting in train a positive feedback, which under appropriate environmental circumstances can concentrate a solitarious locust population and drive it to behavioural gregariousness (see Section 17). Gregarious locust aggregations in turn give rise to mass marching bands of nymphs or flying swarms of adults. Because of the autocatalytic nature of behavioural phase change (gregarization causes aggregation, which stimulates and maintains further gregarization), the entire suite of phase characteristics can change in a coordinated fashion, even though different traits may not share the same underlying mechanisms. Hence, behaviour helps to couple a diverse set of density-dependent phase characteristics into a functionally coordinated threshold trait (Simpson and Sword, 2009; elaborated in Section 17).

11.1 ASSAYING BEHAVIOURAL PHASE STATE

The pioneering studies on locust behavioural phase were carried out by Ellis, who designed elegant laboratory assays for measuring three aspects of behaviour in nymphs of *L. migratoria*, *S. gregaria* and some other species: aggregation (Ellis, 1953a, 1959, 1962, 1963a,b), marching behaviour (Ellis, 1951, 1953b, 1964a,b) and responses to confrontations with other locusts (Ellis, 1962, 1963a,b; Ellis and Pearce, 1962). Wiesel *et al.* (1996) used similar assays to those of Ellis in their study of the effects of JH and JH analogues on the behaviour of solitary- and crowd-reared nymphs of *S. gregaria* and *L. migratoria* (see Section 9.1), and Malual *et al.* (2001) used a similar ring-shaped arena to Ellis (1953a) for measuring the aggregation tendency of *S. gregaria* hatchlings (see Section 16.3), derived by measuring the numbers of locusts found within equal radial segments. Heifetz *et al.* (1994, 1996, 1997, 1998), Heifetz and Applebaum (1995), Bouaïchi *et al.* (1996) and Bouaïchi and Simpson (2003) used a circular arena to derive an ‘associative’ or ‘grouping’ index based on the proportion of locusts found in close proximity to one another.

The most widely employed assay system in recent years has been that devised by Roessingh *et al.* (1993) (described in detail in Simpson *et al.*, 1999), in which a test insect is introduced into the centre of a rectangular arena, at each end of which is a backlit chamber separated from the main body of the arena by a perforated, clear-plastic partition. One of these end chambers contains a group of gregarious locusts (the stimulus group) and the other is empty. The behaviour of the test insect is recorded for 5–10 min onto a lap-top computer. A set of behavioural variables is derived from the raw behavioural records for each test insect. These typically include the position of the test insect with respect to the stimulus group, descriptions of the shape and speed of the path traversed during the assay, the frequency and duration of locomotory events, and various smaller body movements and postural variables.

Having derived individual records for 50–100 solitary-reared and a similar number of crowd-reared insects, multiple logistic regression analysis is used to regresses the dichotomous variable (crowd- versus solitary-reared) against the set of behavioural variables. The logistic algorithm retains those variables that are instructive in distinguishing between the two groups and weights these variables to produce a multi-dimensional model which optimizes the difference between the two rearing groups. By using the logistic regression model, it becomes possible to take the behavioural record of a test insect of unknown phase state and estimate the probability that the insect fits into either the solitary-reared or crowd-reared group. A value for the probability of being considered a member of the solitary-reared group [*p*(solitary)] of 1.0 indicates that the insect behaved in the assay in a manner that was indistinguishable from known solitary-reared insects, whereas a value of 0.0 indicates that the insect could not be distinguished from the crowd-reared group. Intermediate values indicate that the insect is transitional in behaviour. Thus, the value for *p*(solitary) serves to collapse the multiple variables that comprise behavioural phase into a single, linear predictor of phase state.

The Roessingh *et al.* (1993) assay system has been used in studies on *S. gregaria* as hatchlings (Islam *et al.*, 1994a,b; Bouaïchi *et al.*, 1995; McCaffery *et al.*, 1998; Hägele *et al.*, 2000; Elliot *et al.*, 2003), second-instar nymphs (Collett *et al.*, 1998; Despland and Simpson, 2000a; Despland *et al.*, 2000), final-instar nymphs (Roessingh *et al.*, 1993, 1998; Roessingh and Simpson, 1994; Hägele and Simpson, 2000; Sword and Simpson, 2000; Simpson *et al.*, 2001; Hoste *et al.*, 2002b; Lester *et al.*, 2005) and adults (Bouaïchi *et al.*, 1995). Other species studied using the assay include final-instar nymphs of *S. americana* (Sword, 2003) and *L. migratoria* (Ishigaki Island and West African wild-type strains, and Okinawa Island albino strain; Hoste *et al.*, 2002a, 2003), first-instar nymphs of two populations of *L. migratoria* (from Madagascar and Southern France; Chapuis *et al.*, 2008a) and final-instar nymphs of *Chortoicetes terminifera* (Gray *et al.*, 2009).

A simplified version of the Roessingh *et al.* (1993) assay, without a stimulus group, was used by Despland and Simpson (2000b) to study adult *S. gregaria* in the field in Mauritania. Hoste *et al.* (2006) used multiple chambers without stimulus groups to assay activity in *S. gregaria*, manually registering the number of times test locusts contacted the side walls from simultaneous video recordings of 12 chambers for adult and final-instar nymphs, or 48 chambers for hatchlings. Heifetz *et al.* (1996, 1997) used an analysis approach that was similar in conception to that of Roessingh *et al.* (1993). These authors used discriminant analysis (a technique similar to logistic regression, but less forgiving of the non-normal data error distributions that are typical of behavioural variables), based on six behavioural variables related to activity and the tendency to form groups. Data were derived manually from video recordings of groups of *S. gregaria* nymphs in a circular arena.

A drawback of the assay developed by Roessingh *et al.* (1993) is its reliance on manual observation of behaviour. Because individual observers differ subtly in the way in which they record behaviour, it has been necessary for each new experimenter to construct their own logistic regression model and to undertake all their own observations (Simpson *et al.*, 1999). Recently, this problem has been solved by Gray *et al.* (2009), who used automated tracking software to record the behaviour of final-instar nymphs of *C. terminifera* using the Roessingh *et al.* (1993) arena. This system has the extra advantage that two test insects can be observed at once, in side-by-side arenas. Gray *et al.* (2009) optimized the duration of the assay, finding that 8 min provided the shortest duration from which an accurate assessment of behavioural phase state could be made in *C. terminifera*.

11.2 PHASE-RELATED BEHAVIOURAL DIFFERENCES

Logistic regression analysis of the behaviour of solitary- and crowd-reared locusts in the Roessingh *et al.* (1993) assay have invariably produced models which accurately classify the two phases, with models predicting phase state accurately in greater than 80% and usually better than 90% of cases (see references listed earlier). The precise behavioural variables retained in individual models differ somewhat according to the observer (see earlier), but all models share similar features: Crowd-reared locusts typically move towards the stimulus group, are highly active and groom frequently, whereas solitary-reared locusts walk less and more slowly, groom infrequently, freeze when detecting movement in the stimulus group then creep away from the group (Simpson *et al.*, 1999). These behavioural differences typify all ages of *S. gregaria* as well as *L. migratoria* and *C. terminifera*, although in the latter case the automated tracking system used by Gray *et al.* (2009) did not measure grooming behaviour or other small body movements.

Differences in phase-related behaviour within a species have been detected for strains, populations and sub-species (Ellis, 1962; Uvarov, 1966, pp. 372–375; Hoste *et al.*, 2002a, 2003). Uvarov (1966, p. 372) reported that island populations of *L. migratoria* are not known to swarm naturally. A sub-species of *S. gregaria*, *S.g. flaviventris* from south-western Africa, shows less pronounced phase change than more northerly populations of *S. gregaria* (Uvarov, 1966, p. 374) and aggregated less when studied in the laboratory (Ellis, 1962; see also Section 4 for additional strain-dependent differences in phase characteristics of the same locust species). An Okinawa Island albino strain of *L. migratoria*, which lacks [His^7]-corazonin (see Section 7.2.2), was found to show more pronounced behavioural phase change than did a West African wild-type strain (Hoste *et al.*, 2002a) and was thus not obligatorily solitarious as had been suggested previously by Nolte (1967, 1968). There were also phase-independent differences in behaviour between the albino and the wild-type strain, with final-instar albino nymphs being more active but grooming less than wild-type nymphs (Hoste *et al.*, 2002a). Hoste *et al.* (2003) extended their study to another wild-type strain, from Ishigaki Island. This is likely to be similar to the original wild-type of the Okinawa Island albino strain, which has been lost. Crowd-reared Ishigaki Island wild-type nymphs were more similar in behaviour to crowd-reared West African wild-type insects than to crowd-reared Okinawa Island albino nymphs, when classified in a logistic regression model based on crowd-reared nymphs from the West African and albino strains.

The extent to which behavioural difference between locust populations reflect genetic differences in the propensity to phase change, differences in environmental conditions (including parental effects – see Section 16), or an interaction between the two, is not yet clear. Studying the relationship between genotype and environment is important for understanding the ecology and evolution of phase change (Simpson and Sword, 2009). Comparisons based on standard assays offer much in this regard. For example, Sword (2003) used the assay of Roessingh *et al.* (1993) to compare behavioural phase change in first and final-instar nymphs from two populations of *S. americana* (North Carolina and Texas) and with nymphs of *S. gregaria.* Density-dependent changes in behaviour were evident in *S. americana* but were substantially less marked than in *S. gregaria*. Final-instar nymphs of *S. americana* changed behaviour less in response to crowding than did first-instar nymphs, whereas first-instar nymphs from North Carolina changed more than did those from the Texas population.

Recently, Chapuis *et al.* (2008a) compared two population of *L. migratoria*; one from Madagascar with a history of frequently forming outbreaks and the other a non-outbreaking population from Southern France. Insects from both populations were reared in the laboratory under isolated conditions for two generations, after which they were reared in either isolated or crowded conditions for a further two generations. Efforts were made to prevent loss of genetic diversity across the successive generations (see Sections 10.2, 11.4). In the fifth generation, all insects were reared in isolation from soon after emergence from the egg until adulthood, and two phase-related traits were measured: first-instar behaviour (based on the assay of Roessingh *et al.*, 1993) and adult morphometry. The French population expressed far less pronounced behavioural and morphometric gregarization in response to parental and grandparental crowded-rearing than did the Malagasy population. Because behaviour was measured in hatchlings that were isolated on the day of emergence and morphometry assessed in adults that had been solitary-reared throughout their lives, the results indicate that the two locust strains differ genetically in the extent to which behavioural phase state and morphometrical change are transmitted maternally (see Section 16). Whether the French strain is also less likely than the Malagasy strain to gregarize in response to crowding within an insect's lifetime is not known.

11.3 ADULT-SPECIFIC BEHAVIOURAL DIFFERENCES

Adult *S. gregaria* respond similarly to crowding and isolation as do nymphs, demonstrating similar gregarious and solitarious behavioural characteristics as described earlier (Bouaïchi *et al.*, 1995). Adult-specific behaviours that differ between the phases include sexual behaviour and flight. Sexual behaviour has been covered in Section 8.2 and will not be considered further here. There are only limited data on flight behaviour in solitarious locusts, but there are phase-related differences both in the propensity of locusts to take-off and in flight metabolism and performance (Pener *et al.*, 1997b; Fuchs *et al.*, 2003; Ayali *et al.*, 2004). Mass migratory flight is characteristic of gregarious adult locusts. Descriptions of the movement of swarms and the detailed behaviour of locusts within them have been reported in earlier publications (e.g. Waloff, 1962, 1972; Uvarov, 1977; Rainey, 1989) and will not be repeated here. Behavioural aspects of collective behaviour (flight and marching) are covered later (Section 11.5), whereas neural correlates of flight behaviour in solitarious and gregarious phase locusts are considered in Section 11.6 (see also Section 5.3.2). Hormonal aspects of flight are considered in Section 9.3.8.

11.4 THE TIME-COURSE OF BEHAVIOURAL PHASE CHANGE

The detailed time-course of behavioural change within the life of an individual locust has been quantified in nymphs (Roessingh *et al.*, 1993; Islam *et al.*, 1994a; Roessingh and Simpson, 1994) and adults (Bouaïchi *et al.*, 1995) of *S. gregaria.* Accumulation of phase state across generations is considered in Section 16. Behavioural gregarization of solitary-reared locusts, whether first- or fifth-instar nymphs or adults, occurs rapidly, being evident within an hour of crowding and complete after 4–8 hours. If fifth-instar nymphs are re-isolated after a 24-h crowding period, they reassume completely solitarious behaviour within 4h. In contrast, when fifth-instar nymphs that have been reared for many generations in crowded culture are isolated, they initially show rapid behavioural solitarization over the first 4h, but only achieve 40% solitariousness. Thereafter, they remain in a behaviourally transitional state until the end of the stadium. If isolated as first-instar nymphs, however, they are fully behaviourally solitarious by the fifth stadium (Roessingh *et al.*, 1993).

These results indicate that behavioural gregarization and solitarization proceed at different rates (i.e. demonstrate hysteresis) and that solitarization has two phases: an initial rapid phase and a second, slower phase that requires insects to be maintained in isolation across successive moults – or generations

(Roessingh *et al.*, 1993; Section 16). The extent to which complete behavioural solitarization is achieved during the first hours of isolation is a function of the time previously spent crowded. Heifetz *et al.* (1996) also reported that gregarization proceeded more rapidly than behavioural solitarization in nymphs of *S. gregaria*, and earlier studies are also consistent with this conclusion (Ellis, 1962, 1963a; Gillett, 1988).

Such a time-course for behavioural phase change makes adaptive sense (Simpson and Sword, 2009). A gregarious locust that becomes temporarily separated from an aggregation will initially become somewhat more 'cautious' in its behaviour as it partially solitarizes, but continue to be active and attracted by others, which improves its chances of rejoining the safety of the group (see Section 17). In contrast, if isolation occurs only after a brief period of environmentally imposed crowding, a solitarious locust will rapidly reassume cryptic behaviour and reduce the chances of attracting predators (also see Section 17).

Hoste *et al.* (2002b) reported that fifth-instar *S. gregaria* crowded since the first stadium but coming from parents reared in isolation for eight generations in Leuven had not reached complete behavioural gregariousness when tested in the Roessingh *et al.* (1993) assay. The difference between these results and those from the Oxford and other studies (see earlier) might reflect differences between locust strains (also see discussion of Chapuis *et al.*, 2008a in Section 11.2), but another possibility is that the Leuven rearing system for solitarious locusts may have inadvertently selected for loss of gregariousness, given that insects were reared in continuous solitary culture for many generations (see Section 10.2.2).

11.5 COLLECTIVE BEHAVIOUR: MARCHING BANDS AND FLYING SWARMS

The components of gregarious behaviour, especially the tendency to aggregate and exhibit high levels of activity, predispose towards collective behaviour in which juvenile locusts form marching bands and adults formed winged swarms. Uvarov (1977) and Pener (1991) provide detailed descriptions of both these group-level phenomena and review the associated older literature, notably the experimental work of Ellis and colleagues on marching behaviour (see Section 11.1).

Recent work has focused on attempting to understand the transition from aggregation to the onset of marching in hopper bands. Buhl *et al.* (2006) employed models from statistical physics, known as self-propelled particles (SPP) models, to explore the onset of collective behaviour in hoppers of *S. gregaria*. In such simulation models, individuals in a group are represented as particles, with each particle adjusting its speed and/or direction of movement in response to near neighbours. A central prediction from these models is that as the density of hoppers in a group increases, a rapid transition occurs from disordered movement of individuals within the aggregation to highly aligned collective motion. No external stimulus is required to trigger collective movement, nor are there 'group leaders' that initiate group marching. The models also predict that at densities just above the threshold for onset of collective movement there are sudden, unpredictable changes in the direction of movement of the group, but at higher densities, the direction of movement becomes stable for extended periods.

Buhl *et al.* (2006) confirmed the predictions of the SPP model in the laboratory by placing third-instar nymphs in a ring-shaped arena (80 cm diameter) with a central dome to restrict optical flow in the direction opposite to individual motion. The arena was inspired by earlier demonstrations of locusts marching in the laboratory in circular arenas by Ellis (see Section 11.1). Hoppers readily formed highly coordinated marching bands under these conditions, with individuals selecting collectively a randomly determined rotational direction of travel, either clockwise or counter-clockwise, for long periods of time. At any one time about 50% of individuals were moving and the others pausing, as seen in natural bands (Ellis and Ashall, 1957; Stower, 1963). Buhl *et al.* (2006) developed an automated digital tracking system that allowed the simultaneous analysis of group behaviour and the behavioural responses of individual locusts. Experiments were performed on different numbers of locusts in the arena, ranging from 5 to 120 insects, which equated to densities of 13–295 m^{-2}. Coordinated marching behaviour depended strongly on locust density. At low densities (7 or fewer locusts in the arena), there was a low incidence of alignment among individuals; in those trials where alignment did occur it did so only sporadically and after long initial periods of disordered motion. At intermediate densities (10–25 locusts in the arena), there were long periods of collective rotational motion with rapid, spontaneous changes in direction. At densities above 74 m^{-2} (30 or more locusts in the arena), spontaneous changes in direction did not occur and hoppers rapidly adopted a common and persistent rotational direction over the 8 h experimental period. The dynamic instability in motion at intermediate densities may help explain why the direction of bands in the field often changes unpredictably with no obvious relation to external stimuli (Uvarov, 1977).

The critical interaction rule that gave rise to cohesive movement in the SPP model employed by Buhl *et al.* (2006) was that locusts align with their nearby moving neighbours. Optomotor responses are known to be involved in such alignment (see Uvarov, 1966, p. 361), but Ellis (1951, 1962) reported that seeing others marching around an arena was not sufficient to induce alignment in isolated locusts that could see but not contact a marching group. Rather, physical contact seems to be necessary to evoke full marching behaviour. Work on mass marching Mormon crickets, *A. simplex*, suggested why physical contact might be so important. Field studies in NW USA showed that cannibalism provides a powerful aligning force within cricket bands (Simpson *et al.*, 2006). If individuals fail to continue moving when approached from behind by a conspecific, they are likely to be cannibalized.

Cannibalism is also common among locusts and can be a major cause of mortality in the field [see Van Huis *et al.* (2008) and references therein]. Bazazi *et al.* (2008) tested the prediction that cannibalism serves as a mechanism motivating collective motion in locust hopper bands. These authors experimentally manipulated the degree to which fifth-instar *S. gregaria* could detect tactile and visual stimuli from conspecifics approaching from behind and tested behavioural responses of individuals and groups within the same design of arena as used by Buhl *et al.* (2006). Mechanosensory denervation of the abdomen (achieved by sectioning the ventral nerve cord) did not influence the behaviour of isolated gregarious nymphs, but when placed in groups, locusts with severed nerves marched much less than did groups of insects subjected to a sham operation. Detailed video analysis indicated that nerve-sectioned locusts responded in the same manner as controls to being approached by other nymphs from all quarters except from behind; that is, the behavioural and sensory deficit was specific to approaches from the rear. Such insects also experienced significantly higher levels of damage to their abdomen during the experiments as a result of cannibalistic attacks.

The role of vision was explored by Bazazi *et al.* (2008) by painting regions of the compound eyes of otherwise intact nymphs. Groups of individuals with no restriction of visual input showed significantly higher levels of marching than groups comprised of nymphs with a complete restriction of visual input. The proportion of moving locusts in groups in which individuals had only their rearward view occluded was not significantly different from groups of individuals that were totally blind, whereas locusts that could see behind but not ahead exhibited a propensity to march that was intermediate between blind and fully sighted groups. In conjunction with results from the nerve sectioning experiment, these data indicate that stimuli from cannibalistic conspecifics approaching from behind provides an important force serving to motivate and align marching hopper bands.

Attempts have also been made recently to model collective behaviour in flying swarms of adult locusts, although such models have been inspired by, rather than validated with, biological data. Topaz *et al.* (2008) constructed an agent-based model that simulated the behaviour of a rolling migratory locust swarms (Albrecht, 1967; Waloff, 1972; see also Farrow, 1990, as discussed in Section 9.3.8). The model was predicated on the following: (a) flying locusts in a swarm head downwind; (b) locusts land when they reach the front of the swarm, heading slightly upwind as they approach the ground; (c) once landed, locusts remain on the ground until the trailing edge of the swarm passes overhead; (d) at which point they take off and fly slightly upwind before turning to align with the wind; and so on. Terms for social interactions between locusts, gravity, wind and the ground were included in the model. Social interactions included collision avoidance (close-range repulsion) and attraction to neighbours over longer distances. Topaz *et al.* (2008) were stimulated by an earlier, only partially successful, model of locust rolling swarms by

Edelstein-Keshet *et al.* (1998), which was based on coupled non-linear partial differential equations for locusts in the air and on the ground.

In summary, it appears that SPP and other agent-based models and their associated mathematical theory provide a promising quantitative tool for understanding collective behaviour in locusts. The challenge now is to scale such models up to large-scale events in the field.

11.6 NEURAL DIFFERENCES BETWEEN PHASES

Locusts provide an ideal study system for exploring the neuronal bases of behavioural plasticity, given that behavioural phase differences, their time-course and evoking stimuli have been characterized in detail (see earlier) and that a substantial literature exists on gregarious locust neurobiology (reviewed by Burrows, 1996).

To date there have been a small number of published studies in which neural correlates of differences in key behaviours have been quantified in adults of *S. gregaria*. Matheson *et al.* (2004) and Rogers *et al.* (2007) investigated phase-related difference in the visuo-motor pathway, and Fuchs *et al.* (2003) and Ayali *et al.* (2004) explored sensory-motor pathways involved in flight.

Visually evoked behaviours differ substantially between the phases of locusts, notably the switch from repulsion by other locusts to attraction during gregarization. Relating detailed neuronal responses to such behavioural responses is extremely difficult, however, given the multitude and complexity of neural pathways involved. Nevertheless, well-studied visuo-motor pathways offer useful models for exploring the neuronal bases of phase-related behaviour. To this end, Matheson *et al.* (2004) compared responses of the descending contralateral movement detector (DCMD) neurone, which receives input from the compound eye through the lobula giant movement detector (LGMD) neurone and is strongly stimulated by looming objects, signifying an imminent collision or predator attack. DCMD in turn makes synaptic connections with flight and leg motor neurones involved in collision avoidance and escape behaviour. Whereas the response of DCMD to the first presentation of a looming stimulus did not differ between the phases, over 30 repeated stimulus presentations at 1-min intervals the response waned (habituated) substantially (fivefold) more in 3-generation solitarious than in gregarious adults. Once habituated, the peak firing rate and overall number of spikes per stimulus presentation were lower, and the time of peak firing relative to stimulus onset more variable, in solitarious than gregarious insects. The other major phase-related difference was that, in the habituated state, small, fast moving objects evoked briefer periods of higher firing rates in gregarious locusts than did larger, slower moving stimuli. In contrast, habituated solitarious locusts did not show a pronounced tuning in the response of DCMD to object size and speed.

These different responses of DCMD were interpreted by Matheson *et al.* (2004) in relation to the visual environment of the two locust phases. The fact that non-habituated responses did not differ between the phases makes sense in view of both phases needing to respond to imminent collision or attack. Being solitary and relatively sedentary, it might be predicted that solitarious locusts will typically be in a visually non-habituated state. Gregarious locusts, however, are constantly surrounded by small, fast-moving objects; namely, other locusts in flight or in marching bands. Rapid habituation to repeated visual stimulation would be non-adaptive under such circumstances, otherwise collisions in flight and cannibalistic encounters during marching (see previous section) would occur. The pronounced tuning of DCMD responses in gregarious locusts to object size and speed may permit gregarious locusts to respond differently to other locusts than to predators; whereas solitarious locusts would respond more similarly to any looming object, irrespective of size and speed. Perhaps this partly explains how gregarious locusts are attracted to others, whereas solitarious locusts are repelled, although further understanding of the motor consequences of different DCMD responses is needed.

Rogers *et al.* (2007) pursued the visuo-motor pathway from DCMD to a leg motor neurone that receives direct synaptic input from DCMD; the metathoracic fast extensor tibia motor neurone (FETi). As reported by Matheson *et al.* (2004), DCMD in solitarious locusts habituated much more rapidly to repeated stimulation than in gregarious locusts. However, the single post-synaptic potentials (PSPs) evoked in FETi by action potentials in DCMD were twice the amplitude in solitarious than gregarious locusts. This was due to a difference in synaptic transmission strength and served to compensate for the fewer action potentials generated by DCMD in habituated solitarious locusts, such that the size of the summed compound PSP produced in FETi in response to a looming stimulus was the same for both phases. Although the size of the compound PSP was the same, the time of peak depolarization in FETi occurred sooner in gregarious than solitarious locusts in response to small, fast-moving objects. The timing of peak compound PSP amplitude was later and much more variable in solitarious locusts.

Rogers *et al.* (2007) considered possible functions of these phase-related differences in DCMD–FETi coupling. Two possibilities were discussed. The first was that it would be energetically cheaper to generate fewer action potentials in DCMD and to boost the strength of synaptic transmission to FETi than to produce more action potentials each with a weaker synaptic influence. Hence, if widespread within the nervous system, this could represent a means of reducing metabolic costs in solitarious locusts. Second, reducing synaptic strength might serve to compensate for widespread increased in neuronal excitability during the transition to gregariousness, resulting from the substantial increases found in levels of various CNS neuromodulators (Rogers *et al.*, 2004; see Section 10.3) and the enhanced levels of sensory stimulation experienced within a crowd.

The behavioural consequences of the change in DCMD–FETi coupling were also considered by Rogers *et al.* (2007). The main conclusion was that the variable timing of peak response in FETi of solitarious locusts might result in more unpredictable, and possibly more effective, escape responses than in gregarious phase insects; whereas a shorter, more consistent latency of response in FETi to small, fast-moving objects is somehow better suited to life in a crowd.

Fuchs *et al.* (2003) and Ayali *et al.* (2004) chose to study phase-related changes in a well-characterized sensory-motor pathway involved in flight initiation and steering. Young (7 days old) gregarious adult *S. gregaria* were five times more likely to take-off in response to a pulse of air directed at the head than were 1-generation solitary reared adults. Activity was recorded in two descending interneurones that receive sensory input from the head and direct it to the thoracic flight circuitry: the tritocerebral commissure giant (TCG), which receive inputs from head hairs (see Section 5.3.2) and antennal mechanoreceptors, and the tritocerebral commissure dwarf (TCD), which is an inhibitory (GABAergic) interneurone that responds to mechanosensory inputs from the head and other body regions and to reduction in light intensity (Burrows, 1996). Spontaneous activity in TCD was found to be fivefold higher in the dark than in the light in gregarious locusts, but did not differ with light level in solitarious insects. Absolute spontaneous firing rates in TCD were not reported. Presuming that TCD inhibits flight, this result is consistent with reports that gregarious *S. gregaria* fly mainly in the day, whereas solitarious adults have been reported to fly at night.

Spontaneous activity in TCG in the light was significantly higher in gregarious locusts (about twice: ~ 0.5 and $1.0\,Hz$ in solitary- and crowd-reared locusts, respectively), but not measured in the dark. When a wind pulse initiated flight, there was no difference between the phases in TCG firing or flight muscle electrical activity, nor did TCG adapt during the stimulus. However, when a wind stimulus failed to elicit flight, the response of TCG was threefold higher in gregarious than solitarious locusts. In addition, the rate of decline in firing rate throughout the stimulus (adaptation) was threefold greater in solitarious than gregarious insects. The greater responsiveness of TCG in gregarious locusts was shown not to be due to greater numbers of afferent inputs from head hairs, since numbers of hairs were actually 12% higher in solitarious insects (see also Section 5.3.3). Together, the response properties of TCG parallel the differences in flight behaviour between the two phases.

The challenge is to extend these highly promising studies on neuronal plasticity. Thus far, experiments have involved comparison of long-term crowd-reared insects with 1-generation (Fuchs *et al.*, 2003; Ayali *et al.*, 2004) or 3-generation (Matheson *et al.*, 2004; Rogers *et al.*, 2007) solitary-reared insects. It will be important to track neuronal changes throughout the transition from one phase to the other (Rogers *et al.*, 2003) and to relate such changes to levels of neuromodulators (Rogers *et al.*, 2004; Anstey *et al.*, 2009; see Section 10.3) and patterns of gene expression (Section 10.1).

12 Volatiles and volatile pheromones

The volatile emissions of locusts are complex and have been the cause of much research and controversy over the years. Earlier detailed reviews by Loher (1990), Whitman (1990), Byers (1991) and Pener (1991) cover the subject up to the point where modern analytical chemistry techniques began to be applied to locust chemical biology in the early 1990s, notably by the team at the International Centre of Insect Physiology and Ecology (ICIPE) in Kenya, who have worked mainly on *S. gregaria*. The most comprehensive reviews of progress since then are those of Hassanali and Torto (1999), Ferenz and Seidelmann (2003) and, most recently, Hassanali *et al.* (2005a).

A tripartite approach has been adopted in much of the modern research. The first step has been to trap volatile emissions and use gas chromatography (GC) and combined gas chromatography-mass spectrometry (GC-MS) to identify the chemical constituents of trapped volatiles. Presence of a volatile compound does not prove that the substance plays a biological role, however. Accordingly, the second step is to establish which compounds are detected by the sensory system of the locust. The usual initial technique has been to pass volatiles through the GC and to split the effluent such that half goes to the detector and the other half passes across an antennal preparation, with the summed electrical response of the antenna to each compound being recorded. Such coupled gas chromatography–electroantennography (GC–EAD, otherwise known as GC–EAG) shows whether the animal has the capacity to detect a given compound, but not all detected compounds need evoke a behavioural or physiological response, let alone function as pheromones. Thus, the third step is to use bioassays to show that volatiles play a biological role – whether as 'releasers' or 'primers' of a biological process. It is here that a degree of controversy has arisen.

Here we begin by surveying which volatile chemicals have been identified in the atmosphere surrounding locusts (principally *S. gregaria*) and from their faeces. We then consider recent work on neural processing of olfactory inputs in the brain. Finally, we focus on so-called aggregation pheromones, and the particular role of phenylacetonitrile (PAN), which has been the subject of considerable interest and debate in recent years. Detailed discussion may be found elsewhere in the present review about the involvement of volatiles in processes including effect on nymphal colour pattern (Sections 7.2.1 and 7.2.3), adult colouration and yellowing (Section 7.3), sexual maturation (Section 8.1), mating behaviour (Section 8.2), mate location (Section 8.3), oviposition (Section 8.4), control of JH on production of some volatiles (Section 9.1), gregarization (Section 14) and parental transmission of phase state (Section 16).

12.1 THE NATURE OF LOCUST-DETECTED VOLATILE EMISSIONS

12.1.1 *Nymphs*

Fuzeau-Braesch *et al.* (1988) reported guaiacol as the dominant volatile in the atmosphere surrounding fifth-instar, crowd-reared nymphs of *S. gregaria* and two subspecies of *L. migratoria, migratorioides* and *cinerascens*. In each case guaiacol was present with smaller amounts of phenol and veratrole. As discussed in Section 12.2, guaiacol and phenol originate from the faeces (Obeng-Ofori *et al.*, 1994b). Torto *et al.* (1996) used a more efficient technique for trapping volatiles than Fuzeau-Braesch *et al.* (1988) and reported a substantially more diverse list of chemicals in headspace volatiles from mixed-sex crowds of fifth-instar *S. gregaria* nymphs. Compounds that were electroantennogram (EAG)-active included hexanal, octanal, nonanal, decanal, hexanoic acid, octanoic acid, nonanoic acid and decanoic acid. Weakly EAG-active substances included undecanal, dodecanal, pentanoic acid, heptanoic acid, undecanoic acid and dodecanoic acid. Additionally, trace amounts of heptanal, anisole, benzaldehyde, acetophenone, veratrole, guaiacol and phenol were found. The low quantities of guaiacol and phenol may have been due to nymphs being kept without food during the 15-h volatile collection period, which would have resulted in relatively few faecal pellets being produced.

Niassy *et al.* (1999) compared the odour bouquets of crowd-reared fifth-instar *S. gregaria* and *L. m. migratorioides*. Confirming the results of Torto *et al.* (1996), *S. gregaria* were found to emit hexanal, octanal, nonanal, decanal, hexanoic acid, octanoic acid, nonanoic acid and decanoic acid, of which only the acids were EAG-active to *L. m. migratorioides*. In contrast, nymphs of *L. m. migratorioides* emitted just three substances at concentrations that were EAG-active to conspecifics: hexanoic acid, PAN and an unidentified compound. All three of these compounds also stimulated antennae of nymphal *S. gregaria*, although responses to hexanoic acid and PAN were lower than in *L. m. migratorioides*. Faecal volatiles of both species included guaiacol, phenol and indole (see Section 12.2).

12.1.2 *Young adults*

Torto *et al.* (1994) reported only small quantities of volatile compounds released by both sexes of young gregarious adults of *S. gregaria*. Low levels of anisole were evident in the GC trace presented (Torto *et al.*, 1994; Fig. 2C). Faeces from young adults released guaiacol and phenol (Obeng-Ofori *et al.*, 1994b), both of which were found along with small quantities of veratrole in the atmosphere within cages of young adults of *S. gregaria*, *L. m. migratorioides* and *L. m. cinerascens* by Fuzeau-Braesch *et al.* (1988) (see earlier).

12.1.3 *Mature adults*

The complexity of volatile emissions increases with maturity in adult crowd-reared *S. gregaria*. Torto *et al.* (1994) found that both males and females emitted anisole, guaiacol and phenol, the latter two presumably being associated mainly with faeces (Obeng-Ofori *et al.*, 1994b). Mature males, however, produced a distinctively different set of compounds, dominated by PAN but also including benzaldehyde and veratrole. In a brief note, Luber *et al.* (1993) had previously shown that only mature males produce PAN (they used the alternative name benzeneacetonitrile). Fuzeau-Braesch *et al.* (1988) had reported veratrole being produced particularly by cages containing copulating and mature adults. Obeng-Ofori *et al.* (1994b) recorded PAN as the predominant compound in volatiles trapped from faeces of mature males, along with smaller amounts of guaiacol and phenol.

Niassy *et al.* (1999) found veratrole, guaiacol, PAN, phenol and 4-vinylveratrole in trapped volatiles from mature adult males of *S. gregaria*, each of which was EAG-active to male antennae. All of these compounds except phenol also stimulated antennae of mature male *L. m. migratorioides*. The bouquet of volatiles from both male and female *L. m. migratorioides* included (*E*)-2-pentenal, (*Z*)-2-pentenal, (*Z*)-2-hexenal, (*Z*)-2-penten-1-ol, (*Z*)-2-hexen-1-ol, veratrole, hexenoic acid, guaiacol, nonanoic acid and five unknown aliphatic compounds. All were EAG-active, but veratrole had the strongest effect on conspecific male antennae. All compounds except (*Z*)-2-hexen-1-ol also stimulated antennae of male *S. gregaria* at natural concentrations. Yu *et al.* (2007) reported more than 30 compounds in faecal headspace odours from 2–4-day-old gregarious adults of *L. m. manilensis* (see Section 12.2).

Njagi *et al.* (1996) compared volatiles produced by solitarious and gregarious adult male *S. gregaria*. Although PAN was the major compound produced by gregarious males, it was noticeably absent among volatiles produced by solitarious adult males at all ages. Anisole, benzaldehyde, guaiacol and phenol were found in both phase groups, and veratrole ''was present in trace amounts in volatiles of some solitarious adult males'' (Njagi *et al.*, 1996, p. 135). EAG responses were recorded from antennae of male and female adults to concentration series of PAN, benzaldehyde, guaiacol and phenol. Other than some statistically significant, but quantitatively minor, differences at high concentrations, antennal responses were remarkably similar between males and females and between phases, with no indication of any interaction between sex and phase in the size of EAG potentials.

Deng *et al.* (1996) measured emission of PAN by solitary- and crowd-reared adult male *S. gregaria* in response to a change in crowding conditions. Insects from a long-term gregarious stock were either isolated from the first stadium, from fledging as adults, or as mature adults. Controls were continued to be kept crowded throughout their life. In comparison, insects from a long-standing solitarious culture (>19 generations) were either crowded or continued in solitary-rearing for the same developmental periods. PAN production was measured regularly throughout adult life and was found to respond rapidly to a change in population density. Shifting from crowded to solitary rearing or vice versa from the first stadium or later at the adult moult resulted in insects adopting the pattern of PAN production typical of the new rearing regime. Production increased with time after the final moult in crowd-reared locusts, reaching a maximum at reproductive maturity, but no PAN was produced by solitary-reared males of any age. When gregarious mature males were placed in isolation, they ceased producing PAN within two days, whereas crowding mature, solitarious males caused them to start emitting PAN equally rapidly, to a level which after 8–10 days seemed even to exceed that of long-tern gregarious males. When solitarious first-instar nymphs were crowded throughout their lives with either one or three others (i.e. in groups of two or four), the timing of PAN production in adult males was somewhat delayed relative to fully gregarious controls. However, similar maximal release rates to those of gregarious adults were reached in males reared in a group of four. PAN production was an order of magnitude lower when insects were reared in pairs, but the timing of release followed a similar course to insects within groups of four.

Assad *et al.* (1997a) found that exposure of small groups of immature males to fifth-instar nymphs delayed onset of PAN production, but not peak production rates, as part of a general delay in maturation (see Section 8.1 for full discussion). Nymphal faeces had no such effect, but trapped nymphal volatiles or a synthetic blend of the component aldehydes, acids, guaiacol and phenol (Torto *et al.*, 1996; see earlier) were fully effective.

Seidelmann *et al.* (2000) confirmed that PAN (they used the term benzyl cyanide) was produced only by mature adult males and conducted a detailed study of the time-course of release. Daily PAN production was recorded from groups of five adult males. The onset of PAN release occurred on day 15 and was reported to accompany yellowing, indicating sexual maturation. Thereafter, PAN release rose throughout the 35-day period to a maximal rate of 100–140 ng per male. Seidelmann *et al.* (2000) replaced males within the group of five adults with females and found that PAN release was both delayed and lowered progressively as females comprised a greater proportion of the group, such that no PAN was detected when one male was housed with four females. Thus, PAN production was purely a function of the number of other males present.

Seidelmann *et al.* (2000) took groups of four mature males and separated the insects from each other using clear plastic partitions. After 5 days the partitions were removed, then reinstated after a further 5 days. Release of PAN fell rapidly following physical separation, declining to near zero by 4–5 days, and returned equally rapidly upon removal of the partition. Since the insects were still in olfactory and visual contact when partitioned, the conclusion was that physical contact, and hence short-range chemical and/or mechanical cues, is necessary to induce PAN production/release. The fact that crowding with females, which would have provided mechanical stimulation, did not elicit PAN release indicates that short-range chemical cues are responsible. In a later study, Kiessling and Seidelmann (2006) ablated the antennae, mouthpart palps, sensory fields on the legs, or compound eyes of mature males and found that insects produced PAN normally in the presence of other mature males; however, removal of both antennal and eye inputs caused significant reduction in PAN release. Short-range chemical cues are perceived by chemosensory receptors on multiple body parts, including antennae, legs and palps, which could explain why removal of antennae or palps alone had no effect. However, it is not clear why insects separated by transparent partitions in Seidelmann *et al.* (2000) ceased PAN release, if it is presumed that insects could see each other through the partitions. It is possible that severe sensory deprivation occasioned by removal of input from two such important sense organs as the eyes and antennae had more general pathological effects or that both visual and short-range chemosensory inputs converge in their effect on PAN release but with the latter dominating.

In a further study, Seidelmann *et al.* (2003) localized the site of production of PAN in mature adult males. Head, thorax, abdomen, forewings, hindwings, first, second and third pairs of legs were all analysed. The major release sites were found to be the fore and hindwings (two-thirds of total released) and the legs (one-third of total released), particularly the tibia and tarsi of the hind legs. When wings were clipped or amputated, PAN production fell – especially so after forewing amputation – with no evidence of a compensatory increase in release from other body regions. This result suggests that the wings are not only sites of release, but are also sites of production: a conclusion that was confirmed by incubating pieces of body parts with a radiolabelled precursor, phenylalanine. The tibiae and the basal and mid regions of both wings were most active in synthesizing PAN. The suggestion was that vacuolated glandular cells in these regions are responsible for producing PAN. Ferenz and Seidelmann (2003) point out that such an origin of PAN leaves the question as to where the PAN reported by Obeng-Ofori *et al.* (1994b) in mature male faeces came from. Cannibalism is one possibility.

Since incubation of tissue with labelled phenylalanine but not tyrosine resulted in PAN synthesis, the likely pathway of production is L-phenylalanine, to L-*N*-hydroxylphenylalanine to phenylacetaldoxime, to PAN (Ferenz and Seidelmann, 2003). Degradation of PAN produces benzaldehyde, which is another component of mature male volatile emissions (see earlier), but was also reported in small amounts in volatiles from solitarious males, which do not produce PAN (Fig. 1 in Njagi *et al.*, 1996). Sites of release of another mature male produced volatile, veratrole, were also studied by Seidelmann *et al.* (2003). In contrast to PAN, the conclusion was that all body regions emit veratrole, suggesting a different glandular origin.

In their review article, Ferenz and Seidelmann (2003) state that PAN synthesis appears to be under hormonal control and that raw or partly HPLC-purified methanolic extracts from brains of brains of mature adult males stimulate release of PAN in solitarious males after 3–5 days, 'suggesting that a putative neuropeptide controls the synthesis and not the release of the pheromone'. Kiessling and Seidelmann (2006) provided data supporting the statement in Ferenz and Seidelmann (2003). They concluded from injections of methanolic extracts of various tissues into young adult males that both synthesis and release of PAN are controlled by a circulating biosynthesis activating neuropeptide (PAN-BAN) originating from the brain and sub-oesophageal ganglion, the chemical identity of which remains unknown. Capacity to produce PAN was found to be unaffected by removing testes and/or reproductive accessory glands from young adult males, whereas transplanting corpora allata from mature into immature gregarious males rapidly initiated PAN production, implying a role of JH in the regulation of PAN synthesis – as in other maturation processes (see Section 8.1).

Three sources of volatiles produced by mature adult females warrant only brief mention since they are covered in other sections. The first of these is egg pod foam and associated secretions. Rai *et al.* (1997) identified veratrole and acetophenone in egg pod foam as behaviourally active substances responsible for oviposition aggregation in *S. gregaria* (see Section 8.4 for discussion). Malual *et al.* (2001) claimed that the volatile unsaturated ketones, (*Z*)-6-octen-2-one, (*E*,*E*)-3,5-octadien-2-one and (*E*,*Z*)-3,5-octadien-2-one, which were identified by Torto *et al.* (1999b) and are associated with eggs and to a lesser extent egg foam, are responsible for maternal transfer of phase state to developing embryos (see Section 16.3 for a discussion of this claim). Second, Njagi and Torto (2002) report that dichloromethane extracts of Comstock-Kellog glands (eversible glands located in the genital chamber) from recently sexually mature female *S. gregaria* contain pentanoic acid, hexanoic acid, heptanoic acid, octanoic acid, nonanoic acid and decanoic acid. Pentanoic acid was EAG-active and was suggested to elicit sexual behaviour in mature males (see Section 8.2 for further discussion). Finally, as discussed in Section 8.3, there have been claims that solitarious females emit a sex attractant pheromone (Inayatullah *et al.*, 1994), although no distinct chemical emission from solitarious females or specialized release organ has yet to be identified.

12.2 VOLATILES ASSOCIATED WITH GUT BACTERIA AND FAECES

That faecal volatiles may play a role in locust phase change was suggested by Nolte (Nolte *et al.*, 1973; Nolte, 1976), who identified 5-ethylguaiacol or 'locustol' as the active compound. As has been documented most recently by Hassanali *et al.* (2005a) and in several earlier reviews (Loher, 1990; Whitman, 1990; Byers, 1991; Pener, 1991), Nolte's findings have not been replicated. Although it is now considered that locustol is not a bioactive, locust-produced volatile, related phenolic compounds are released from the faeces of *S. gregaria.* Guaiacol (2-methoxyphenol) and phenol are the predominant compounds emitted from faeces of hoppers and young adults. Both are found in faecal odours of mature, crowd-reared males, with PAN also recorded in substantial amounts (Obeng-Ofori *et al.*, 1994b). Guaiacol and phenol were reported in headspace vapour from crowds of two subspecies of *L. migratoria* as nymphs, young adults and mature adults (Fuzeau-Braesch *et al.*, 1988), and presumably were also of faecal origin. This was proven to be the case by Niassy *et al.* (1999), who additionally reported indole from faeces of fifth-instar nymphs in both *S. gregaria* and *L. migratoria*. The absence of indole in the earlier report from Obeng-Ofori *et al.* (1994b) was attributed to its relatively low volatility.

Yu *et al.* (2007) collected faecal volatiles from young adult *L. m. manilensis* and analysed them by GC–EAD, using antennae from male and female adults that were 5–15 days beyond the final moult. Of the more than 30 compounds evident in GC traces, 11 elicited an electrophysiological response from antennae. The data shown are for female faeces and female antennal responses, but it is stated that both sexes responded similarly. Of the 11 electrophysiologically active compounds, nine were identified as 2-hexenal, 2,5-dimethylpyrazine, cyclohexanol, hexanal, benzaldehyde, benzyl alcohol, nonanal, 2,6,6-trimethylcyclohex-2-en-1,4-dione and β-ionone. Of these, 2-hexenal was the most active. The absence of guaiacol and phenol among the reported components of faecal odour is noteworthy, but hard to explain.

Dillon *et al.* (2000) showed that guaiacol and phenol are produced by bacteria found in the gut, notably associated with the lining of the hindgut. Germ-free (axenic), crowd-reared nymphs and adults of *S. gregaria* produced no guaiacol and very small amounts of phenol. Infection of first-instar, axenic nymphs with a single species of bacterium, *Pantoea* (formerly *Enterobacter*) *agglomerans*, which is a dominant member of the normal gut biota, instated guaiacol and phenol production in fifth-instar nymphs and adults under otherwise germ-free conditions. Incubation of undigested wheat with bacteria did not result in guaiacol production, indicating that digestion of food in the gut is necessary for bacteria to produce more than trace amounts of phenolics. The likely precursor for guaiacol production by bacteria was shown to be vanillic acid, which is derived from the diet and found in the faeces of both normal and axenic locusts. Feeding locusts with filter paper treated with vanillic acid solution resulted in the production of large amounts of guaiacol.

Dillon *et al.* (2002) extended their investigation by using germ-free faecal pellets to culture *P. agglomerans* and two other bacteria from the locust gut biota, *Klebsiella pneuomoniae* and *Enterobacter cloacae*. All three produced substantial amounts of guaiacol and smaller quantities of phenol. Another species of gut bacterium, *Enterococcus casseliflavus*, produced no phenol and only small amounts of guaiacol, whereas an opportunistic pathogenic species, *Serratia marcescens*, produced neither compound. The production of guaiacol from faeces closely matched the ability of the different bacterial species to produce guaiacol by decarboxylation when cultured in a nutrient broth containing vanillic acid, confirming that vanillic acid is the precursor.

Dillon *et al.* (2008) have recently explored differences in the bacterial communities in the guts of wild-caught specimens of three locust species (brown locust, *Locustana pardalina*, Moroccan locust, *Dociostaurus maroccanus*, and Italian locust, *Calliptamus italicus*), using denaturing gradient gel electrophoresis (DGGE) analysis of 16S rRNA gene fragments. In the case of *L. pardalina*, individuals connoted as 'solitarious', 'transiens' and 'gregarious' were collected from 'mixed phase groups' in a 1–2 ha area 'from a population that comprised a small swarm of gregarious individuals, which had been settled in the area for several days, integrated with a resident cohort of transiens and solitaries' (Dillon *et al.*, 2008). Substantial variation in bacterial gut communities was found both within and between these phase categories, and between the locust species (and the meadow grasshopper, *Chorthippus parallelus*, which was also studied). However, within *L. pardalina* solitarious individuals possessed fewer bacterial species in their guts than did transiens or gregarious individuals. It is not possible to place much emphasis on this result, given the small number of individuals tested ($n = 4-7$) from only a single location. Nevertheless, the authors considered possible reasons for a phase-related difference in the diversity of gut biota. They discounted differences in diet breadth (cf. Despland, 2005 for *S. gregaria*; see Section 17.2), on the grounds that brown locust is graminivorous and that specimens came from the same location, and suggested instead that locusts have the ability to regulate their gut microbiota through immune reactions. Differences between phases in activity levels and their propensity to interact could also have accounted for the result, however, by altering encounter frequency with potential sources of bacterial inoculation. Dillon *et al.* (2008) proposed that differences between the phases in the diversity of their gut microbiota might reflect a trade-off between the costs and benefits of maintaining a high bacterial diversity in the gut. Costs could include increased pathogenic effects; benefits might involve increased stability within the ecological milieu of the gut against invasion by virulent pathogens, and hence serve to protect against an elevated risk of disease in high-density populations (Wilson *et al.*, 2002; see Section 17.4). This is an intriguing idea, but much more extensive data are required to test its veracity.

Whether phase-related differences occur in bacterial communities within the guts of other locust species is not yet known, nor whether such differences would translate into quantitative or qualitative differences in faecal volatile production.

12.3 PROCESSING OF ODOUR STIMULI IN THE CENTRAL NERVOUS SYSTEM

As discussed in Section 12.1, much of the preliminary screening for electrophysiological activity of locust-produced volatiles has come from use of electroantennography, which measures the summed electrical potential evoked by an odour stimulus from an entire antenna. Hansson and colleagues have conducted a series of detailed studies in which physiological responses to various volatiles were recorded from receptor neurones (RNs) within single olfactory sensilla, and from interneurones within the antennal lobes of the deutocerebrum.

In an initial study, Hansson *et al.* (1996) recorded from antennal RNs in gregarious adult *S. gregaria* of undisclosed age. Recordings were made from two types of sensillum, which they termed 'trichoid/basiconic' and 'coeloconic' (see Section 5.3.1 for a discussion of terminology in relation to other workers). In a more detailed study, Ochieng' and Hansson (1999) recorded from RNs in sensilla categorized as basiconic, trichoid and coeloconic in gregarious and solitarious young adults. Hence, in contrast to Hansson *et al.* (1996), basiconic and thrichoid sensilla were separated. All RNs (or more accurately, assemblages of RNs, since single unit recording was difficult) recorded from basiconic sensilla in both males and females were excited by compounds produced by mature males (PAN, benzaldehyde, guaiacol, phenol, and a 100:15:4:4 blend of these, with sensitivity being greatest to PAN and guaiacol among the individual compounds) and a blend of nymphal volatiles (hexanal, octanal, nonanal, decanal, hexanoic acid, octanoic acid, nonanoic acid and decanoic acid in a ratio of 4:9:20:20:14:50:100:16, respectively). Additionally, RNs were excited by (*E,Z*)-2,6-nonadienal and acetophenone. The former of these compounds has been recorded as a component of volatiles from the host plant *Tribulus terrestris* (Njagi and Torto, 1996) and was (and is often) stated to be a potential sex pheromone component, although no concrete data exist on this (see Section 8.3). Acetophenone was reported by Rai *et al.* (1997) to be a component of an oviposition aggregation pheromone (see Section 8.4).

Coeloconic RNs, were excited by the plant odours (*E*)-2-hexenal and butyric acid and by the nymphal odour blend, but were inhibited at high concentrations by all adult-produced compounds other than benzaldehyde, which was excitatory. RNs in trichoid sensilla were excited by (*E,Z*)-2,6-nonadienal, but inhibited by high doses of butyric acid. Phase-related differences were few and difficult to interpret: none was seen for responses of RNs in coeloconic and trichoid sensilla, and only for the nymphal odour blend, acetophenone and (*E,Z*)-2,6-nonadienal, there was a consistent dose-related difference between the phases in responses of RNs in basiconic sensilla. In each of these cases, RNs in solitarious insects had a higher responsiveness than those in gregarious individuals, with the difference being most marked in males (Ochieng' and Hansson, 1999).

Hansson *et al.* (1996) mapped the projections of RNs in the antennal lobes of the deutocerebrum. It was found that RNs branch to enter multiple glomerular structures, of which there were estimated to be approximately 1000 in each antennal lobe. Later estimates of numbers of glomeruli in adults were about half that number, and it was subsequently reported than a minority of RNs project to a single glomerulus rather than to multiple glomeruli (Anton *et al.*, 2002). Such glomeruli comprise tangles of neuropile containing (1) the terminal arborisations of antennal RNs, (2) local interneurones that make synaptic connections within and between glomeruli, (3) projection interneurones that receive inputs within multiple glomeruli and leave the antennal lobe to terminate at the lateral lobes of the protocerebrum and the calyces of the mushroom bodies and (4) the terminals of centrifugal interneurones that originate in the protocerebrum, suboesophageal ganglion, ventral nerve cord and perhaps elsewhere in the antennal lobes (Hansson and Anton, 2000).

Anton and Hansson (1996) recorded from projection interneurones (PNs) in the antennal lobes of young (1–6 days after fledging) and older (>25-day-old) crowd-reared adult *S. gregaria* in response to a range of locust-produced and food-related odorants. Subsequently, Ignell *et al.* (1998) recorded from PNs in gregarious and solitarious fifth-instar nymphs, and Ignell *et al.* (1999) carried out a similar study on third-instar nymphs. In a culminating paper in the series, Anton *et al.* (2002) quantified developmental changes in the size and structure of the antennal lobes and in electrophysiological responses of PNs from gregarious and solitarious locusts across development (first-, third- and fifth-instar nymphs and young adults).

The gross structure and organization of the antennal lobe was found to be consistent between solitarious and gregarious *S. gregaria* and across development. An approximate doubling occurred from first instar to the adult in the size of the antennal lobes, the maximal diameter of glomeruli, and the number of glomeruli (Anton *et al.*, 2002). The overall morphology of PNs remained constant during development, but the number of glomeruli innervated by each PN increased from first-instar nymphs to adults, from an average of 10 glomeruli per PN for first-instar nymphs to 18 for adults, with no difference between phases (Anton *et al.*, 2002).

The electrophysiological response characteristics of PNs were established during intracellular recording by stimulating the antenna with PAN, guaiacol, phenol, benzaldehyde, veratrole, acetophenone, (*E*,*Z*)-2,6-nonadienal, (*E*)-2-hexenal, a 60:40 blend of guaiacol and phenol (termed 'nymph blend', designed to match the composition of nymphal faecal emissions, but, as discussed in Section 12.2, could equally be argued to match immature adult faeces), and an 80:3:3:5:4:5 blend of PAN, guaiacol, phenol, benzaldehyde, veratrole and anisole ('adult blend', mimicking the mixture of compounds emitted by mature adult males). PNs were categorized into five basic types according to their responses to these stimuli (Anton *et al.*, 2002): (1) pheromone component specific, which responded to a single locust-produced compound or any blend containing that compound; (2) blend specific, which responded to the nymphal odour blend and/or the mature adult male blend (see earlier), but not to single compounds; (3) pheromone generalist, which responded to at least two individual locust-produced compounds and the blends in which they occurred; (4) pheromone-plant generalist, which responded to both locust-produced and plant volatiles and (5) plant-specific PNs, which only responded to plant volatiles.

When summed across these five categories and across sexes, phases and ages, about 40% of all responding PNs were stimulated by PAN, and similar percentages responded to guaiacol and to phenol. Approximately 20% of PNs responded to benzaldehyde and a similar percentage to (*E*)-2-hexenal. There were phase-specific differences across development in response to PAN, guaiacol and phenol. Whereas there were no significant changes in response to these compounds in solitarious insects across development, in gregarious individuals there was a doubling in the percentage of PNs responding to PAN in young adults. There was also some evidence of increases in the percentage of PNs responding to guaiacol and phenol from first- to third-instar, followed by a decline in the fifth stadium, and a further increase in response to guaiaicol but not phenol in gregarious adults. Both phases showed an increase in the number of PNs responding to benzaldehyde with age.

In an earlier study of third-instar nymphs, Ignell *et al.* (1999) found significant differences between the phases in the proportion of PNs responding to the adult male odour blend, PAN (solitarious > gregarious in both cases), phenol, veratrole, acetophenone, nonadienal and hexenal (gregarious > solitarious for each of these compounds), but no difference to a blend of non-faecal nymphal volatiles (the eight aldehydes and acids detailed in Section 12.2), the nymphal faecal blend, guaiacol, benzaldehyde, hexanoic acid or hexenal. The only difference that reached a high level of statistical significance ($P < 0.001$) was that for acetophenone. In their parallel study on fifth-instar nymphs, Ignell *et al.* (1998) only found significant differences (all indicated as $P < 0.05$) for the nymphal faecal blend, PAN, guaiacol and phenol, with solitarious values being greater than gregarious values in each case. Dose–response studies indicated a generally higher sensitivity in PNs of solitarious nymphs, a result which was opposite to that reported for third-instar nymphs by Ignell *et al.* (1999) (see further discussion later on this point).

Pheromone generalists were the most frequently encountered type of PN, comprising around 40%–50% of PNs recorded in first-instar nymphs, 60%–75% in later-instar nymphs, and 50%–60% in young adults (Anton *et al.*, 2002). The only striking difference in response between the phases among pheromone generalist PNs was seen in adult females, in which solitarious females had double the percent of PNs responding to veratrole and acetophenone than did gregarious females (70% versus 40% for veratrole, 75% versus 30% for acetophenone). Males did not differ between the phases (~50%–60% of PNs responded in each phase to veratrole or acetophenone).

Blend-specific neurones comprised only a few percent of PNs in all age groups and in both phases, with numbers so low as to be difficult to draw any conclusions. About 20% of PNs were categorized as component-specific, with no evidence of statistical differences with phase or age. Anton *et al.* (2002) commented that those responding to PAN and nonadienal occurred less frequently with age, most notably in gregarious locusts, whereas those responding to guaiacol and acetophenone increased in number with age in both phases, but numbers were too low for statistical analysis. Plant-specific PNs were very rare in both phases and at all developmental stages.

Young adults were used in the experiments of Anton *et al.* (2002). Puzzlingly, in an earlier study, Anton and Hansson (1996) found that none of the PNs recorded from in mature (> 25-day old) adults responded to any of the locust-produced or plant compounds, even though mature adults are behaviourally responsive to such odours, and antennal RNs were functional. Ignell *et al.* (2001) explored this issue in a detailed study in young (8-day-old) and mature (29-day-old) gregarious adults of both sexes. One treatment group was allatectomized at day 2 after adult ecdysis; another group was injected with JH III on days 1, 2 and 3; a third group was sham operated; and a fourth group was left untouched. Electroantennogram responses to locust-produced volatiles were identical among the treatment groups and ages, indicating that mature adults retain sensory responsiveness to odours and that manipulating JH had no effect. The proportion of locust preparations with responding PNs in their antennal lobes, however, diminished with age from 90% at day 8 to 60% at day 29. Of the total number of PNs tested, there was a decline from 80% responding at day 8 to around 30% by day 29. These age-related changes are consistent with but not nearly as marked as reported by Anton and Hansson (1996) and occurred in both sexes. The changes in PN responsiveness were found to be related to JH. Allatectomized insects at day 29 were similar in PN responsiveness to controls on day 8, whereas JH-injected insects at day 8 were comparable to controls on day 29, implying that elevated titres of JH are linked to deterioration in the electrophysiological competence of the antennal lobes. Behavioural responses to PAN were recorded in the various experimental groups at weekly intervals until day 29. Changes accompanying those occurring in PN responsiveness were found and will be considered in more detail in the next section and are also critically discussed in Section 9.1.

What can be concluded from these studies of antennal odour processing in relation to developmental and phase-specific behavioural differences? Evidently locusts have olfactory receptors that respond to locust-emitted compounds, and such receptor inputs converge and are integrated within the antennal lobe to produce a variety of outputs, which are coded in the responses of projection interneurones. Sensory receptors and projection interneurones are, in the main, generalist in their action spectra, with few RNs and PNs being restricted in their response to specific compounds, and only a minority of PNs responding exclusively to particular blends of compounds. Beyond that, however, it is difficult to draw compelling conclusions. The risk is that the complexity of chemical emissions and the multiplicity of their reported functions are such that *post hoc* explanations can be generated to explain just about any difference that might appear with age and phase. There were remarkably few phase-specific differences apparent among the large number of paired comparisons made between the phases in studies on RNs, PNs and in associated EAG investigations; and those that achieved statistical significance were in the main relatively subtle.

Some of the general trends seen throughout development are likely to reflect increases in the numbers of olfactory sensilla with age. Sensilla numbers increase throughout development at a substantially greater rate than do the number and size of antennal lobe glomeruli. For example, basiconic sensilla increase in number 13–15-fold from first instar to adulthood (Ochieng *et al.*, 1998), yet the number and size of glomeruli in the antennal lobes only double (Anton *et al.*, 2002). As a result, there must be greater convergence of RNs onto primary interneurones, with potential consequences for both the sensitivity and response spectra of antennal lobe interneurones. The peculiar reversal of PN sensitivity between the phases seen from the third to the fifth stadium is probably explicable in these terms, as was concluded by Ignell *et al.* (1999). Hence, PNs from gregarious insects were found to be more sensitive then those of solitarious insects in the third stadium, but the opposite was the case in the fifth stadium. As discussed in Section 5.3.1, Ochieng *et al.* (1998) found a correlated reversal in the numbers of antennal sensilla, with gregarious nymphs having more sensilla than gregarious nymphs in the third stadium but fewer in the fifth; perhaps indicating that a proportion of solitarious insects had inserted an extra instar during development.

12.4 AGGREGATION PHEROMONES

The past 15 years have seen considerable attention paid to the behavioural roles of locust-emitted volatiles in *S. gregaria*, in particular in the context of aggregation behaviour. Fuzeau-Braesch *et al.* (1988) proposed that locust-emitted volatiles (phenol, guaiacol and the mixture of these with veratrole) in both *S. gregaria* and *L. migratoria* serve as 'cohesion pheromones', encouraging gregarious nymphs and adults to remain together, but not attracting them. These results were based on an assay with four arms and a central crossing region, with locusts released as groups in one arm and the odour source positioned in another. Locusts tended to aggregate within the central zone at the juncture of the four arms. It should, therefore, be borne in mind that visual stimuli provided a confounding source of stimuli promoting aggregation. Inspired by these experiments, the group at ICIPE developed a behavioural assay system particularly designed to measure whether locust volatiles arrest locusts within regions suffused with odour. Their assay system is a single-chambered, two-choice olfactometer. Typically, a clean, vertical airstream is paired with an adjacent vertical stream enriched with volatiles. Locusts are introduced into the chamber either alone or in groups and their positions measured some time later (variously after 10, 15, 20 or 30 min). The number of locusts within the control and test regions of the chamber are recorded and used to derive an Aggregation (or Retention) Index, which is $100 \times (T-C)/N$, where T is the number of locusts found on the test side, C the number on the control side and N the total number of locusts tested. Insects in the middle between the test and the control zones are omitted from analysis.

The first published study using the vertical airflow olfactometer was that of Obeng-Ofori *et al.* (1993). These authors tested gregarious *S. gregaria* hoppers of mixed sexes (second, third, fourth and fifth instar), young and mature adults. Locusts were tested in the olfactometer either singly and in groups of 10, with similar results. Hoppers responded to nymphal odours by remaining within the odour-saturated zone of the arena, but did not respond to adult odours. In contrast, mature adults (sexes combined) only responded by remaining within the test zone when presented with their own odour; not when the test airstream contained the odour of young adults or nymphs. Young adults only responded to volatiles from mature adults. Hence, the conclusion was that juveniles have a separate aggregation pheromone system to adults.

In a subsequent study, Torto *et al.* (1994) studied the chemical constituents of the adult odour blend on mixed-sex groups of five fifth-instar nymphs, young adults, or mature adults. Crude volatile extracts from mature adults and al blend of six constituent compounds in mature male odour (a 100:15:5:5:4:4 mix of PAN, benzaldehyde, veratrole, anisole, guaiacol and phenol, respectively) gave high values for the Aggregation Index in both sexes of young and mature adults, but not in nymphs. When chemical were presented singly, PAN elicited the strongest aggregation response from young and mature adults. Guaiacol and phenol were moderately effective and benzaldehyde less so, whereas anisole and veratrole alone or together were inactive. When guaiacol, phenol and benzaldehyde were combined, the effect was as strong as the complete blend, even though PAN was absent. The conclusion was that PAN, guaiacol, phenol and benzaldehyde comprise the aggregation pheromone of mature adults.

A worrying feature of the study by Torto *et al.* (1994), first pointed out in the review of Ferenz and Seidelmann (2003), was the statement that 'If this calculation [of the Aggregation Index, AI, for a given trial] resulted in a negative number, the aggregation index was assumed to be zero'. The problem with this adjustment is that it shifts the null expectation of no response to a test odour from a mean AI value of zero, to one near 25 (Seidelmann *et al.*, 2005), with potentially serious implications for data interpretation. Rather than an AI of zero indicating no effect of a test volatile, this value would indicate moderate avoidance. In a subsequent paper, Torto *et al.* (1996) also set negative values of AI to zero. Whether this was the case for other studies using the vertical airflow olfactometer assay (Obeng-Ofori *et al.*, 1994a,b; Njagi *et al.*, 1996; Niassy *et al.*, 1999; Ignell *et al.*, 2001) is not stated; but in all these publications, including those of Torto *et al.* (1994, 1996), the clear implication is that a mean value for AI of zero indicates no behavioural effect of a test odour, not avoidance of the odour.

In relation to the results of Torto *et al.* (1994), changes in the interpretation are required as a result of the authors truncating AI. First, benzaldehyde alone was ineffective on adults (giving AI values near 25). Second, anisole and veratrole alone and in combination were probably repellent. These are listed as 'inactive' in data tables, without an accompanying value for AI, so presumably had values near zero. Third, since AI values for nymphs are not reported, other than the statement that adult odour blends 'Evoked no aggregation response from fifth instar nymphs' (Torto *et al.*, 1994; caption to Table 1), it may be concluded that mean AI values were near zero. This would indicate active avoidance by nymphs of mature adult odours.

Obeng-Ofori *et al.* (1994a) tested the separate responses of male and female *S. gregaria* as hoppers (third, fourth and fifth instar), young and mature adults to male and female volatiles. Groups of five insects were introduced into the olfactometer and offered a choice between a clean airstream and one permeated with volatiles from live locusts. Both male and female nymphs responded to nymphal odours from both sexes, but not to adult volatiles. Young and mature adults of both sexes responded only to mature male volatiles. The fact that some mean AI values were slightly negative (as was also the case for Obeng-Ofori *et al.*, 1993) indicates that negative AI values were not truncated to zero.

Obeng-Ofori *et al.* (1994b) studied behavioural responses to faecal volatiles and found that groups of five nymphs of both sexes had elevated values for AI in response to both nymphal and young adult faeces, but were reported to be indifferent to mature adult faeces. Young adult and mature adults of both sexes responded to each other's faecal odours and to those from nymphal faeces. None of the mean values reported for AI were negative (even slightly so, as would be expected for ineffective test odours), and values for the lowest concentrations tested in dose–response studies, for all stages except nymphs tested with mature adult volatiles, were approximately 25; both of which suggest that negative AI values were truncated to zero. If this were the case, then the interpretation would change in one important respect: nymphs were not indifferent to faecal volatiles from mature adults, but rather repelled by them, having AI values close to zero. Such a result would be in accordance with the reinterpretation of the results of Torto *et al.* (1994) discussed earlier. Such an interpretation would, in turn, imply that PAN is repellent to nymphs, given that other faecal volatiles (guaiacol and phenol) are shared between nymphs and mature adult males (a possibility that was raised by Pener and Yerushalmi, 1998, p. 373 and is considered further later).

Torto *et al.* (1996) state that they truncated the AI values in their study on groups of five, fifth-instar *S. gregari*a subjected to crude nymphal odours and their components. Crude nymphal volatiles and synthetic blends of the constituent aldehydes (hexanal, octanal, nonanal and decanal) and acids (hexanoic acid, octanoic acid, nonanoic acid and decanoic acid) yielded AI values well in excess of 25 (the null expectation for no response) at concentrations higher than 75 'locust-hour equivalents' (LH, volatiles released by one locust over one hour). At 37.5 LH, the mean value was near 20. Mature adults, however, had AI values close to zero when exposed to nymphal volatiles (Table 1 in Torto *et al.*, 1996): which would indicate avoidance. When guaiacol and phenol (faecal volatiles) were presented together at various concentrations to nymphs, AI values of only 30–40 resulted, with no evidence of a dosage effect, suggesting little response. These results are in contrast with the dose-dependent responses to faecal volatiles reported by Obeng-Ofori *et al.* (1994b). Adding these two compounds to the full blend of aldehydes and acids perhaps slightly increased the AI values for nymphs. It was claimed that individual acids were moderately active but that none of the individual aldehydes showed significant activity. However, in view of the null expectation of mean AI near 25, data in Fig. 3 of the paper indicate no, or only weak, behavioural responses to individual acids (values are 20–30, reaching around 40 at the highest concentrations of 300 LH). Values for the aldehydes are not given, but presumably they were close to zero, which would suggest avoidance; not indifference.

Njagi *et al.* (1996) extended the study of mature adult volatiles to consider gregarious and solitarious phases of *S. gregaria*. Single mature adults were tested in the vertical airflow olfactometer. An AI equivalent (termed a Retention Index, RI since it was based on individual locusts rather than groups) was calculated and odours that produced values less than 10 were listed as 'inactive' and their mean values not reported. It is not stated whether such 'inactive' RI values were in fact negative. Both sexes of both phases were retained on the test side of the arena (RI up to 70–80) by the mature adult male blend (a 100:15:6:5:4:4 mixture of PAN, benzaldehyde, veratrole, anisole, guaiacol and phenol, respectively) at the tested concentrations, which ranged from 25 to 250 LH. Anisole and veratrole together or separately were classed as 'inactive' (hence perhaps repellent, which would coincide with reinterpretation of data from Torto *et al.*, 1994, discussed earlier). Removing these two compounds from the complete blend had no effect; nor did further removal of PAN, followed by removal of benzaldehyde, leaving residual activity in the blend of only guaiacol and phenol. In contrast to responses to blends of compounds, single application of PAN was fully effective, whereas responses to guaiacol, phenol and benzaldehyde alone in that order were decreasingly effective in retaining test locusts in the odour-suffused region.

Ignell *et al.* (2001) tested the responses of individual adult male and female gregarious *S. gregaria* to PAN at weekly intervals from ecdysis until 29 days old. In contrast to results from Njagi *et al.* (1996), values for AI were around 30 for males and females until 8–15 days old, thereafter falling to zero. Allatectomized insects maintained AI values around 40 (males) or 20 (females) throughout adult life. Detailed behavioural observations were undertaken at the different ages, including time individuals spent inactive, walking, moving their legs, and moving their antennae. At 15 days or older the elevated AI values in allatectomized insects were accompanied by insects spending more time inactive on the control side than the PAN-suffused side of the olfactometer, and walking and moving their legs and antennae more on the PAN side than the control side, suggesting an arousing effect of the compound. However, behavioural differences between the sides of the arena were small at younger ages in control and allatectomized locusts, even though they had elevated AI values, making the relationship between detailed behaviour and the resulting distribution of the test insects in the arena difficult to interpret. See Section 9.1 for further discussion on difficulties in the interpretation of these behavioural data from Ignell *et al.* (2001).

Interspecific responses to locust-emitted odours were explored by Niassy *et al.* (1999) in their study of gregarious *L. migratoria* and *S. gregaria*. The vertical airflow olfactometer was used, with groups of five locusts being introduced into the arena. Whether negative values for AI were set to zero is not indicated. Nymphs of both species responded to the volatiles emitted by nymphs of the other species, as well as to their own smell; albeit the dose–response characteristics differed, with *L. migratoria* nymphs showing a flat dose–response to *S. gregaria* nymphal odours, rather than the inverted U-shaped response to their own species' volatiles, and *S. gregaria* nymphs having inverted U-shaped responses to both species' odours, but peaking at a lower concentration for conspecific than for heterospecific odour. The only common chemical reported in the nymphal odours of the two species was hexanoic acid, and *L. migratoria* nymphs released PAN (Niassy *et al.*, 1999; see Section 12.1.3). Earlier results indicated that PAN (only found in odours from mature gregarious male *S. gregaria*) should inhibit aggregation in *S. gregaria* nymphs (see earlier), which might explain the differences between the species in dose–response curves.

Mature adult *L. migratoria* responded to odours from mature adults of both species. However, adult *S. gregaria* were more responsive to odours of adult conspecifics than to heterospecific volatiles, which might have been due to absence of PAN in volatile emissions from mature adult *L. migratoria* (Niassy *et al.*, 1999). Mature adult *L. migratoria* had AI values near zero for conspecific nymphal volatiles, but high values for their own odour, whereas nymphs responded to both nymphal and adult conspecific odours. There were no differences between the sexes of *L. migratoria* either as sources of, or responders to, conspecific nymphal odour. Both sexes of adults elicited similar responses as odour sources, in keeping with their similar odour profiles (see Section 12.1.3), although there was a tendency for the sexes to respond more strongly to the smell of the opposite sex than to their own sex.

In summary, the following conclusions can be drawn from the extensive work undertaken using the vertical airflow olfactometer:

First, gregarious juveniles of *S. gregaria* (hoppers of both sexes, from second to fifth stadium) remain in areas suffused with nymphal volatiles, a response that can be mimicked with synthetic blends of constituent acids and aldehydes.

Second, nymphs are retained in areas permeated with faecal odours from all stages except mature adult gregarious males. Since faeces of all stages and sexes emit phenol and guaiacol, but mature males faeces also emit PAN, the implication is that the former two compounds are responsible for retaining nymphs, whereas PAN is avoided. Consistent with this conclusion, juveniles are indifferent to the smell of immature adults, and either indifferent to or avoid volatiles from mature adult males, which release PAN. Indeed, a recent study by Seidelmann *et al.* (2005), which is discussed in detail in the next section, also indicates an avoidance of PAN by gregarious nymphs. A repellent or other dispersive effect of PAN might help explain the unpublished reports that spraying PAN dispersed nymphal bands in the field (Hassanali and Torto, 1999, pp. 313, 322).

Third, sexually immature adults of both sexes, and mature adults of both sexes and both phases, are induced to remain in areas permeated with mature, gregarious adult male odour or synthetic blends containing components such as PAN, guaiacol, phenol and benzaldehyde; or by faecal odours of any developmental stage, that is, phenol and guaiacol, with PAN in the case of mature males.

A gap remaining in behavioural studies using the vertical airflow olfactometer is a comparison of solitarious and gregarious nymphs in relation to their response to nymphal and adult volatiles. As discussed in Section 12.3, differences in electrophysiological responses of antennal receptors and antennal lobe interneurones are not especially marked between the phases in *S. gregaria* juveniles, making it difficult to predict or interpret behavioural responses. Heifetz *et al.* (1996) used a Y-tube olfactometer to test attraction to odour sources in solitary-reared fourth-instar nymphs of *S. gregaria* and found attraction to faecal extracts of gregarious hoppers but not to cuticular extracts or airborne volatiles. The results of Heifetz *et al.* (1996) do not allow predictions to be made about behavioural responses in the vertical airflow olfactometer of Obeng-Ofori *et al.* (1993), since retention within a vertical airstream need not involve attraction, but could instead be achieved by an arrestment effect of volatiles (*sensu* Kennedy, 1978). Despland (2001), in contrast, reported that solitarious nymphs moved away from the smell of a visually obscured group of gregarious nymphs. These experiments were undertaken in the assay of Roessingh *et al.* (1993) (see Section 11.1), modified such that the group of stimulus insects (including their faeces) was confined within a white-painted, perforated box through which air was pumped into the arena. Unlike solitarious nymphs, gregarious test nymphs remained close to the smell of the stimulus group. These results could have involved directed and/or undirected behavioural responses (Kennedy, 1978), and once more are difficult to relate to behaviour in a vertical airflow olfactometer.

12.5 THE PHENYLACETONITRILE PARADOX

A very different interpretation of the behavioural role of phenylacetonitrile (PAN) in *S. gregaria* arose from a series of papers by the group of Seidelmann and Ferenz (Seidelman *et al.*, 2000, 2003, 2005; Seidelmann and Ferenz, 2002; Seidelmann, 2006). Seidelmann *et al.* (2000, 2005) employed what they termed a 'dynamic combined Y and T olfactometer' and concluded that PAN is a powerful repellent, released by mature gregarious males to help prevent competition from rival males during courtship and copulation and to prevent homosexual mating attempts. The olfactometer consisted of a Y-shaped arrangement, through which warm, humidified air was drawn. One of the arms contained the test compound and the other was odour-free. Rather than test insects being released into the outflow of the olfactometer as in a traditional Y-olfactometer, they were inserted through a T-arm connected to the middle of the stem of the Y. Insects were introduced individually and the assay terminated after 3 min. Locusts found at the exit (outflow end of the Y) were considered to have avoided the odour; those that moved upwind and entered the Y-arm containing the odour were classified as being attracted; those that moved upwind but then entered the odour-free Y-arm were interpreted as showing no attraction; and locusts that remained in the middle section of the olfactometer (mostly upwind of the entry point) were considered to be aroused by the odour, but neither attracted nor repelled. Seidelmann et al. (2000) used a single concentration of synthetic PAN and found that, relative to control trials in which PAN was not present in either Y-arm, mature gregarious males had a reduced probability of entering the arm that contained PAN, were less likely to stay in the middle of the arena, and more likely to travel either to the exit or to the PAN-free Y-arm. Immature adult males were less likely to avoid the PAN Y-arm than the mature males, but showed increased tendencies to enter the clean arm and exit.

Seidelmann *et al.* (2005) conducted a more extensive study using the Y-T olfactometer, in which they tested different stages of gregarious *S. gregaria*, various concentrations of PAN, and other components of mature male odour bouquet. They also measured concentrations of PAN at the outflow of the olfactometer. Relative to control trials, fifth-instar nymphs, immature and mature adults of both sexes were less likely to move into the PAN-emitting Y-arm or to remain in the middle section of the olfactometer, and more likely to end up at the exit or in the PAN-free Y-arm. The strongest response was seen in mature males, and the weakest in male nymphs. There was an interesting trend in a dose–response study on mature adult males, in which PAN doses of 43 ng l^{-1} to 88 μg l^{-1} were tested. The main response at concentrations of 4.7 μg l^{-1} or lower was an increased tendency to stay in the middle section of the arena, rather than move to the exit. At 38 μg l^{-1} or higher concentrations, there was strong avoidance of the odour-containing Y-arm and a substantial increase in locusts collected at the exit. In a final experiment, PAN was tested with and without associated compounds in the mature male odour blend (anisole, benzaldehyde, guaiacol, phenol and veratrole in a 5:7:1:4:10 blend; alone or mixed in a 1 to 10 ratio with PAN). The addition of extra compounds did not affect the response to PAN. When the blend of other compounds was tested alone, locusts (male nymphs and mature adults of both sexes) were more likely to remain in the middle section of the olfactometer and less likely to move towards the exit, but unchanged in their probability of entering either of the Y-arms.

Seidelmann and Ferenz (2002) confirmed their conclusion that PAN serves as a courtship inhibition pheromone in experiments in which they manipulated PAN emission and observed courtship and mate guarding behaviour. They first forcibly separated a gregarious mating pair of *S. gregaria* and placed the female with a fresh gregarious male. Once the new male had coupled with the female, the original male was reintroduced to the arena. In none of 11 trials did the original male attack the pair. In a second experiment, after separation from its gregarious mate, the female was placed with a male that had been isolated for 7 days (and hence ceased producing PAN: see Section 12.1.3). After they had paired, the original male was introduced – and in all 12 trials leapt onto the back of the mating pair and attempted to couple with the female. This same experiment was repeated, but 1 μl of PAN solution was applied to the pronotum of the previously isolated male. In no case of the 12 trials did the original gregarious male attempt to mount the couple: PAN treatment of the solitary-reared male had prevented usurping by the competitor. In contrast, couples were mounted by the rival male in all 12 control trials in which only solvent was applied to the pronotum of the previously isolated male. Next, after separation of the initial mating pair, the female was treated with either PAN or solvent and the original male placed back with her. Control-treated females were remounted within 30 min in 80% of cases, but none of the PAN-treated females were remounted. Finally, males were offered two females: PAN-treated and control; PAN-treated females were strongly avoided in preference for controls.

Seidelmann (2006) explored the relationship between inhibition of mating by PAN and the time that mature gregarious males had been deprived of females. Test males that had been kept in groups of 10 without females for 0, 1, 3 or 3 days were placed with mating couples. Takeover attempts by the test males increased progressively in frequency with time of deprivation, beginning after 1 day. These results indicate that males were decreasingly influenced by PAN released by the mating male as time without females increased.

How might the stark difference in interpretation between the behavioural effects of PAN as an aggregation pheromone on the one hand, and a courtship-inhibiting pheromone on the other, be reconciled? Hassanali *et al.* (2005a) proposed one possibility that PAN serves both roles but at different concentrations. Seidelmann *et al.* (2005) estimated that 48 ng l^{-1} was the maximally active concentration of PAN for male aggregation in the vertical airflow olfactometer (based on data in Torto *et al.*, 1994). This was at the low end of concentrations used by Seidelmann *et al.* (2005). In their Y-T olfactometer such a concentration resulted in locusts tending to remain in the middle zone of the arena and showing substantially less avoidance of the odour source than at higher PAN concentrations (especially 38 μl l^{-1} or more). Remaining in the middle zone of the arena could plausibly translate into remaining within a PAN-suffused column of air in the vertical airflow olfactometer.

Rono *et al.* (2008) set out explicitly to test the hypothesis that PAN serves to retain locusts in the local area at low concentrations and acts as a courtship inhibition pheromone at the higher concentrations that occur in the near vicinity of mature males, which are the sources of PAN emission. Two experimental regimes were employed. In the first, the vertical airflow olfactometer was used. Mature gregarious males were tested in groups of six to a range of PAN concentrations, extending well beyond those used in earlier studies. Concentrations ranged from 0 to 128 μl, which equated to 0–1600 LH (locust emission hours), whereas Torto *et al.* (1994) tested up to 250 LH. Negative values for RI (equivalent to AI) were not set to zero. Values for RI increased from zero when no PAN was present to around 60 at concentrations of 8–16 μl (100–200 LH), then fell progressively as concentration increased further. At 80 μl (1000 LH), RI reached zero, and thereafter became negative at higher concentrations. Hence, PAN caused mature males to remain in the column of air suffused with PAN at low concentrations and was avoided at high concentrations.

It is difficult to equate concentrations used by Rono *et al.* (2008) with those in the Y-T olfactometer, but we will attempt to do so here. Seidelmann *et al.* (2005) estimated from their own measurements and the conditions used by Torto *et al.* (1994) that 48 ng l^{-1} was the maximally active concentration of PAN in the vertical airflow olfactometer. With reference to Fig. 4 in Torto *et al.* (1994), in which maximal responses by mature males to PAN are at around 100–150 LH, and the statement in Rono *et al.* (2008) that 1 μl of PAN applied in their olfactometer represents 12.5 LH, the concentrations tested by Rono *et al.* (0–128 μl) would equate to a range from 0 to 1 μg l^{-1} in the Y-T olfactometer. Seidelmann *et al.* (2005) tested a much greater range of concentrations, from 43 ng l^{-1} to 88 μg l^{-1}.

Rono *et al.* (2008) conducted a second experiment in which they measured the preferred position of individual mature males in a PAN gradient. Locusts were introduced into the middle of an 84-cm long, 7-cm wide arena, with a central PAN source. Because the ends of the arena were covered with mesh, gradients of PAN would have been established away from the source in both directions. Different PAN doses were used at the central source and the position and orientation of test locusts were assessed after 15 min. As the concentration of PAN was increased, locusts moved further away from the source towards either end of the arena. At low PAN concentrations, locusts were located closer to the source than if no PAN was present.

The paradox seems, therefore, to have been resolved: PAN has two very different behavioural effects, depending on its concentration. At low concentrations, it retains adults, but not nymphs, in the vicinity; but at high concentrations it provokes avoidance. Since mature males are the source of PAN, the latter effect is to inhibit homosexual encounters and to inhibit competition from rival males during mating.

There are additional problems concerning PAN. Studying *S. gregaria*, Mahamat *et al.* (1993, 2000) concluded that PAN is the dominant factor in the ‘maturation accelerating pheromone’ emitted by crowded mature males. These authors also concluded that this pheromone enhances sexual maturation of adults (see also Mahamat *et al.*, 1997b) and it induces the bright yellow colouration in sexually fully mature crowded adult males. In contrast, Schmidt and Albütz (2002) concluded that PAN (under the term benzyl cyanide) has no yellow colour–inducing effect, and neither volatiles from crowded adult males, nor PAN alone, accelerate maturation of immature males in *S. gregaria*. These issues are thoroughly discussed in Sections 7.3 and 8.1. It may be added that according to Niassy *et al.* (1999), crowded adults of *L. m. migratorioides* do not emit PAN, nevertheless, in this species, crowded mature males accelerate maturation of immature adults and crowded males become bright yellow with sexual maturation (Sections 8.1 and 7.3, respectively).

13 Cuticular substances and contact pheromones

Cuticular substances have been implicated as cues in stimulating development of yellow colouration and accelerating maturation in adult *S. gregaria* (Schmidt and Albütz, 2002; see Sections 7.3 and 8.1, respectively); development of yellow background colour in crowd-reared nymphs of *S. gregaria* (Lester *et al.*, 2005; see Section 7.2.3); acceleration of maturation and stimulation of production/release of phenylacetonitrile (PAN) in male *S. gregaria* (Amerasinghe, 1978a; Seidelmann *et al.*, 2000; see Sections 8.1 and 12.1.3); and stimulation of behavioural gregarization in nymphs of *S. gregaria* (Heifetz *et al.*, 1996, 1997, 1998; see Section 14).

The most abundant components of cuticular lipids are long-chain linear and branched hydrocarbons, which are known to play roles in chemical communication in a wide range of other insect species (Howard and Blomquist, 2005). The most detailed analysis of differences between the phases of locusts remains that of Genin *et al.* (1986), who studied nymphs and young and mature adults of *L. migratoria cinerascens*. Hydrocarbons comprised 52%–78% of cuticular lipids, with the smallest percentage being found in 1-generation solitary-reared individuals. The profiles of hydrocarbons were reported to differ with phase, with a general trend of more condensed hydrocarbons in solitary-reared than gregarious insects of all ages and both sexes. Génin *et al.* (1987) also reported differences with phase in the profile of cuticular aliphatic ethers in *L. migratoria cinerascens,* with heavier ethers being associated more with solitary-reared insects, and differences in the frequencies of different structural isomers.

The precise roles in locust biology of cuticular hydrocarbons and other compounds in cuticular lipids remain largely obscure: as can be seen by reference to sections elsewhere in the present review.

14 Factors that induce gregarious phase characteristics

Progress has been made in recent years towards understanding the stimuli associated with crowding that evoke gregarious phase characteristics in *S. gregaria*. The main focus has been on induction of gregarious behaviour, and this will provide the main substance of the current section. Factors stimulating other gregarious phase characteristics are less well understood and have been considered in earlier sections, notably black patterning in Section 7.2.1 and yellow colouration in Section 7.2.3, both in hoppers.

Roessingh *et al.* (1998) investigated the single and interactive effects of visual, olfactory and contact stimuli associated with crowding on the behaviour of final-instar solitarious (at least 2-generation solitary-reared) nymphs of *S. gregaria*. Test insects were confined for 4h within a clear plastic container, which sat within a larger box. Olfactory stimulation of the test insect was provided by piping air that had passed across a crowd of gregarious nymphs into the inner chamber. A group of gregarious nymphs, contained between the outer and inner boxes, provided visual but no olfactory stimulation. The effects of physical buffeting within a crowd were simulated by placing papier maché spheres inside the inner box and rocking the entire contraption gently with an electric motor and eccentric. After 4 h the behaviour of the test insect was assessed, using the behavioural assay of Roessingh *et al.* (1993; see Section 11.1).

Visual or olfactory stimuli alone had no effect on the behaviour of solitarious test insects, but together they caused significant behavioural gregarization (we return to this finding later). The most striking result, however, was that persistent buffeting with papier maché spheres caused total behavioural gregarization, whether presented alone or in combination with visual and/or olfactory stimulation (Roessingh *et al.*, 1998).

Buffeting with paper balls provided two possible gregarizing stimuli: mechanical and contact chemical. The latter might be provided by the balls picking up chemical cues from the test insect's cuticle, such that the insect is stimulated by its own surface chemicals. That contact chemical cues might play a role was shown by Heifetz *et al.* (1996), who exposed fourth-instar nymphs that had been solitary-reared since the second stadium for 5 h to dichloromethane extracts of nymphal cuticle, airborne nymphal volatiles or faecal extracts. These authors found that only cuticular extracts caused a shift towards gregarious behaviour, as measured using discriminant analysis based on measures of grouping and activity (see Section 11.1). The shift was partial, with the overall discriminant function remaining marginally solitarious. Heifetz *et al.* (1997) next fractionated cuticular extracts into wax esters, triglycerides, fatty acids and hydrocarbons. The only fraction to induce a behavioural shift (albeit, overall the locusts again remained within the region classified as solitarious), was the hydrocarbons. Antennectomy inhibited any shift towards gregarious behaviour (Heifetz *et al.*, 1996), and antennal levels of the second messenger inositol triphosphate (IP_3) but not cAMP, were rapidly and transiently elevated by exposure to cuticular hydrocarbons (Heifetz *et al.*, 1997). Thus, the conclusion was drawn that cuticular hydrocarbons are detected by antennal sensory receptors and elicit behavioural gregarization.

In a further publication, Heifetz *et al.* (1998) subjected fourth-instar nymphs that had been solitary-reared since the second stadium to 5 h access to mesh screens coated with one of a series of concentrations of cuticular surface extracts of gregarious or solitary-reared conspecific nymphs or gregarious nymphs of *L. migratoria*. Test insects were then grouped and their activity and degree of association were measured (see Section 11.1). There was no significant effect of treatment on the extent to which nymphs aggregated (actual data are not presented). However, activity levels increased in a dose-dependent manner upon exposure to surface waxes from gregarious conspecifics, but not to surfaces extracts of solitary-reared nymphs or *L. migratoria*. Heifetz *et al.* (1998) compared the chemical profiles of cuticular hydrocarbons in fourth-instar nymphs of *L. migratoria* (crowd-reared), and in *S. gregaria* that had been solitary-reared for two stadia or long-term crowd-reared. Statistical testing of differences was not done and the extent of replication is unclear. The claim is made that the quantity of cuticular hydrocarbons was greater in solitary-reared than crowd-reared *S. gregaria*, although differences in body size were not accounted for. The types of compounds present did not differ between species or phases, but it appears that the ratio of compounds may have been different. The point is appreciated by the authors that there is a logical difficulty in their claim that cuticular hydrocarbons stimulate behavioural gregarization (or, more accurately, increased activity levels). If cuticular extracts of solitary-reared locusts do not stimulate increased activity, yet placing solitarious locusts together induces rapid behavioural gregarization, for cuticular hydrocarbons to be responsible it must be the case that the profile of cuticular hydrocarbons changes to the behaviourally efficacious gregarious formulation very rapidly indeed upon exposure to a crowd. To this end, the authors compared the profiles of cuticular hydrocarbons in crowd-reared nymphs that were isolated for 3 days then recrowded for 1 day, with locusts that were crowded throughout and then isolated for 1 day. Again, there are no statistical results presented, but it seems that both of these treatment groups had hydrocarbon profiles more like 2-stadium solitary-reared nymphs than their fully crowd-reared counterparts [compare Figs 3 and 6 in Heifetz *et al.* (1998)]. Hence, whereas solitary rearing induced a change over 24 h towards a solitary-reared cuticular hydrocarbon profile, the reverse transition did not occur. It may be the case that there are particular cuticular compounds in solitary-reared insects that do change in their concentrations or ratios exceedingly rapidly upon exposure to crowding, but there is no evidence that this is the case.

Hägele and Simpson (2000) disentangled the relative effects of mechanical and contact chemical stimulation by subjecting 3-generation solitary-reared fifth-instar *S. gregaria* nymphs to 4-h stimulation with cuticular extracts from gregarious nymphs. Extracts were prepared and presented according to the protocols developed by Heifetz *et al.* (1996), with or without mechanical stimulation in the form of a constant shower of millet seeds. Behavioural gregarization was complete following mechanical stimulation, but not apparent at all following exposure to cuticular hydrocarbons. It remains possible that contact chemical cues play a role in behavioural re-gregarization of recently isolated gregarious nymphs, as used by Heifetz *et al.* (1996, 1997), but there is no doubt that their role in eliciting behavioural gregarization is substantially weaker than that of mechanical stimulation. However, cuticular chemical cues appear to be important for production of other gregarious characteristics such as yellow colouration in nymphs of *S. gregaria* (Lester *et al.*, 2005; discussed further in Section 7.2.3). Under crowded conditions, of course, mechanical, chemical and visual cues will be present together.

Simpson *et al.* (2001) localized the site of mechanosensory input responsible for gregarization, by stroking one of 11 body regions (antenna, face, mouthparts, prothorax, wingpad, abdomen, foreleg, midleg, hind femur, hind tibia or hind tarsus) with a paintbrush for 5s each minute over a 4-h period. Full behavioural gregarization was induced by stimulating the outer face of a single hind femur, whereas no other body region induced significant gregarization.

Rogers *et al.* (2003) further localized the mechanosensory receptors involved. Significant behavioural gregarization was induced when just one quarter of the femur was stimulated; unless that was the lower, distal quartile, which contains fewer mechanosensilla than other regions (see Section 5.3.3). Stimulating half of the femur produced a stronger behavioural effect than stroking only a quarter. However, if locusts were immobilized ventral surface uppermost with legs fixed in position during stimulation, gregarization failed to occur in response to leg stroking. In unrestrained locusts, leg stroking causes medial displacement of the femur towards the body. Thus, both mechanosensilla on the outer femoral surface (exteroreceptors) and proprioceptors associated with the basal leg joints are involved in behavioural gregarization: stimulation of either alone is insufficient.

Rogers *et al.* (2003) used a combination of nerve sectioning and patterned electrical stimulation of metathoracic leg nerve 5 and its main branches to discover where the afferents from relevant exteroreceptors and proprioceptors travel. Electrical stimulation of the entire nerve 5 in restrained locusts elicited complete behavioural gregarization, as did stimulation of nerve branch 5B, but not 5A. Within nerve 5B, stimulation of branches 5B1 and 5B2 together was necessary to induce behavioural gregarization, with neither alone being effective. Afferent from mechanosensilla on the outer femur and proprioceptors at the leg base are known to travel in both 5B1 and 5B2 [see Fig. 9 in Rogers *et al.* (2003)].

A sub-population of exteroreceptors on the hind femur, the basiconic sensilla, contains both mechanosensory and chemosensory neurones. These two modalities of sensory neurone converge onto a common population of thoracic spiking interneurones, which also receive mechanosensory inputs from the other purely tactile hairs on the femur. If these local interneuronal circuits are involved in gregarization, then stimulation of the chemosensory afferents alone, in conjunction with proprioceptive input from the leg base, should cause gregarization. Rogers *et al.* (2003) tested this hypothesis by selectively stimulating the chemosensory neurones by puffing acetic acid vapour onto the femur. Even though this caused medial movement of the femur and elicited strong reflex avoidance responses, locusts remained fully solitarious in behaviour. Therefore, either there is a separate circuitry for mechanosensory inputs involved in behavioural gregarization, or convergence with other mechanosensory signals occurs further along the integrative pathway in the CNS.

Localising the source of gregarizing input to particular femoral receptors and nerves has opened the way to a detailed analysis of the neurochemical bases of phase change, further aspects of which are covered in Sections 10.3 and 11.6. It is fortuitous indeed that the hindleg of the locust, its motor control and reflex responses, sensory receptors and integrative circuitry, is among the best known models in neurobiology (Burrows, 1996).

Although mechanosensory cues are powerful inducers of gregarious behaviour, Roessingh *et al.* (1998) showed that a combination of olfactory and visual cues can also cause a degree of behavioural gregarization, but either alone was largely ineffective. Hence, visual stimulation alone caused no change in behaviour over 4 h in nymphs, but had some effect after 24 h. When solitarious nymphs were subjected to either natural odour from gregarious nymphs or a synthetic adult male odour blend (Torto *et al.*, 1994), there was no evidence of behavioural gregarization over 4 or 24 h. Nor did male or female pre-reproductive or reproductive adults behaviourally gregarize in response to a 4-h period of exposure to adult male synthetic odour. Similarly, Heifetz *et al.* (1996) reported no change towards gregarious behaviour in fourth-instar nymphs of *S. gregaria* that had been reared in isolation since the second stadium and subjected for 5 h to odour collected from gregarious nymphs, faecal extracts, or synthetic phenol derivatives (guaiacol, veretrole or the two combined).

Anstey *et al.* (2009) found that a combination of visual and olfactory stimulation led to a transient rise in levels of serotonin (5HT) in the thoracic ganglia of solitarious nymphal *S. gregaria*, as did mechanosensory or electrical stimulation of hindleg mechanoreceptor neurones (also see Section 10.3). These results indicate convergence of the two gregarizing pathways (mechanosensory and visual/olfactory) onto a single neuromodulatory effector system for behavioural phase change, which is located within the thoracic ganglia. This conclusion was confirmed by pharmacological manipulations of serotonin levels in the thoracic nervous system. Injecting a mixture of two 5HT antagonists (ketaserin and methiothepin) into the meso- and metathoracic ganglia of solitarious locusts blocked the initiation of behavioural gregarization in response to a 1-h period of mechanosensory or olfactory and visual stimulation, whereas saline-injected controls gregarized as expected. When 5HT synthesis was inhibited by injection of α-methyltryptophan (AMTP), a competitive antagonist of tryptophan hydroxylase, behavioural gregarization was blocked in response to 2-h leg stroking. In contrast, behavioural gregarization was enhanced by topical application onto the thoracic ganglia of serotonin, or injection into the ganglia of a mixture of the 5HT agonists, α-methylserotonin and 5-carboxamidotryptamine. Finally, when endogenous serotonin synthesis was enhanced by injection into the haemocoel of the 5HT precursor, 5-hydroxytryptophan (5HTP), the gregarizing effect of a brief period of crowding was amplified relative to saline-injected controls. It should be noted (see Section 10.3) that the spike in serotonin in the thoracic ganglia that causes behavioural gregarization is transitory and that Rogers *et al.* (2004) showed that levels of serotonin are actually elevated by approximately 20%–35% in long-term solitary reared nymphs as compared to gregarious ones. Serotonin was also substantially elevated during the initial stages of the reverse transition from long-term gregariousness to solitariousness; 24 h newly isolated gregarious hoppers showed eightfold higher amounts of serotonin, but only in the optic lobes of the brain and not in the thoracic ganglia. Therefore, the locality where serotonin levels change seems to be very important.

Lester *et al.* (2005) explored longer-term effects of sensory cues in solitarious nymphs that were reared from the second until the fifth stadium in the presence of stimuli from conspecific gregarious nymphs or those of *L. migratoria*. Odour alone did not cause behavioural gregarization; however, it did elicit a degree of black patterning, even if the odour came from *L. migratoria* (see also Section 7.2.1). Visual and olfactory stimulation together caused behavioural gregarization, extensive blackening and a degree of yellowing of the cuticle; once again, even if the smell was heterospecific (see Section 7.2.3). Actual physical contact with other locusts during rearing did not produce more extensive blackening or behavioural gregarization than did the combination of odour and sight without contact, but rearing in contact with conspecifics was required for full yellowing (Section 7.2.3).

In conclusion, different gregarious phase characteristics are triggered in *S. gregaria* nymphs by different combinations of sensory cues emanating from other locusts. Rapid and complete behavioural gregarization is evoked by touching the hind femur. The neural pathways responsible are becoming known and are mediated, at least initially, by levels of serotonin in the thoracic ganglia. In combination, visual and odour cues are also gregarizing, producing evidence of behavioural phase change within hours, and extensive development of black patterning and partial yellowing over successive stadia. Odour alone, even prolonged over stadia, does not induce gregarious behaviour, or yellowing, but does cause partial blackening. It is not known if prolonged mechanostimulation of the hind leg, without other cues, affects black patterning, yellowing or other phase characters.

As a final word, it is important to note that all experiments discussed in this section have concerned the *initiation* of gregarization, as distinct from its maintenance (see Roessingh *et al.*, 1993; Hassanali *et al.*, 2005a). Whether odour or other cues serve to maintain gregarious characteristics once acquired, thus inhibiting a return to the solitarious condition when a gregarized locust becomes isolated (see Section 11.4), is not yet known. Results from a Ph.D. thesis from 1999 by Malual as cited by Hassanali *et al.* (2005a, p. 230), suggest that olfactory cues slow solitarization in nymphs and adults of *S. gregaria*.

Another possible example is the loss of gregarious characteristics in antennectomized crowd-reared nymphs (Mordue, 1977; Gillett, 1983; Heifetz *et al.*, 1996), which is discussed further in the next section.

15 Factors that induce solitarious phase characteristics

There is no incontrovertible evidence, in *S. gregaria* at least, that solitarious locusts release any biologically active, phase-typical volatiles (see Section 12; Hassanali *et al.*, 2005a; Rono *et al.*, 2008). It is now generally accepted that solitarization results from the absence of crowding. In other words, gregariousness is the derived phenotype, with solitariousness being the default condition. Accordingly, induction of solitariousness will occur in response to (a) the absence or diminution of crowding cues or (b) as a result of a change in a gregarious insect's ability to respond to such cues. One caveat to this conclusion comes from the recent report that solitarious adult females release higher levels of a compound in their egg foam than do gregarized females, yet the two phases have similar levels of a second compound (an alkylated L-dopa analogue) that was shown to have a strong gregarizing effect on the behaviour of developing hatchlings (Miller *et al.*, 2008). It was hypothesized that the former compound inhibits the gregarizing action of the second material: if so, then the former compound may be the first instance of a pheromone that is released by solitarious insects to cause a solitarizing effect (Miller *et al.*, 2008; see Section 16.3).

Reduction in the rates of interaction among locusts in a group will occur if individuals are either dispersed or become less mobile for some reason. Regarding the first of these, the most obvious scenario is when a locust becomes separated from others in the group. Another example of dispersal among nymphs of *S. gregaria* might be behavioural repellence by phenylacetonitrile (PAN), which is released by mature adult males and appears to be avoided by nymphs (see Section 12.4 for detailed discussion). This avoidance may be considered as the effect of a solitarizing pheromone, reported by Gillett and Phillips (1977) and Gillett (1983) as a 'solitarizing stimulus'.

Factors causing changes in mobility include temperature, nutritional state, photoperiod and disease. For example, Roessingh and Simpson (1994) reported a degree of behavioural solitarization overnight, when locusts in crowds are quiescent and unable to see each other. Rogers *et al.* (2003) found accompanying changes in levels of neuromodulators in the CNS (Section 10.3). Despland and Simpson (2000a) showed that the likelihood that locusts would move and contact each other at or between feeding sites was a function of the nutritional quality and distribution foods in the environment, with nutritional state mediating these effects (see Section 17).

A possible example of a solitarizing effect of disease comes from the work of Elliot *et al.* (2003). These authors infected gregarious adults of *S. gregaria* with the fungal pathogen *Metarhizium anisopliae* and allowed the insects to survive to reproduction by exhibiting behavioural fever, by which infected locusts select higher body temperatures and inhibit rates of fungal growth. The first-instar offspring from these parents were found to be slightly but significantly more likely to be green than the offspring of control parents (65% of nymphs were black and 15% all green from infected parents, as compared to 83% and 1%, respectively, for controls). The behaviour of nymphs that were all green (colour score 1 on a scale of 1–5, where 5 is all black) was compared with a cohort of hatchlings that were more darkly pigmented (reported to be mainly colour score 2), using a modification of the Roessingh *et al.* (1993) behaviour assay (Section 11.1). Green-coloured hatchlings behaved more solitariously than did those darker in colour, implying that not only were the offspring of infected parents more likely to be green, they were also more solitarious in behaviour.

There are two possible explanations of the solitarizing effect of parental infection. First, the infection itself reduced mobility of the adults, which limited contacts between locusts and hence led to less maternal gregarization of embryos (see Section 16). Alternatively, infection induced behavioural fever, and it was the elevated body temperature that had the effect, either directly or via reduced activity levels. Accordingly, Elliot *et al.* (2003) kept uninfected parents under one of two thermal regimes. Both treatment groups were kept at 20°C at night (15 h), whereas one group was at 38–39°C during the day (9 h) and the other had an additional period of 44°C for 5 h during the middle of the day to represent a typical behavioural fever temperature. Adults kept for 20 days under the fever regime produced 14% green hatchlings, as had infected parents. Hence, the conclusion was that solitarization was a side-effect of behavioural fever, not a direct effect of fungal infection. Since adult behaviour was not measured, it is not possible to say whether the solitarizing effect of fever related to the frequency or intensity of interaction among adults during the pretreatment period, but this seems a likely explanation. The solitarizing effect of fungal pathogens, behavioural fever and heat treatment resembling behavioural fever are also discussed in Section 8.5.

Another example of disruption of aggregation behaviour due to disease comes from the work of Shi and Njagi (2004) on *Locusta migratoria manilensis* infected with *Nosema locustae*. Third-instar nymphs were infected with fungal spores and behaviourally tested as fourth- and fifth-instar nymphs (and, it is stated, as young adults, although no data are presented) to volatiles trapped from nymphs and young adults and their faeces, using the vertical olfactometer of Obeng-Ofori *et al.* (1993) (see Section 12.4). Infected female (but not infected male) fourth-instar nymphs, and infected male and female fifth-instar nymphs, showed lower 'aggregation' responses to volatiles than did uninfected controls (i.e. individually tested infected insects were less likely to remain in the region of the arena suffused with volatiles than were uninfected nymphs). There were also differences with infection in the electrophysiological responsiveness of the antennae to stimulation with volatiles. These results are difficult to interpret, but show that fungal infection could encourage solitarization by interfering with the ability of locusts to respond to volatile cues and interact with one another.

Heifetz *et al.* (1996) removed the antennae from fourth-instar crowd-reared nymphs and found that they were behaviourally solitarized 24 h later when kept in groups of four, whereas control insects with hind tarsi amputated remained gregarious. The conclusion was drawn that antennal receptors responding to cuticular hydrocarbons are necessary for sustaining behavioural gregariousness, even if visual and contact cues are present (see Section 14). However, removal of antennae will have rendered insects less active due to major loss of sensory input; and less activity would result in fewer interactions between locusts and hence partial loss of gregarious behaviour. A similar explanation could account for earlier reports of colour (Mordue, 1977) and behavioural (Gillett, 1983) changes following antennectomy. Alternatively, chemical cues received by the antennae may be necessary for maintaining gregariousness once acquired, even if those cues alone are not responsible for initiation of gregarious behaviour (see end of Section 14).

16 Transmission of phase from parents to progeny

Numerous components of phase state have been found to accumulate epigenetically across generations in both *S. gregaria* and *L. migratoria* (Faure, 1932; Gunn and Hunter-Jones, 1952; Albrecht *et al.*, 1958; Hunter-Jones, 1958; Uvarov, 1966; Injeyan and Tobe, 1981a; Islam *et al.*, 1994a,b; Islam, 1997; McCaffery *et al.*, 1998; Rahman *et al.*, 2002b). Traits including hatchling behaviour, colouration, mass, ovariole number, morphometry and development time depend on parental rearing density; with crowded parents producing hatchlings with gregarious features and isolated adults producing hatchlings with solitarious characteristics.

Simpson and Miller (2007) reviewed in detail the 45 published experiments and associated controls that have investigated parental transmission of hatchling behaviour and colouration in *S. gregaria* (originally published in Islam *et al.*, 1994a,b; Bouaïchi *et al.*, 1995; Islam, 1997; McCaffery *et al.*, 1998; Hägele *et al.*, 2000; Malual *et al.*, 2001; Rahman *et al.*, 2002b; Bouaïchi and Simpson, 2003; Tanaka and Maeno, 2006). Key results and conclusions will be emphasized here.

16.1 TRANSGENERATIONAL ACCUMULATION OF GREGARIOUS CHARACTERISTICS

Bouaïchi *et al.* (1995) found that the extent to which solitary-reared mothers produced hatchlings that behaved gregariously depended on how recently they were crowded relative to the time of oviposition. In effect, the mother has a memory of how recently she experienced crowding and predisposes the phase state of her hatchlings accordingly. It would appear that the mother gains further evidence of population density through the phase of her mate, as indicated by the fact that mating with a crowd-reared male resulted in a majority of behaviourally gregarized hatchlings (Islam *et al.*, 1994a). Despland and Simpson (2000b) found that these parental effects were not confined to laboratory studies. They kept small populations of solitary-reared adults in field enclosures in Mauritania (10 locusts per 2×2 m arena). The enclosures contained the same density of food plants (13% ground cover of *Hyoscyamus muticus*), but these were distributed at one of four levels of clumpiness, from completely clumped to highly dispersed. Adults were allowed to feed, mate and lay eggs, with boxes of moistened sand being provided as oviposition sites. The behavioural state of hatchlings was assessed using a simplified version of the assay designed by Roessingh *et al.* (1993) (see Section 11.1). The degree of behavioural gregarization in hatchlings was positively related to the extent to which parental food plants were aggregated. It was already known that clumped resource distribution at small spatial scales encourages congregation and subsequent gregarization of solitarious *S. gregaria* through high levels of contact between individuals (Collett *et al.*, 1998; Despland *et al.*, 2000; see Section 17): Despland and Simpson's (2000b) results show that such an effect translates to the next generation.

Experimental treatments that induce behavioural gregarization tend to produce darker coloured hatchlings. The relationship between hatchling colour and behaviour is weak, however, with colour explaining only 10% of the variance in behaviour (Islam *et al.*, 1994a). Having assessed the published data, Simpson and Miller (2007) concluded that behaviour precedes colouration as an indicator of gregarious phase transformation in hatchlings. In only one case of the 23 experiments surveyed – that in which solitary-reared adults laid into oviposition tubes containing more than three recently-laid gregarious egg pods (McCaffery *et al.*, 1998) – did colour change appear to precede behavioural gregarization.

16.2 TRANSGENERATIONAL ACCUMULATION OF SOLITARIOUS CHARACTERISTICS

Just as crowding of solitary-reared parents induces development of gregarious characteristics in their hatchlings, the reverse is also the case, with isolation of crowd-reared adults as late as at the time of oviposition resulting in a degree of solitarization of hatchlings (Islam *et al.*, 1994a,b; Simpson and Miller, 2007). However, rather than behaviour being more responsive than colouration, as for acquisition of gregarious features, solitarious behaviour and colouration are transmitted across generations less readily and at approximately the same rate (Simpson and Miller, 2007). This finding provides another example of hysteresis in locust phase transition, in which rates of gregarization differ from rates of solitarization (see Sections 9.3.8, 10.2, 11.4).

In one of the 11 relevant experiments surveyed by Simpson and Miller (Hägele *et al.*, 2000), there was a clear disconnection between colour and behaviour, indicating once again that, although these traits are correlated, they do not necessarily share the same underlying mechanisms. In that experiment, the accessory glands were ligatured in gregarious females, with the result that hatchlings emerged behaviourally solitarized but black rather than green in colour (see later in this review for further discussion).

16.3 THE NATURE OF THE MATERNAL GREGARIZING AGENT

A detailed series of experiments by McCaffery *et al.* (1998) showed that gregarizing activity is found in the foam deposited with the eggs during oviposition. The active material was found to be present in aqueous extracts of egg foam and to degrade rapidly, losing its effect if stored for more than 24 h. Eggs from solitary-reared females only remained sensitive to application of the material during the first day after laying, after which they became insensitive to topical application of the material, either as a result of the chorion becoming impermeable or the developing embryo becoming refractory (McCaffery *et al.*, 1998).

When eggs from pods laid by crowd-reared females were separated and incubated singly, either following washing in saline or not, a successively greater proportion of hatchlings exhibited green colouration as the time between oviposition and egg separation decreased (McCaffery *et al.*, 1998). No solitarizing effect of egg separation was apparent if the interval exceeded 10–24 h (McCaffery *et al.*, 1998). This result accords with treatments in which hatchling gregarization occurred when aqueous extracts of fresh gregarious egg foam were applied to newly solitary-laid eggs (see earlier). McCaffery *et al.* (1998) noted that early separation of eggs appeared to be less effective at solitarizing hatchlings when the eggs were not freshly ovulated, indicating that eggs could become committed to gregarious development if held within the oviducts.

Hägele *et al.* (2000) reported that behavioural gregarization could be reinstated if early-separated eggs from crowd-reared parents were topically treated with aqueous extracts of female accessory glands. Ligaturing of the accessory glands in crowd-reared females resulted in behaviourally solitarized hatchlings, suggesting that the accessory glands are a source of the gregarizing agent (Hägele *et al.*, 2000). Such hatchlings were black not green, however. This result indicates either that colour determination is under separate control from behaviour or that another source of the gregarizing material exists within the reproductive tract, which was present in sufficient amounts to prevent loss of black colouration. The walls of the oviduct may provide such a source, since they possess glandular cells that appear identical to those of the accessory glands and (Szopa, 1982; Hägele *et al.*, 2000).

It was further demonstrated by McCaffery *et al.* (1998) and Islam (1997) that the capacity of egg foam extracts from crowd-reared females to induce dark colouration in freshly laid eggs from solitary-reared females relates strongly to the polarity of the solvent: hexane, chloroform and acetone extracts had no effect, ethanol had a minor effect, and saline extracts induced the greatest degree of darkening (but not full black colouration) (see Simpson and Miller, 2007). Behavioural gregarizing activity was, furthermore, found to be present in the filtrate after aqueous foam extracts were passed through a 3-kD filter. Heat treatment at 100 °C for 10 min caused partial inactivation (McCaffery *et al.*, 1998).

Miller *et al.* (2008) confirmed that behavioural gregarizing activity was found in aqueous extracts of gregarious egg foam. They pursued the chemical identity of the gregarizing agent by applying successively more specific HPLC fractions of aqueous egg foam extracts to eggs laid by solitary-reared females and assaying the behaviour of hatchlings. Gregarizing activity was tracked to a specific fraction, which was subjected to nuclear magnetic resonance (NMR) spectroscopy. The bioactive compound was found to have characteristics consistent with an alkylated L-dopa analogue. Intriguingly, comparison of the egg foam of gregarious and solitarious females showed similar levels of this compound in the two phases; however, another closely similar peak was found to be in higher concentrations in solitarious foam. It was postulated that this second peak, which does not have gregarizing activity, is an inhibitor of the active compound and that gregarization results from reduced production of the inhibitor by adult females subjected to crowding. This possibility remains to be tested.

Tanaka and Maeno (2006) did not find that hatchlings became green upon early separation of eggs from crowd-reared female *S. gregaria* and as result concluded that hatchling colouration is committed before oviposition and questioned the presence of the maternal gregarizing agent in egg foam. Simpson and Miller (2007) offered two possible explanations for Tanaka and Maeno's results. First, high egg mortality, both when eggs were incubated singly, and even in control egg pods (60% or more; see Fig. 2 in Tanaka and Maeno, 2006; contrasted with 15% mortality in, e.g., Hägele *et al.*, 2000) perhaps indicated problems with infection or some other stressor. It is well known that dark hatchlings are larger and more robust than green ones (Hunter-Jones, 1958; Uvarov, 1966; Tanaka and Maeno, 2006), which makes it likely that eggs destined to become green hatchlings would have been disadvantaged. The second possible reason for the results concerned the way in which eggs were collected. Tanaka and Maeno (2006) provided crowd-reared females with oviposition medium for 7 h each day, and oviposition tubes were shallower than is usually preferred by *S. gregaria*. For the rest of the day, locusts had no access to an oviposition site. Restricted availability of a suitable oviposition site would have increased the incidence of females retaining eggs within the oviduct after ovulation. McCaffery *et al.* (1998) suggested that the failure to solitarize hatchlings in some of their experiments in which eggs were separated soon after laying ''may lie in the degree of exposure of the eggs in the oviducts to the gregarizing factor before oviposition. Possible changes in the length of time that eggs remain in the oviducts before oviposition could affect their degree of prior commitment to the gregarious phase and may need to be controlled in future experiments'' (McCaffery *et al.*, 1998, p. 361). The failure of separated eggs that had been held back in the reproductive tract to produce green hatchlings is consistent with observations that the gregarizing agent is released into the reproductive tract from the accessory glands and possibly also the epithelial lining of the oviducts (Hägele *et al.*, 2000) and that responsiveness of eggs to topically applied egg foam extracts lasts for only a brief period, suggesting that they are then committed to a particular developmental trajectory (see earlier; McCaffery *et al.*, 1998).

Tanaka and Maeno (2008) responded to Simpson and Miller's (2007) suggestions in a subsequent study, in which they convincingly demonstrated that egg mortality or delayed oviposition were not responsible for the failure to find evidence of green colouration in early-separated eggs in their culture of *S. gregaria*. Nor did they find evidence that solitarious eggs were more likely to produce melanized first-instar nymphs if oviposited into sand that had previously been used as an oviposition site by crowd-reared females. Tanaka and Maeno (2008) also did not find that eggs from the lower part of egg pods were more likely to produce green hatchlings than those from the upper regions, as reported by Papillon (1960) and interpreted by Simpson *et al.* (1999) as perhaps indicating a gradient of the maternal gregarizing agent with distance from the upper foam pod. The effects of parental density on hatchling colouration are discussed also in Section 7.2.1, and the putative effect of bursicon on darkening of *S. gregaria* hatchlings is outlined in Section 9.3.5.

Tanaka and Maeno's (2008) results leave another explanation for the discrepancy with other published results, which was raised by both Tanaka and Maeno (2006) and Simpson and Miller (2007): genetic and/or culturing differences between laboratory strains. Long-term culturing, coupled with high mortality rates in the culture may well have either resulted in loss of the trait, or shifted the timing of its action. The work of Chapuis *et al.* (2008a), discussed in Section 11.2, indicates that strains of *L. migratoria* differ genetically in the extent to which they demonstrate maternal transmission of phase. Additionally, recently published microsatellite markers for *S. gregaria* indicate the potential for loss of genetic diversity in laboratory cultures. Hence, Yassin *et al.* (2006) reported a high degree of heterozygote deficiency relative to a population in Hardy–Weinberg equilibrium for most microsatellite loci derived from African field populations. The same was true for microsatellites for both *L. migratoria* and *Chortoicetes terminifera* (see Chapuis *et al.*, 2008b). However, Kaatz *et al.* (2007) found no significant heterozygote deficiency in markers derived from a laboratory culture of *S. gregaria* originally from Niger, which would seem to indicate substantially reduced genetic variation in laboratory culture. High levels of heterozygote deficiency are to be expected in field populations of locusts, partly because of the large effective population size, and also because of a high incidence of null alleles in Orthoptera (Chapuis *et al.*, 2008b). Another example of a possible strain differences is the different rates of behavioural gregarization measured in solitary-reared insects from the Leuven and Oxford cultures, measured by the same workers using the Roessingh *et al.* (1993) assay system (Roessingh and Simpson, 1994; Hoste *et al.*, 2002b; see Section 11.4). For additional strain-dependent differences in phase characteristics, see Section 4.

Maeno and Tanaka (2008c) recently provided data confirming the earlier demonstration that hatchling colour in *S. gregaria* responds rapidly to a change in the population density experienced by the mother prior to egg laying (Islam *et al.*, 1994a,b). Crowding of solitary-reared females resulted in the production of an increased proportion of dark-coloured hatchlings from the next egg pod, whereas isolation of crowd-reared parents induced more green-coloured hatchlings in the next pod. Hence, the laboratory strain used in Tanaka and Maeno's earlier experiments clearly does demonstrate maternal inheritance of hatchling colour form. Maeno and Tanaka (2008c) furthermore found that crowding females with males of a different species, *L. migratoria*, was as effective as conspecific crowding in inducing an increase in egg length. Although colour of hatchlings was not reported in that experiment, increased egg length is typical of a gregarizing influence. Hence, mechanical or non-specific chemical cues are involved, both of which have previously been found to induce gregarious behavioural characteristics and black patterning in older nymphs (Simpson *et al*, 2001; Lester *et al.*, 2005; see Section 14).

Given that such stimuli experienced by the mother mediate a rapid gregarizing effect on the terminal oocytes, involvement of a chemical agent in determining hatchling colour seems inescapable. The results of Tanaka and Maeno (2006) and Maeno and Tanaka (2007) indicate that, in their experimental locusts, the site of action of maternal gregarizing agent on terminal oocytes is within the ovary, rather than in the egg pod immediately after laying. In other words, their results are consistent with a difference in the timing of action of the maternal factor across studies, as opposed to constituting evidence for the lack of such an agent in the first place, as implied in their papers.

As discussed earlier, a plausible reason for the discrepancy in timing of action of the maternal gregarizing agent between Tanaka and Maeno's and earlier experiments (McCaffery *et al.*, 1998) is a genetic difference in the experimental locusts. However, another intriguing possibility emerges from the data in Maeno and Tanaka (2008c). Some of their data (Fig. 4) indicate that the first egg pod laid by a female is more likely to contain a small proportion of hatchlings of the 'wrong' colour (green in the case of crowd-reared females, and black in the case of solitary-reared mothers) than are subsequent egg pods (see also Section 7.2.1). The authors specify that in their previous experiments (Tanaka and Maeno, 2006; Maeno and Tanaka, 2007) they used 'only egg pods deposited later in the adult stage…' (Maeno and Tanaka, 2008c, p. 1079). The implication of this result is not articulated by the authors, but it may offer a clue to explain their failure to obtain development of green colouration in hatchlings from separated gregarious eggs and the inconsistency in colour effects seen in some of the earlier experiments.

If the maternal agent is secreted from the accessory glands and oviduct epithelia at around the time of ovulation when egg foam is being secreted (Hägele *et al.*, 2000), then it will act on the first batch of eggs while they are within the oviducts and in the egg pod after oviposition. Accordingly, the first batch of eggs laid by crowd-reared females will be sensitive to early separation and washing. The next, previtellogenic oocytes in line, however, will receive a burst of gregarizing agent while still in the ovaries. To the extent that these oocytes are receptive and prepared to commit (both of which may be genetically labile and strain-dependent), hatchling colour will be determined to some degree by the time of laying. Removal of the gregarizing agent due to egg separation and washing may be too late to generate green hatchlings. Thus, the inability to produce green hatchlings from separated and washed eggs from pods other than the first is entirely predicted based on previous findings about the source (Hägele *et al.*, 2000) and activity of the maternal agent (McCaffery *et al.*, 1998; Hägele *et al.*, 2000; Miller *et al.*, 2008).

The extent to which varying exposure of oocytes to maternal gregarizing agent in the reproductive tract or genetic differences between the locust cultures can explain these colour observations remains to be seen. What is not at issue is that a maternal agent (or agents) exists and acts upon eggs to determine the development of phase-related behavioural characteristics. Whether the same compound(s) affect behaviour and the development of black hatchling colouration is not known (Simpson and Miller, 2007; Miller *et al.*, 2008) and elucidation of these gregarizing pathways remains an outstanding research topic.

Maeno and Tanaka (2008c) also found that the first egg pod laid by crowded females of *S. gregaria* contained significantly smaller eggs and more numerous eggs per egg pod than egg pods laid later. With ageing of the crowded females, the eggs were larger and their number per egg pod decreased (see further references on the effect of the female's age on the number of eggs per egg pod in Section 8.5). In contrast, the first egg pod for isolated females contained fewer, larger eggs than those laid later. The trends of declining egg size and increasing number in isolated females were weak (see Fig. 2 in Maeno and Tanaka, 2008c) and the authors acknowledged that the pattern may have been due to adults dying progressively throughout the period. Accordingly, another experiment was conducted to control for this possibility, in which pods were collected from identified crowded- and isolated-reared females. Although the hatchling colour effect discussed earlier remained (Fig. 4C and 4F are similar to Fig. 3A and 3B), the egg size and number effects were lost for isolated females (Fig. 4A and 3B) and the change over progressive egg pods for crowd-reared females looks much less marked (compare Fig. 4D and 4E with Fig. 2A and 2B).

Malual *et al.* (2001), using cultures of *S. gregaria* housed at the International Centre of Insect Physiology and Ecology (ICIPE), confirmed McCaffery *et al.*'s (1998) finding that egg pods from gregarious females contains a behaviourally gregarizing agent. *S. gregaria* females were provided with 24-h access to oviposition tubes filled with moistened sand. The deposited egg pods were then removed and the sand was sifted and allowed to dry overnight, before being remoistened and used to incubate eggs from solitary-reared females. Eggs incubated in this contaminated sand emerged into hatchlings which had an elevated tendency to aggregate within a ring-shaped arena (a feature of behavioural gregariousness). The time delay between egg laying by the crowd-reared females and exposure of solitary-reared eggs to the contaminated sand was between 12 and 36 h, which means that these results are consistent with earlier work in which McCaffery *et al.* (1998, Figs 1, 2) reported partial gregarization of hatchlings when solitary-reared females laid eggs into sand containing multiple pods from crowd-reared females.

Malual *et al.* (2001) concluded that the active agents are C8 unsaturated ketones, not a low-molecular weight, aqueous material as suggested by McCaffery *et al.* (1998) and proven by Miller *et al.* (2008). Malual *et al.*'s conclusion was based on three lines of indirect evidence, each of which is also consistent with existence of a water-soluble, short-lived agent. The first observation was that hatchlings from eggs laid by solitary-reared females showed no elevated tendency to aggregate if incubated in contaminated sand that had been washed with methanol, acetone and dichloromethane then dried for 24 h before being remoistened and used as an incubation substrate. Malual *et al.* (2001) argued that solvent washing would have removed C8 unsaturated ketones and that this explained the loss of gregarizing activity. However, methanol extraction will also have removed low-molecular weight polar compounds, and the solvent washing and drying process extended the 12–36 h pretreatment period by a further 24 h, after which the gregarizing agent would have lost activity (McCaffery *et al.*, 1998, Fig. 11). No control for this extra 24-h period was provided by Malual *et al.* (2001). This 24-h extended pretreatment period could also explain why contaminated sand did not gregarize hatchlings from solitary-laid eggs when flushed for 24 h with nitrogen gas; a finding that was used as the second line of evidence to support the conclusion that C8 unsaturated ketones were involved.

The third source of evidence was that gas chromatography (GC) showed the presence of C8 unsaturated ketones in contaminated sand and crushed crowd-laid eggs, which had disappeared after nitrogen flushing. These compounds were also detected in the accessory glands of gravid, crowd-reared females (Malual *et al.*, 2001). These results are consistent with earlier findings on the site of production of the maternal gregarizing agent (Hägele *et al.*, 2000), but given the limited solubility in water of the C8 ketones, it is hard to explain how ketones could account for the fact that highest levels of bioactivity were found in aqueous extracts of egg foam (Islam, 1997; McCaffery *et al.*, 1998; Simpson and Miller, 2007).

Hägele *et al.* (2004) investigated total ecdysteroid levels in eggs laid by solitary- and crowd-reared *S. gregaria* and of solitary-reared females that had been crowded for a period (2–48 h) immediately before egg laying (long enough to evoke development of gregarious behaviour in hatchlings; see earlier). As expected from previous studies (Tawfik *et al.*, 1999b), there were marked difference in total ecdysteroid content of eggs from solitary- and crowd-reared females, with gregarious eggs having substantially higher total ecdysteroid content. This difference could not be accounted for simply in terms of differences in egg mass. Ecdysteroid content in gregarious eggs rose to a peak in early embryogenesis (day 3) and a second pronounced peak occurred later in embryogenesis (day 10), whereas eggs from solitary-reared locusts had low total contents of ecdysteroids, which rose gradually throughout embryonic development (see Section 9.2 for more detailed findings of Tawfik *et al.*, 1999b). If ecdysteroids were responsible for development of gregarious behavioural traits, either directly or as mediators of the action of the aforementioned maternally produced gregarizing agent, it would be expected that a period of maternal crowding before oviposition would have resulted in a shift in ecdysteroid levels in the eggs towards values typical of gregarious females. However this did not occur; total ecdysteroid contents of eggs were indistinguishable from solitary-reared females that had not been crowded prior to oviposition.

Rahman *et al.* (2002a) (see also De Loof *et al.*, 2006) reported that a 6.08-kD peptide that was found to occur at higher levels in the haemolymph of gregarious than solitarious adults was also present in the freshly laid eggs of crowd-reared females at higher concentrations than in solitary-reared females (see Section 10.2.). Clearly the peptide is not the same as McCaffery *et al.*'s (1998) small molecular weight, aqueous gregarizing agent, or Malual *et al.*'s (2001) C8 ketones, but it could conceivably play a role in determining the phase state of hatchlings.

In summary, the balance of available data indicates that the maternal gregarizing influence comes from small molecular weight, polar compound(s) originating from the reproductive accessory glands, and perhaps also from glandular tissue lining the oviduct. The most active compound isolated to date has features in common with an alkylated N-dopa analogue. Exposure to the gregarizing agent soon after ovulation, first in the oviduct and then after oviposition, causes hatchlings to develop gregarious behaviour and colouration. The exact chemical identity of the gregarizing compound(s) is yet to be established, and interactions with other compounds such as a postulated inhibitor found in elevated concentration in solitarious egg foam, peptides and steroid hormones are unknown. It is also not known how the gregarizing agent has its influence over embryonic development. One intriguing possibility is that DNA methylation might be involved, as has recently been demonstrated in honey bee gyne development following feeding larvae royal jelly (Kucharski *et al.*, 2008). Other possible mechanisms include chemical modifications of histones, alternative splicing producing different transcripts from a single gene, and modulation of gene expression by small non-coding RNAs (e.g. Brennecke *et al.*, 2008; Wei *et al.*, 2009).

17 Ecology and ecophysiology

The population dynamics of phase change can briefly be summarized as follows: crowding triggers behavioural gregarization, which promotes aggregation and coalescence of locally scattered solitarious populations, and drives a shift towards the gregarious condition in other, slower-responding phase characteristics. The resulting gregarious phenotype is well adapted to life in a crowd, which promotes local population growth and further spread of gregarization. The potential for population growth is higher in gregarious than solitarious populations under favourable conditions, even though fecundity is lower in gregarious individuals (Cheke, 1978; Holt and Cheke, 1996; see Section 8.5). The process of population gregarization will continue until environmental influences cause a decline in population density and a return to the solitarious condition (see Roffey and Magor, 2003; Simpson and Sword, 2009). Crowding initiates this sequence of events, but how is it that solitarious populations become crowded in the first place, if solitarious locusts have a powerful tendency to avoid one another (Section 11)?

Two processes are important in bringing solitarious locusts together. The first of these is an increase in population size due to improved conditions for survival and reproduction, generally following rain. The second is concentration of locusts within the habitat. Convergent weather systems can concentrate adult populations over large areas, and smaller-scale features of the environment also play an important role. Roffey and Magor (2003) provide a detailed review of the older literature; here, we concentrate on more recent experimental and modelling results.

17.1 EFFECTS OF RESOURCE DISTRIBUTION AT FINE SPATIAL SCALES

Even though solitarious locusts avoid others except to mate, requirements for resources such as food, areas of suitable microclimate, and overnight roosting sites, might overcome mutual repulsion and potentially encourage the switch to behavioural gregarization (e.g. Kennedy, 1939; Chapman, 1955; Roffey and Popov, 1968).

Bouaïchi *et al.* (1996) conducted experiments in the laboratory and the field to test whether resource distribution at fine spatial scales is important in determining the extent to which phase change occurs in a local population of solitarious *S. gregaria* nymphs. Ten third-generation solitary-reared nymphs were placed into an arena (130 × 70 cm) with either a single or multiple resource sites. The resources used included food, warm spots and perches. The positions of nymphs were video-taped throughout the 4-h trial and their individual behavioural phase state assessed both prior to entry into the arena and at the end of the trial. When a single resources site was present, locusts were behaviourally gregarized after four hours, whereas for multiple sites they remained more solitarious. A single variable explained the extent of behavioural gregarization across all treatments: the time during the assay that each individual had spent in close contact (within a single body length) of other locusts. Hence, resource distribution determined the degree of mutual contact, which in turn translated into the extent of behavioural gregarization.

Bouaïchi *et al.* (1996) found similar results to those from their laboratory studies in a parallel experiment conducted under field conditions in Morocco. Nymphs from local low- and high-density subpopulations of *S. gregaria* were placed in groups of 10 within 4.5-m diameter circular enclosures, containing either a single or multiple food plants (*Lotus* sp) or plants used for roosting (*Emex* sp). The extent of grouping among locusts (measured by an association index; see Section 11.1) over a 2-h test period was greatest when single resource sites were available and did not differ between subpopulations. However, when multiple food or perching sites were present, low-density subpopulation nymphs associated substantially less than did high-density subpopulation nymphs, typically occupying different food or perching sites. Nymphs taken from the high-density subpopulation, however, tended to remain aggregated across the 2-h period when multiple resource sites were present.

In a related field study, Despland and Simpson (2000b) showed that the effect of the fine-scale distribution of food plants extends across generations: manipulating the distribution of plants in the parental environment of *S. gregaria* in field cages in Mauritania translated into changes in behavioural phase state of hatchlings. This study is discussed further in Section 16.1.

The three-way relationships between resource distribution, resource abundance, and locust population size were explored by Collett *et al.* (1998) in agent-based computer simulations, parameterized according to biological measures of the time-course and mechanisms of behavioural phase change. The extent of gregarization within a simulated population was shown to increases both as population size increased and with increasing clumping of food resources. A simulated locust population would suddenly gregarize across particular combinations of resource abundance, resource distribution and population size. For example, if a fixed number of resource sites were distributed in a highly dispersed manner, increasing population size in the model had a gradual effect on the level of gregarization within the population, whereas when the same number of resource sites were clumpily distributed, increasing population size had no effect until a threshold was reached, beyond which the entire population rapidly gregarized.

The predictions of the Collett *et al.* (1998) model were tested in laboratory experiments, in which 2-generation solitary-reared, second-instar *S. gregaria* nymphs were placed into arenas (70 × 70 cm) in which the number of locusts and the number and spatial distribution of pots of wheat seedlings were systematically varied (Collett *et al.*, 1998; Despland *et al.*, 2000). Resource distribution was calculated and manipulated according to a fractal algorithm. The positions of all locusts were video-taped throughout the 8-h trial, and the behavioural phase state of each locust was assessed at the end. Experimental results verified the prediction of the agent-based model. Locusts were more likely to come into contact on patchily distributed food clumps and this translated into increased behavioural gregarization for higher degrees of resource clumping.

Collett *et al.* (1998) explored the contribution of the shift within individuals from repulsion to attraction with behavioural gregarization to phase change at the population level. When the tendency to avoid or be attracted by other locusts is removed from the Collett *et al.* model, population gregarization in response to locusts coming into close contact with one another becomes a much more gradual function of resource distribution. Including social interactions both inhibits gregarization when resources are dispersed, as locusts successfully avoid one another, and enhances gregarization above a critical degree of resource clumping, as locusts are forced together and become mutually attracted. As a result, behavioural phase change serves to couple together the suite of different phase characteristics into a single threshold trait at the population level (Simpson and Sword, 2009).

17.2 EFFECTS OF NUTRITION AND HOST-PLANT QUALITY

In the model of Collett *et al.* (1998), it was assumed that all resource sites are of equal quality. However, it is well known that the chemical properties of food resources play an important role in determining patterns of movement, and hence the likelihood that locusts will interact with one another (Simpson and Raubenheimer, 2000). Despland and Simpson (2000a) used chemically defined artificial diets within the same experimental setup as employed by Despland *et al.* (2000) to show that the nutritional quality of foods interacts with their fine-scale distribution pattern to influence behavioural phase state. Twenty food pots were distributed within a 70 × 70-cm arena in either a scattered or a clumped distribution. When all foods were of uniformly high nutritional quality (containing 21% protein and 21% digestible carbohydrate), gregarization among second-instar nymphs was inhibited, irrespective of food distribution pattern. However, when all foods were of low nutritional quality (7% protein and 7% carbohydrate), significant gregarization occurred even when foods were widely dispersed, with the effect being even more marked when foods were clumped. The strongest influence of resource distribution was seen when half of the food pots contained high protein but low carbohydrate (35%:7%) and the other half were nutritionally complementary (7% protein : 35% carbohydrate). In this treatment, locusts remained behaviourally solitarious when foods were scattered and became fully gregarious when foods were clumped.

The results from Despland and Simpson (2000a) were explicable in terms of known effects of nutritional state on locomotion and food selection behaviour (reviewed by Simpson and Raubenheimer, 2000). Thus, when locusts have access to nutritionally optimal food (21% protein : 21% carbohydrate is near optimal), they do not move far from a feeding site after a meal, resulting in limited interactions between insects irrespective of whether foods are clumped or scattered. In contrast, locusts with only low-quality foods are much more mobile and therefore contact each other more often, even when foods are scattered in distribution. When locusts have access to nutritionally complementary foods, they alternate between food types to balance their nutrient intake. When foods were scattered, insects did not need to move far to locate a complementary food and could avoid others, but high levels of movement between complementary foods within a clumped resource distribution resulted in gregarization. A related issue is that inadequate nutrition – in particular protein shortage – translates into an increased likelihood of cannibalistic interactions, with implications for group size (Section 17.4) and collective movement (Section 11.5).

Simpson *et al.* (2002) compared the macronutrient-balancing strategies of singly housed final-instar nymphs of *S. gregaria* that had been solitary- or crowd-reared. Locusts were either allowed to select a balanced diet from nutritionally complementary foods or confined throughout the stadium to a single food containing one of five protein to digestible carbohydrate ratios. Crowd- and solitary-reared nymphs selected a similar ratio of protein to carbohydrate in their diet when offered a choice of complementary foods. However, the two phases differed when restricted to a single, nutritionally unbalanced food, with crowd-reared insects eating more of unbalanced foods than did solitarious nymphs. On low-protein, high-carbohydrate diets, this resulted in crowd-reared insects gaining high levels of body fat and suffering increased mortality. Both phases regulated protein growth across the stadium, but crowd-reared nymphs converted ingested protein to growth less efficiently and maintained a lower adult body protein content.

These phase-related differences in nutrient balancing and utilization were interpreted by Simpson *et al.* (2002) in relation to the different nutritional ecologies of the two phases. Across the species' recession zone, solitarious phase *S. gregaria* have a narrower host-plant range than do gregarious conspecifics [data from various publications are collated by Despland (2005), Table 1]. Additionally, solitarious locusts tend to associate with specific host plants in a given region (Louveaux *et al.*, 1998; Bashir *et al.*, 2000; Woldewahid *et al.*, 2004; Despland, 2005; Van der Werf *et al.*, 2005; Hassanali *et al.*, 2005b; Van Huis *et al.*, 2008). Being less mobile and having a more restricted host-plant range presumably means that solitarious phase *S. gregaria* experience less nutritional heterogeneity than their more mobile gregarious counterparts. Hence, when gregarious locusts feed upon a nutritionally imbalanced food, they have a higher likelihood of redressing the imbalance in the near future by locating a nutritionally complementary food. Since there are fitness costs of ingesting nutrients in excess of requirements (Simpson *et al.*, 2004; Raubenheimer *et al.*, 2005; Lee *et al.*, 2008), solitarious *S. gregaria* would therefore be expected to be less prone to consuming nutritionally unbalanced foods than would gregarious phase insects, which can discount the short-term costs of nutrient excesses against the higher likelihood of converting the excesses in balanced nutriment by encountering a complementary food in the near future. Supporting evidence for this 'nutritional heterogeneity' hypothesis has come from several studies on pairs of related species of insect differing in host-plant range, including a comparison between gregarious phase *S. gregaria* and the grass-specialist *L. migratoria* (Raubenheimer and Simpson, 2003).

The food switching behaviour of solitarious and gregarious *S. gregaria* nymphs was further explored by van der Zee *et al.* (2002), who placed individual final-instar nymphs within arenas containing either two dishes of the same nutritionally balanced food or two nutritionally complementary foods that were separated by varying distances, up to 2m apart. As predicted, solitarious locusts were less likely to move between food dishes than were previously crowd-reared locusts. This difference was especially marked when foods were close together and was even evident when both foods were nutritionally optimal, and switching was therefore unnecessary for nutrient balancing. As food distance increased, both phases tended to switch less.

In addition to nutritional quality, responses to plant secondary metabolites differ with experience of crowding in *S. gregaria*. A full discussion of this aspect may be found in Section 7.2.3, with a further mention in Section 17.4. The effect of food on maturation is discussed in Section 8.1 and on fecundity and fertility in Section 8.5.

17.3 TEMPORAL SYNCHRONIZATION

For populations to remain gregarious rather than disperse and solitarize, locusts not only need to be spatially aggregated but also must become synchronized in their temporal patterns of activity and development. On a daily schedule, temporal synchronicity is imposed by diurnal cycles of daylight and temperature (Uvarov, 1977 and references therein). Thus, for example, the early morning period of high-density basking that follows overnight roosting might help re-establish behavioural gregarization after an overnight decline caused by locusts being inactive and hence not interacting (Roessingh and Simpson, 1994; see Section 15). At a longer time-scale, environmental conditions such as rain events synchronize lifecycles, in conjunction with synchronous egg laying and egg hatching, and pheromonal coordination of development and maturation within populations of gregarious locusts (Uvarov, 1977; Hassanali *et al.*, 2005a,b; see Sections 8.1 and 8.4).

An additional nuance is that synchronization of activity can be mediated by resource distribution, as shown by Despland and Simpson (2006). Experimental data came from the same study on nymphal *S. gregaria* as in Despland *et al.* (2000), with additional inferences and tests of predictions derived from a realistically parameterized agent-based simulation. Results showed that activity, including feeding, was stimulated in quiescent nymphs by active neighbours, but only within a radius of approximately 5 body lengths. This mutual excitation causes phase-setting of short-term endogenous activity rhythms (Simpson and Raubenheimer, 2000) and consequently synchronization of activity between individuals. Because the probability of having an active neighbour is greater when foods are clumped rather than scattered, the strength of entrainment and phase-coupling of individual activity cycles is also greater, which results in emergence of synchronized movement and feeding within groups. Synchronization of activity will increase the intensity of mutual stimulation during periods of activity and hence accelerate population gregarization. In contrast, when resources are scattered, activity in one locust stimulates activity of any solitarious near neighbours, but usually the response is to move away, eliciting a wave of dispersal among scattered solitarious locusts and inhibiting gregarization. Hence temporal synchronization may serve to further sharpen the transition from one behavioural phase state to the other induced by increased resource clumping or an increase in population density (Collett *et al.*, 1998; see Section 17.1).

17.4 AVOIDING NATURAL ENEMIES

Recently, Reynolds *et al.* (2009) hypothesized that selection pressure from predators has been an important factor underlying the initial evolution of the switch to active aggregation under high population densities, which is a central component of phase polyphenism in locusts. They used a model from 'percolation theory' to show that as population densities increase, the switch to an extremely clumped distribution disrupts the connectivity of the predators' food-patch network and hence provides protection against predation. Forming aggregations of itself is thought to confer protection against predators under conditions of high population density through the safety of numbers, as demonstrated by Sword *et al.* (2005) in radio-tracking experiments on individuals from bands of Mormon crickets.

Forming high-density aggregations may serve to disrupt predator foraging, but it brings its own problems, and certain other phase-related phenotypic traits would appear to serve as secondary adaptations for reducing these negative effects under crowded conditions. Solitarious behavioural responses and colouration are consistent with reduced visibility to predators. The work of Sword and colleagues, considered in detail in Section 7.2.3, indicates that gregarious colouration, at least in *S. gregaria* and some congeners, acts as a warning to predators through association with toxic gut contents. Also discussed in Section 7.2.3, experiments and simulations by Despland and Simpson (2005a,b) indicate that crowding of solitarious *S. gregaria* nymphs elicits a switch towards readily accepting rather than rejecting foods containing chemicals that are toxic to vertebrate predators.

Forming aggregations presents another source of threat: cannibalism. As discussed in Section 11.5, field experiments on Mormon crickets and laboratory studies and agent-based simulations on *S. gregaria* show that the threat of cannibalism, detected by tactile and visual cues, provides an explanation for why individuals align within marching bands (Simpson *et al.*, 2006; Bazazi *et al.*, 2008). Marching therefore not only serves to increase the likelihood that fresh resources will be located but also reduces the risk that individuals are eaten by conspecifics approaching from behind.

Another consequence of being in high density, interactive populations is that spread of pathogens is facilitated. Wilson *et al.* (2002) found that gregarious adult *S. gregaria* survived topical application of spores of the fungal pathogen, *Metarhizium anisopliae* var. *acridum* better than solitarious adults. Gregarious locusts had higher antibacterial activity and somewhat higher haemocyte counts, but there was no difference between the phases in phenoloxidase activity, encapsulation or behavioural fever responses. Whether such 'density-dependent prophylaxis' is a true phase-related character or simply reflects differences in exposure to pathogens as a result of rearing history is not yet clear. Teasing these two possibilities apart would require rearing individuals under simulated crowding, without altering exposure to pathogens.

17.5 LARGER SPATIAL SCALES

Small-scale features of the habitat such as resource abundance, quality and distribution are therefore important in either promoting or inhibiting phase transition within local populations, at least of *S. gregaria*. Babah and Sword (2004) tested whether similar effects of resource distribution are apparent at larger spatial scales: kilometres rather than metres. They quantified the distribution patterns of two dominant tussock grasses between adjacent regions in Mauritania that differ in their history of supporting gregarious populations. As predicted, the tussock grasses were aggregated to a greater extent along 2 km^2 belt transects in the region with the higher historical frequency of locust gregarization.

At a yet larger spatial scale, Woldewahid *et al.* (2004) recorded densities of solitarious locust populations on the Red Sea coastal plains of Sudan and measured the extent to which population densities were spatially correlated. It was found that populations were well correlated over scales of 5–24 km, but not at larger distances, and that such distribution patterns reflected a strong association between locusts and plant communities dominated by *Heliotropium* sp and millet cultivation (*Pennisetum* sp).

Despland *et al.* (2004) attempted to link small-to-midscale resource processes to landscape-scale patterns, using satellite imagery to estimate resource abundance and distribution in Mauritania and the Red Sea coast, and compared values between years in which desert locust outbreaks occurred or were absent. They found that whereas at small spatial scales a fragmented habitat with many dispersed vegetation patches inhibits gregarization (see Section 17.1), at the landscape scale the opposite occurs, with habitat fragmentation bringing migrating locusts together into localized regions and encouraging outbreaks. On the Red Sea coast, which is habitually fragmented in its vegetation distribution (Woldewahid *et al.*, 2004), locust outbreak years were those with high resource abundance, but in Mauritania both increased resource abundance and fragmentation were features of outbreak years.

Understanding patterns of movement of locusts between resource patches at the landscape scale requires knowledge of the movement responses of individuals as well as of hopper bands and flying swarms. In Section 11.5 we discussed how agent-based models have been used recently to study collective movement in bands and swarms. At a higher scale, numerous studies have concerned swarm migration in relation to wind and weather patterns (reviewed by Farrow, 1990; Cheke and Tratalos, 2007), but there is a need to link these two and to integrate them with ecological models of phase change. Time series analyses offer one means of inferring connectivity between populations of locusts across regions (Cheke and Tratalos, 2007); another is to use population genetics techniques.

17.6 POPULATION GENETICS

Population genetics offers a powerful tool for linking population structure across different spatial scales, allowing the extent of genetic mixing between populations to be estimated and patterns of population movement and interbreeding to be inferred. Ibrahim and coworkers (Ibrahim *et al.*, 2000; Ibrahim, 2001) pioneered these techniques on locusts, in their study of solitarious populations of *S. gregaria* in Eritrea. Ibrahim *et al.* (2000) used a highly variable DNA marker (the noncoding 3′ end of the *antennapedia*-class homeobox gene) to screen locust samples and concluded that significant genetic divergence existed among solitarious population, implying that solitarious populations were not the remnants of a single homogenous gregarious population. Ibrahim (2001) next used Monte Carlo simulation to infer population dynamics within a metapopulation structure, based on known frequencies of outbreaks and recessions. The model indicated that to develop genetic divergence among local solitarious populations during recession periods founder effects (due to small numbers of colonists coming from different populations) and genetic drift must exceed the rate of genetic mixing during outbreak periods.

Highly variable microsatellite loci provide a particularly powerful set of genetic markers for population genetics research; although there are severe technical challenges in developing suitable PCR primers for locusts, due in particular to the abundance of null alleles in Orthoptera. Yassin *et al.* (2006) published a set of nine pairs of PCR primers for polymorphic microsatellite loci in *S. gregaria*, based on a sample of locusts collected in Eritrea (Ibrahim *et al.*, 2000). Kaatz *et al.* (2007) published a further nine pairs of primers, based on locusts from a longstanding laboratory culture from Niger and shown also to be effective on *S. gregaria* from Senegal and with *S. g. flaviventris* (see also Section 16.3). Recently, Chapuis *et al.* (2008b) have characterized eight polymorphic microsatellite loci from field-collected *Chortoicetes terminifera*. In conjunction with new techniques for dealing with the complicating effects of null alleles on statistical analysis of microsatellite data (Chapuis and Estoup, 2007), these new markers open the possibility to begin detailed population genetic analysis in locusts.

The most detailed study to date employing microsatellites to study population genetics while controlling for null allele effects is that of Chapuis *et al.* (2008c, 2009). These authors used 14 microsatellite loci, developed by Zhang *et al.* (2003) and Chapuis *et al.* (2005) to assess 25 populations of *L. migratoria* from Africa, Europe, Asia and Australia, attributed to nine subspecies (*L. m. capito*, *cinerascens*, *manilensis*, *migratoria*, *migratorioides*, *gallica*, the Australian subspecies, the Arabian subspecies, and the Palavas form from Southern France). The data showed no evidence of a distinct genetic signature associated with a population's propensity to outbreak. Nor did the genetic structure apparent across populations align with the sub-specific classifications based on morphometric data. Instead, genetic structure was explicable by a combination of geographical and ecological barriers (mountains and oceans) and the homogenizing effects of outbreaks. Populations on Indian and Pacific Ocean islands had a higher level of genetic differentiation that did continental populations from Africa, Asia and Europe, probably reflecting isolation effects on islands. Similarly, the Northern Hemisphere mountain belt (stretching from the European Alps across the Himalayas) appeared to provide a barrier to gene flow between populations from Eurasian and Eurafrican populations, with a zone of mixing apparent in Western Europe. Unfortunately, samples were not included from Indian and Tibetan subspecies, which would have helped confirm this hypothesis. Intriguingly, Australian and Indonesian populations were more closely related to African than to Chinese populations, as had previously been proposed based on morphometry by Farrow and Colless (1980). Chapuis *et al.* (2008c) suggested that this pattern reflects climatic effects, with populations dispersing and colonizing more effectively within a given climatic region (tropical or temperate) than between regions. See Zhang and Kang (2005) for additional discussion of Chinese representatives of *L. migratoria* as outlined in Section 1.2.

In a more detailed regional study, Chapuis *et al.* (2009) analysed genetic variation within and between 24 populations in Western Europe (historically non-outbreaking), Madagascar and Northern China (outbreaking populations). Gene flow was found to be greater among outbreaking than non-outbreaking populations, but the effective population size and degree of genetic variation was similar between the two types of population, confirming the findings of Chapuis *et al.* (2008c).

18 Concluding remarks

As should be abundantly clear from this review, substantial progress has been made in the study of locust phase polyphenism over the past 10 years. The topic has well and truly emerged from the realms of applied entomology to assume a prominent position in the modern study of phenotypic plasticity, whereby adaptive phenotypes arise during development as a result of plastic interactions between genes and the environment.

The study of behavioural gregarization has seen some of the most far-reaching progress. We now have a quantitative understanding in *S. gregaria* of the environmental conditions that favour crowding, the stimuli and neurochemical mediators of behavioural phase change and the link from individual behaviour to population-level events that lead to mass movement (Fig. 3). There is still much of interest to be done at the level of the neural circuitry underlying behavioural phase change, but extremely promising advances have already been made, and will no doubt continue.

The involvement of locust-emitted volatiles in aspects of phase biology has generated substantial research, and its fair share of productive controversy. It has been especially heartening to see that some of the major areas of controversy, such as the role of phenylacetonitrile, appear to have been at least partially resolved during the past year.

Substantial progress has also been made during the past 10 years in cataloguing molecular markers of phase. No integrated picture has yet to emerge, but the growing availability of genetic resources will have a substantial impact during the coming decade, with genomics and proteomics techniques helping to uncover the molecular genetic and epigenetic mechanisms underlying locust phase change. Having a sequenced genome for one of the major locust species would be of enormous benefit, and it is to be hoped that this will happen within the next few years.

New genetic resources also offer a solution to a more prosaic problem in locust research. There has been a growing realization that some of the differences reported between laboratories in aspects of phase polyphenism most likely reflect effects of rearing locusts in long-term culture in the laboratory. Now that suitable genetic tools are becoming available, such a microsatellite markers, it is imperative that assessment be made of the genetic status of laboratory cultures.

As a result of advances over the past 10 years, there is now the possibility for understanding how gene transcription, translation, and regulation during phase change relate to the physiological, behavioural and ecological interactions that result in outbreaks, mass migration, phylogenetic and biogeographical patterns at continental scales. Such an integrated understanding would be unique in the biological sciences. However, many questions remain to be answered. It is generally accepted that transformation from the solitarious to gregarious phase occurs due to a local increase in population density and the first effect of crowding is behavioural change. However, we still know little, or anything, of the physiological and especially the molecular mechanisms that induce many other gregarious characteristics in morphology, anatomy, colour (including pigments that underlie colour changes) and reproductive variables. The same is true for transformation from the gregarious to the solitarious phase, which seems

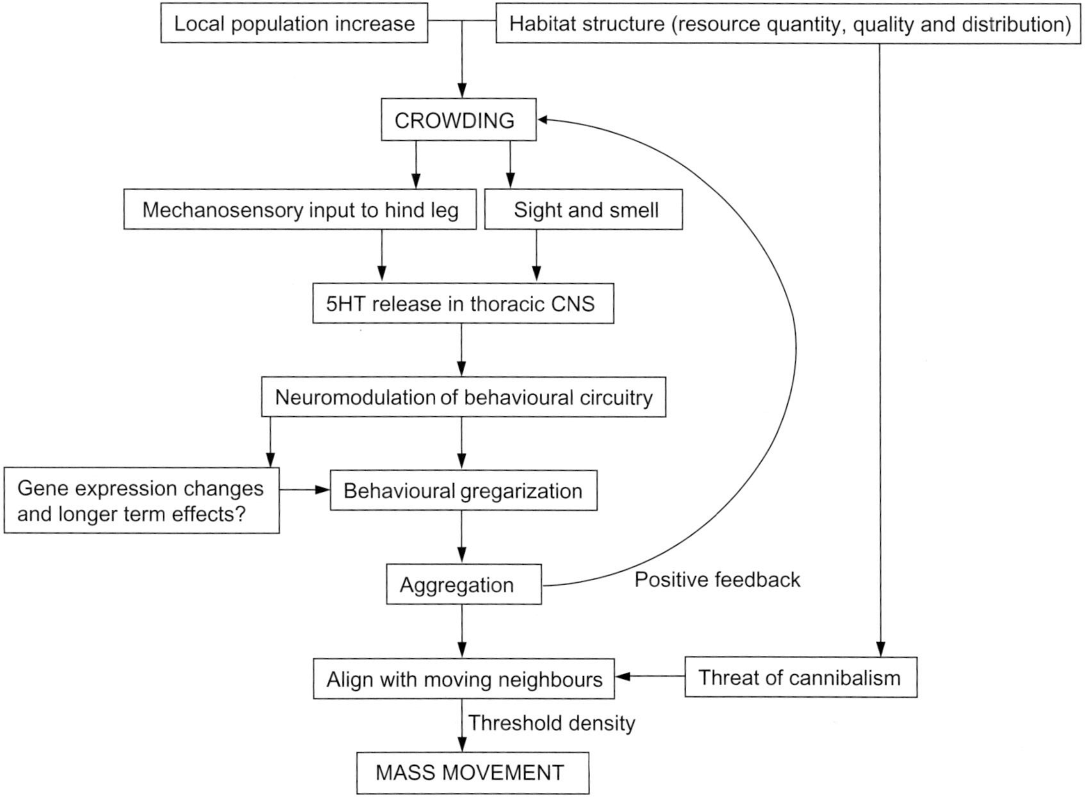

FIG. 3 A schematic setting out the key stages in behavioural gregarization in *Schistocerca gregaria*, beginning with initial crowding of solitarious populations and culminating in mass movement. See text for detailed discussion of each point.

to be less abrupt than the opposite transformation. These mechanisms, as well as the roles of various neuropeptides (see Section 9.3.9) need to be revealed for an integrated understanding of locust phase polyphenism.

Notes added to the proofs

After completion of this review, we found additional studies, published meanwhile, in relation to locusts, locust phases or locust phase characteristics. The subjects of these studies are briefly summarized as follows:

- Yamagishi and Tanaka (2009) investigated morphometrics and adult colouration after an outbreak of *L. migratoria* in Iheya Island, Okinawa, Japan. Maeno and Tanaka (2008a) described the colour of a reddish-brown mutant of *S. gregaria* and phase-dependent expression of this mutant.
- Tanaka and Zhu (2008) found that the width of the egg pod, clutch size and hatchling weight tend to decrease with decreasing latitude at various geographic localities of *L. migratoria* in China. Maeno and Tanaka (2008b) studied phase-specific developmental and reproductive strategies in *S. gregaria*. Lange (2009) reviewed neural mechanisms coordinating female reproductive system in locusts, as based mainly on *L. migratoria*, without paying attention to possible phase-dependent differences.
- Clynen and Schoofs (2009) published a peptidomic survey of the neuroendocrine system of *L. migratoria* and *S. gregaria*, presenting the amino acid sequences, molecular mass and related literature. They characterized biochemically approximately 50 (neuro)peptides. Manière *et al.* (2009) reported that a neurohormone found in *L. migratoria*, different from known ovary maturating parsins and neuroparsins of locusts (see Section 9.3.7), induces steroidogenesis, *in vitro*, in blowfly ovaries. Liu *et al.* (2008) investigated phase-dependent differences in the ultrastructure of flight muscles of *L. m. manilensis*. Jiang *et al.* (2009) studied the binding specificity of an odorant binding protein identified in the antenna of (presumably crowded) *L. migratoria*.
- Maeno and Tanaka (2009) reported that artificial miniaturization of eggs laid by crowded *S. gregaria* produces green hatchlings characteristic to isolated conspecifics (see Sections 7.2 and 16). Ben Hamouda *et al.* (2009) investigated the role of the egg pod foam and rearing density in *L. m. migratoria*, to reveal vertical transmission of phase characteristics, such as colouration, morphometrics and number of mature oocytes, to the next generation of *L. m. migratoria*.

Unfortunately, we overlooked an older publication by Islam (1995), who reported that application of a JH analogue, fenoxycarb, to crowded *S. gregaria* resulted in decreased egg size. Maeno and Tanaka (2009) reported similar results. Such reduction in egg size is in strong contrast to the report by Schneider *et al.* (1995) that application of fenoxycarb leads to oversized oocytes in *S. gregaria* (see Section 9.1).

Acknowledgements

The authors thank Dr. Alexandre V. Latchininsky for information, interpretation and references of the Russian literature, cited mostly in Section 1.2. We also thank Dr. Greg Sword for reading parts of the manuscript and for comments and discussion. SJS was supported on an Australian Research Council Federation Fellowship and MPP by a Hebrew University grant to retired scientists. Finally, we would like to thank Pedro Teixeira for his invaluable help in formatting and editing the manuscript.

References

Abashidze, E. (2000). Locust outbreaks in Georgia. *Zaschita i Karantin Rastenii (Plant Protection and Quarantine)* (10), 16–17, [in Russian].

Abashidze, E., Tsakadze, T. and Shavliashvili, I. (1998). An epizootics of the Italian locust in Georgia. *Zaschita Rastenii v Kazakhstane (Plant Protection in Kazakhstan)* (4), 27–28, [in Russian].

Abbassi, K., Atay-Kadiri, Z. and Ghaout, S. (2003a). Biological effects of alkaloïds extracted from three plants of Moroccan arid areas on the desert locust. *Physiol. Entomol.* **28**, 232–236.

Abbassi, K., Atay-Kadiri, Z. and Ghaout, S. (2003b). Caractérisation des populations de *Schistocerca gregaria* (Forskål 1775), durant la recrudescens de 1995 au Sud du Maroc. *J. Orthop. Res.* **12**, 63–69.

Albrecht, F. O. (1953). *The Anatomy of the Migratory Locust*. London: The Athlone Press.

Albrecht, F. O. (1955). La densité des populations et la croissance chez *Schistocerca gregaria* (Forsk.) et *Nomadacris septemfasciata* (Serv.); la mue d'ajustement. *J. Agr. Trop. Bot. Appl.* **2**, 109–192.

Albrecht, F. O. (1959). Facteurs internes et fluctuations des effectifs chez *Nomadacris septemfasciata* (Serv.). *Bull. Biol. Fr. Belg.* **93**, 414–461.

Albrecht, F. O. (1965). Influence du groupement, de l'état hygrométrique et de la photopériode sur la résistance au jeune de *Locusta migratoria migratorioides* R. et F. (Orthoptère acridien). *Bull. Biol. Fr. Belg.* **99**, 287–339.

Albrecht, F. O. (1967). *Polymorphisme Phasaire et Biologie des Acridiens Migrateurs*. Paris: Mason et Cie Éditeurs.

Albrecht, F. O., Verdier, M. and Blackith, R. E. (1958). Détermination de la fertilité par l'effet de groupe chez le criquet migrateur (*Locusta migratoria migratorioides* R. et F.). *Bull. Biol. Fr. Belg.* **92**, 349–427.

Albrecht, F. O., Verdier, M. and Blackith, R. E. (1959). Maternal control of ovariole number in the progeny of the migratory locust. *Nature* **184**, 103–104.

Altner, H. (1977). Insect sensillum specificity and structure: an approach to a new typology. In: *Olfaction and Taste VI (Proceedings of the 6th International Symposium* *on Olfaction and Taste, Paris)* (eds LeMagnen, J. and MacLeod, P.), pp. 295–303. London: Information Retrieval.

Amar, Z. (2004). *The Locust in Jewish Tradition.* Ramat Gan, Israel: Bar Ilan University, [in Hebrew].

Amerasinghe, F. P. (1978a). Pheromonal effects on sexual maturation, yellowing, and the vibration reaction in immature male desert locusts (*Schistocerca gregaria*). *J. Insect Physiol.* **24**, 309–314.

Amerasinghe, F. P. (1978b). Effects of J.H.I and J.H.III on yellowing, sexual activity and pheromone production in allatectomized male *Schistocerca gregaria. J. Insect Physiol.* **24**, 603–611.

Anarbaev, A. (2001). Locust management in the Republic of Kyrgyzstan. *Zaschita i Karantin Rastenii v Kazakhstane (Plant Protection and Quarantine in Kazakhstan)* (1), 21–22, [in Russian].

Anstey, M. L., Rogers, S. M., Ott, S. R., Burrows, M. and Simpson, S. J. (2009). Serotonin mediates behavioral gregarization underlying swarm formation in the desert locusts. *Science* **323**, 627–630.

Anton, S. and Hansson, B. S. (1996). Antennal lobe interneurons in the desert locust *Schistocerca gregaria* (Forskal): processing of aggregation pheromones in adult males and females. *J. Comp. Neurol.* **370**, 85–96.

Anton, S., Ignell, R. and Hansson, B. S. (2002). Developmental changes in the structure and function of central olfactory system in gregarious and solitary desert locusts. *Microsc. Res. Tech.* **56**, 281–291.

Applebaum, S. W. and Heifetz, Y. (1999). Density-dependent physiological phase in insects. *Annu. Rev. Entomol.* **44**, 317–341.

Applebaum, S. W., Avisar, E. and Heifetz, Y. (1997). Juvenile hormone and locust phase. *Arch. Insect Biochem. Physiol.* **35**, 375–391.

Arthurs, S. and Thomas, M. B. (2000). Effects of a mycoinsecticide on feeding and fecundity of the brown locust *Locustana pardalina. Biocontrol Sci. Techn.* **10**, 321–329.

Ashall, C. and Ellis, P. E. (1962). Studies on numbers and mortality in field populations of the desert locust, (*Schistocerca gregaria* Forskål). *Anti-Locust Bull.* **38**, 1–59.

Assad, Y. O. H., Hassanali, A., Torto, B., Mahamat, H., Bashir, N. H. H. and El Bashir, S. (1997a). Effects of fifth-instar volatiles on sexual maturation of adult desert locust, *Schistocerca gregaria. J. Chem. Ecol.* **23**, 1373–1388.

Assad, Y. O. H., Torto, B., Hassanali, A., Njagi, P. G. N., Bashir, N. H. H. and Mahamat, H. (1997b). Seasonal variation in the essential oil composition of *Commiphora quadricincta* and its effect on the maturation of immature adults of the desert locust, *Schistocerca gregaria. Phytochem.* **44**, 833–841.

Audsley, N., McIntosh, C. and Phillips, J. E. (1992a). Isolation of a neuropeptide from locust corpus cardiacum which influences ileal transport. *J. Exp. Biol.* **173**, 261–274.

Audsley, N., McIntosh, C. and Phillips, J. E. (1992b). Actions of ion-transport peptide from locust corpus cardiacum on several hindgut transport processes. *J. Exp. Biol.* **173**, 275–288.

Ayali, A. and Pener, M. P. (1992). Density-dependent phase polymorphism affects response to adipokinetic hormone in *Locusta. Comp. Biochem. Physiol. A* **101**, 549–552.

Ayali, A. and Pener, M. P. (1995). The relations of adipokinetic response and body lipid content in locusts (*Locusta migratoria migratorioides*) with special reference to phase polymorphism. *J. Insect Physiol.* **41**, 85–89.

Ayali, A., Golenser, E. and Pener, M. P. (1994). Differences in response to adipokinetic hormones between gregarious and solitary locusts. In: *Insect Neurochemistry and Neurophysiology 1993* (eds Bořkovec, A. B. and Loeb, M. J.), pp. 177–180. Boca Raton, FL: CRC Press.

Ayali, A., Pener, M. P. and Girardie, J. (1996a). Comparative study of neuropeptides from the corpora cardiaca of solitary and gregarious *Locusta*. *Arch. Insect Biochem. Physiol.* **31**, 439–450.

Ayali, A., Golenser, E. and Pener, M. P. (1996b). Flight fuel related differences between solitary and gregarious locusts (*Locusta migratoria migratorioides*). *Physiol. Entomol.* **21**, 1–6.

Ayali, A., Pener, M. P., Sowa, S. M. and Keeley, L. L. (1996c). Adipokinetic hormone content of the corpora cardiaca in gregarious and solitary migratory locusts. *Physiol. Entomol.* **21**, 167–172.

Ayali, A., Fuchs, E. and Kutsch, W. (2004). Neurophysiological studies on flight-related density-dependent phase characteristics in locusts. *Acta Biol. Hungarica* **55**, 137–141.

Babah, M. A. O. and Sword, G. A. (2004). Linking locust gregarization to local resource distribution patterns across a large spatial scale. *Environ. Entomol.* **33**, 1577–1583.

Bachem, C. W., van der Hoeven, R. S., de Bruijn, S. M., Vreugdenhil, D., Zabeau, M. and Visser, R. G. (1996). Visualization of differential gene expression using a novel method of RNA fingerprinting based on AFLP: analysis of gene expression during potato tuber development. *Plant J.* **9**, 745–753.

Badisco, L., Claeys, I., Van Loy, T., Van Hiel, M., Franssens, V., Simonet, G. and Vanden Broeck, J. (2007). Neuroparsins, a family of conserved arthropod neuropeptides. *Gen. Comp. Endocrinol.* **153**, 64–71.

Badisco, L., Claeys, I., Van Hiel, M., Clynen, E., Huybrechts, J., Vandersmissen, T., Van Soest, S., Vanden Bosch, L., Simonet, G. and Vanden Broeck, J. (2008). Purification and characterization of an insulin-related peptide in the desert locust, *Schistocerca gregaria*: immunolocalization, cDNA cloning, transcript profiling and interaction with neuroparsin. *J. Mol. Endocrinol.* **40**, 137–150.

Baggerman, G., Clynen, E., Mazibur, R., Veelaert, D., Breuer, M., De Loof, A., Tanaka, S. and Schoofs, L. (2001). Mass spectrometric evidence for the deficiency in the dark-color-inducing hormone, [His^7]-corazonin in an albino strain of *Locusta migratoria* as well as for its presence in solitary *Schistocerca gregaria*. *Arch. Insect Biochem. Physiol.* **47**, 150–160.

Baggerman, G., Huybrechts, J., Clynen, E., Hens, K., Harthoorn, L., Van der Horst, D., Poulos, C., De Loof, A. and Schoofs, L. (2002). New insights in adipokinetic hormone (AKH) precursor processing in *Locusta migratoria* obtained by capillary liquid chromatogtraphy – tandem mass spectrometry. *Peptides* **23**, 635–644.

Bailey, N. W., Gwynne, D. T. and Ritchie, M. G. (2005). Are solitary and gregarious Mormon crickets (*Anabrus simplex*, Orthoptera, Tettigoniidae) genetically distinct? *Heredity* **95**, 166–173.

Baker, G. L. (1993). *Locusts and Grasshoppers of the Australian Region.* The Orthopterists' Society Series of Field Guides, No. D9E. Québec: NSW Agriculture/Orthopterists' Society.

Ballan-Dufrançais, C. (1978). Biaccumulation minérale, purique et pigmentaire dans le tégument de *Schistocerca gregaria* (Forsk.). Dimorphisme sexuel et control endocrine. *Acrida* **7**, 85–99.

Bar-Ilan, A. R., Shulov, A. and Pener, M. P. (1969). The stridulation of sexually mature males of Locusta migratoria migratorioides (R. and F.) in relation to temperature and some other factors. *Physiol. Zool.* **42**, 417–428.

Barth, R. H. and Lester, L. J. (1973). Neuro-hormonal control of sexual behavior in insects. *Annu. Rev. Entomol.* **18**, 445–472.

Bashir, M. O., Hassanali, A., Rai, M. M. and Saini, R. K. (2000). Changing oviposition preferences of the desert locust, *Schistocerca gregaria*, suggest a strong species predisposition for gregarization. *J. Chem. Ecol.* **26**, 1721–1733.

Bazazi, S., Buhl, J., Hale, J. J., Anstey, M. L., Sword, G. A., Simpson, S. J. and Couzin, I. D. (2008). Collective motion and cannibalism in locust migratory bands. *Curr. Biol.* **18**, 735–739.

Beenakkers, A. M. Th. (1969). The influence of corpus allatum and corpus cardiacum on lipid metabolism in *Locusta migratoria*. *Gen. Comp. Endocrinol.* **13**, 492.

Beenakkers, A. M. Th., Bloemen, R. E. B., De Vlieger, T. A., Van der Horst, D. J. and Van Marrewijk, W. J. A. (1985a). Insect adipokinetic hormones. *Peptides* **6** (Suppl 3), 437–444.

Beenakkers, A. M. Th., Van der Horst, D. J. and Van Marrewijk, W. J. A. (1985b). Biochemical processes directed to flight muscle metabolism. In: *Comprehensive Insect Physiology Biochemistry and Pharmacology, Vol. 10, Biochemistry* (eds Kerkut, G. A. and Gilbert, L. I.), pp. 451–486. Oxford: Pergamon Press.

Bell, M. (2005). The 2004 desert locust outbreak. *Bull. Am. Meteor. Soc.* **86**, S60, special section on ''State of the climate in 2004.''

Ben-Aziz, O., Zeltser, I. and Altstein, M. (2005). PBAN selective antagonists: inhibition of PBAN induced cuticular melanization and sex pheromone biosynthesis in moths. *J. Insect Physiol.* **51**, 305–314.

Ben Hamouda, A., Ammar, M., Ben Hamouda, M. H. and Bouain, A. (2009). The role of egg pod foam and rearing conditions on the phase state of the Asian migratory locust *Locusta migratoria migratoria* (*Orthoptera, Acrididae*). *J. Insect Physiol.* **55**, 617–623.

Ben-Shahar, Y. (2005). The *foraging* gene, behavioral plasticity, and honeybee division of labor. *J. Comp. Physiol. A* **191**, 987–994.

Benichou-Redouane, K. and Fuzeau-Braesch, S. (1982). Comparaison des taux d'octopamine chez *Locusta migratoria cinerascens* grégaire, solitaire et solitarisé au gaz carbonique, dans différents organes nerveux. *C. R. Hebd. Séanc. Acad. Sci. Paris, Série III,* **294**, 385–388.

Bensch, S. and Akesson, M. (2005). Ten years of AFLP in ecology and evolution: why so few animals? *Mol. Ecol.* **14**, 2899–2914.

Bergot, B. J., Schooley, D. A. and de Kort, C. A. D. (1981). Identification of JH III as the principal juvenile hormone in *Locusta migratoria*. *Experientia* **37**, 909–910.

Bermudez, I., Beadle, D. J., Trifilieff, E., Luu, B. and Hietter, H. (1991). Electrophysiological activity of the C-peptide of the *Locusta* insulin-related peptide. Effect on the membrane conductance of *Locusta* neurons in vitro. *FEBS Letters* **293**, 137–141.

Bernays, E. A. (1972a). The muscles of newly hatched *Schistocerca gregaria* larvae and their possible functions in hatching, digging and ecdysial movements (Insecta: Acrididae). *J. Zool.* **166**, 141–158.

Bernays, E. A. (1972b). The intermediate moult (first ecdysis) of *Schistocerca gregaria* (Forskål) (Insecta, Orthoptera). *Z. Morph. Tiere* **71**, 160–179.

Bernays, E. A. (1972c). Changes in the first instar cuticle of *Schistocerca gregaria* before and associated with hatching. *J. Insect Physiol.* **18**, 897–912.

Bernays, E. A. and Chapman, R. F. (1998). Phenotypic plasticity in numbers of antennal chemoreceptors in a grasshopper: effects of food. *J. Comp. Physiol. A* **183**, 69–76.

Blanford, S. and Thomas, M. B. (2001). Adult survival, maturation and reproduction of the desert locust *Schistocerca gregaria* infected with the fungus *Metarhizium anisopliae* var *acridum*. *J. Invertebr. Pathol.* **78**, 1–8.

Bodenheimer, F. S. and Shulov, A. (1951). Egg-development and diapause in the Moroccan locust (*Dociosaturus maroccanus* Thnb.). *Bull. Res. Counc. Israel* **1**, 59–75.

Bogerd, J., Kooiman, F. P., Pijnenburg, M. A. P., Hekking, L. H. P., Oudejans, R. C. H. M. and Van der Horst, D. J. (1995). Molecular cloning of three distinct cDNAs, each encoding a different adipokinetic hormone precursor, of the migratory locust, *Locusta migratoria*. *J. Biol. Chem.* **270**, 23038–23043.

Boigegrain, R.-A., Mattras, H., Brehélin, M., Paroutaud, P. and Coletti-Previero, M.-A. (1992). Insect immunity: two proteinase inhibitors from hemolymph of *Locusta migratoria*. *Biochem. Biophys. Res. Commun.* **189**, 790–793.

Botens, F. F. W., Rembold, H. and Dorn, A. (1997). Phase-related juvenile hormone determinations in field catches and laboratory strains of different *Locusta migratoria* subspecies. In: *Advances in Comparative Endocrinology: Proceedings of the XIIIth International Congress of Comparative Endocrinology, Yokahama (Japan), November* (eds Kawashima, S. and Kikuyama, S.), pp. 197–203. Bologna: Monduzzi Editore.

Bouaïchi, A. and Simpson, S. J. (2003). Density-dependent accumulation of phase characteristics in a natural population of the desert locust *Schistocerca gregaria*. *Physiol. Entomol.* **28**, 25–31.

Bouaïchi, A., Roessingh, P. and Simpson, S. J. (1995). An analysis of the behavioural effects of crowding and re-isolation on solitary-reared adult desert locusts (*Schistocerca gregaria*) and their offspring. *Physiol. Entomol.* **20**, 199–208.

Bouaïchi, A., Simpson, S. J. and Roessingh, P. (1996). The influence of environmental microstructure on the behavioural phase state and distribution of the desert locust *Schistocerca gregaria*. *Physiol. Entomol.* **21**, 247–256.

Boureme, D., Tamarelle, M. and Girardie, J. (1987). Production and characterization of antibodies to neuroparsins A and B isolated from the corpora cardiaca of the locust. *Gen. Comp. Endocrinol.* **67**, 169–177.

Bouthier, A. (1966). Modifications des pigments (ommochromes et ptérines) en relation avec la mutation *albinos* chez *Locusta migratoria cinerascens* Fabr. (Orthoptères, Acrididae). *C. R. Hebd. Séanc. Acad. Sci. Paris, Série D* **262**, 1480–1483.

Bouthier, A. (1975). Nature and biosynthesis of ommochromes in *Locusta migratoria cinerascens* Fabr. (Orthoptera, Acrididae) and its *albino* mutant. *Acta Vitaminol. Enzymol. Milano* **29**, 210–218.

Bouthier, A. and Lhonoré, J. (1984). Developmental changes in the amount of pigments, inorganic material and uric acid in *Locusta migratoria cinerascens* Fabr. (Orthoptera, Acrididae) epidermis, during the last larval instar and the imaginal life. *J. Comp. Physiol. B* **154**, 549–560.

Brehélin, M., Boigegrain, R.-A., Drif, L. and Coletti-Previero, M. A. (1991). Purification of a protease inhibitor which controls prophenoloxidase activation in hemolymph of *Locusta migratoria* (insecta). *Biochem. Biophys. Res. Commun.* **179**, 841–846.

Brennecke, J., Malone, C. D., Aravin, A. A., Sachidanandam, R., Stark, A. and Hannon, G. J. (2008). An epigenetic role for maternally inherited piRNAs in transposon silencing. *Science* **322**, 1387–1392.

Breuer, M., Hoste, B. and De Loof, A. (2003). The endocrine control of phase transition: some new aspects. *Physiol. Entomol.* **28**, 3–10.

Breugelmans, B., Simonet, G., van Hoef, V., Claeys, I., Van Soest, S. and Vanden Broeck, J. (2008). Quantitative RT-PCR analysis of pacifastin-related precursor transcripts during the reproductive cycle of solitarious and gregarious desert locusts. *Insect Mol. Biol.* **17**, 137–145.

Breyne, P., Dreesen, R., Cannoot, B., Rombaut, D., Vandepoele, K., Rombauts, S., Vanderhaeghen, R., Inze, D. and Zabeau, M. (2003). Quantitative cDNA-AFLP analysis for genome-wide expression studies. *Mol. Genet. Genomics* **269**, 173–179.

Broom, M., Speed, M. P. and Ruxton, G. D. (2006). Evolutionarily stable defence and signalling of that defence. *J. Theor. Biol.* **242**, 32–43.

Broza, M. and Pener, M. P. (1969). Hormonal control of the reproductive diapause in the grasshopper, *Oedipoda miniata*. *Experientia* **25**, 414–415.

Buhl, J., Sumpter, D. J. T., Couzin, I. D., Hale, J. J., Despland, E., Miller, E. R. and Simpson, S. J. (2006). From disorder to order in marching locusts. *Science* **312**, 1402–1406.

Burrows, M. (1996). *The Neurobiology of an Insect Brain*. Oxford, UK: Oxford University Press.

Burtt, E. (1951). The ability of adult grasshoppers to change colour on burnt ground. *Proc. R. Entomol. Soc. Lond. A* **26**, 45–48.

Byers, J. A. (1991). Pheromones and chemical ecology of locusts. *Biol. Rev.* **66**, 347–378.

Canavoso, L. E., Jouni, Z. E., Karnas, K. J., Pennington, J. E. and Wells, M. A. (2001). Fat metabolism in insects. *Annu. Rev. Nutr.* **21**, 23–46.

Candy, D. J., Becker, A. and Wegener, G. (1997). Coordination and integration of metabolism in insect flight. *Comp. Biochem. Physiol. B* **117**, 497–512.

Cantacuzène, A.-M. (1967). Effets comparés de l'allatectomie sur l'activité des glandes annexes mâles et le comportement sexuel de deux Acridiens: *Schistocerca gregaria* et *Locusta migratoria* (souches *migratorioides* et 'Kazalinsk'. *C. R. Hebd. Séanc. Acad. Sci. Paris, Série D* **265**, 224–227.

Cantacuzène, A.-M. (1968). Recherches morphologiques et physiologiques sur les glandes annexes mâles des orthoptères. II – Les glandes annexes mâles des acridiens après l'ablation des corps allates et au cours de la diapause imaginale. *Bull. Soc. Zool. Fr.* **93**, 545–557.

Carlisle, D. B. and Ellis, P. E. (1959). La persistance des glandes ventrales céphaliques chez les criquets solitaires. *C. R. Hebd. Séanc. Acad. Sci. Paris* **249**, 1059–1060.

Carlisle, D. B. and Ellis, P. E. (1962). Endocrine glands and phase in locusts. *Sym. Gen. Biol. Italica* **10**, 219–224.

Carlisle, D. B. and Ellis, P. E. (1963). Prothoracic gland and gregarious behaviour in locusts. *Nature* **200**, 603–604.

Carlisle, D. B. and Ellis, P. E. (1965). Synchronisation of sexual maturation in desert locust swarms. *Compt. Rend. Ve Congr. Un. Int. Étude Insectes Soc., Toulouse*, 63–67.

Carlisle, D. B., Ellis, P. E. and Betts, E. (1965). The influence of aromatic shrubs on sexual maturation in the desert locust *Schistocerca gregaria*. *J. Insect Physiol.* **11**, 1541–1558.

Cassier, P. and Fain-Maurel, M.-A. (1968). Influence de la photopériode sur la persistance ou la dégénérescence des glandes de mue chez les imagos grégaires du criquet migrateur. Etude expérimentale et infrastructurale. *C. R. Hebd. Séanc. Acad. Sci. Paris, Série D* **267**, 646–648.

Cassier, P. and Fain-Maurel, M.-A. (1969). Étude infrastructurale des glandes de mue de *Locusta migratoria migratorioides* (R. et F.) III. Sur la persistance ou la dégénérescence des glandes ventrales chez les imagos solitaires. *Arch. Zool. Exp. Gen.* **110**, 203–224, + 6 plates.

Cayre, M., Strambi, C., Charpin, P., Augier, R. and Strambi, A. (1997). Specific requirement of putrescine for the mitogenic action of juvenile hormone on adult insect neuroblasts. *Proc. Natl. Acad. Sci. U. S. A.* **94**, 8238–8242.

Ceccato, P., Cressman, K., Giannini, A. and Trzaska, S. (2007). The desert locust upsurge in West Africa (2003–2005): Information on the desert locust early warning system and the prospects for seasonal climate forecasting. *Int. J. Pest Manag.* **53**, 7–13.

Chamberlain, D. J. (1980). The effects of pod crowding on the viability of eggs of *Locusta migratoria migratorioides* (R. & F.) and *Schistocerca gregaria* (Forsk) (Orthoptera: Acrididae). *Acrida* **9**, 51–62.

Chapman, R. F. (1955). A laboratory study of roosting behaviour in hoppers of the African migratory locust (*Locusta migratoria migratorioides* R. & F.). *Anti-Locust Bull.* **19**, 1–40.

Chapman, R. F. (2002). Development of phenotypic differences in sensillum populations on the antennae of a grasshopper, *Schistocerca americana*. *J. Morph.* **254**, 186–194.
Chapman, R. F. and Lee, J. C. (1991). Environmental effects on numbers of peripheral chemoreceptors on the antennae of a grasshopper. *Chem. Senses* **16**, 607–616.
Chapuis, M.-P. and Estoup, A. (2007). Microsatellite null alleles and estimation of population differentiation. *Mol. Biol. Evol.* **24**, 621–631.
Chapuis, M.-P., Loiseau, A., Michalakis, Y., Lecoq, M. and Estoup, A. (2005). Characterization and PCR multiplexing of polymorphic microsatellite loci for the locust *Locusta migratoria*. *Mol. Ecol. Notes* **5**, 554–557.
Chapuis, M.-P., Estoup, A., Augé-Sabatier, A., Foucart, A., Lecoq, M. and Michalakis, Y. (2008a). Genetic variation for parental effects on the propensity to gregarise in *Locusta migratoria*. *BMC Evol. Biol.* **8**, 37.
Chapuis, M.-P., Popple, J.-A., Simpson, S. J., Estoup, A., Martin, J.-F., Steinbauer, M., McCulloch, L. and Sword, G. A. (2008b). Eight polymorphic microsatellite loci for the Australian plague locust, *Chortoicetes terminifera*. *Mol. Ecol. Resources* **8**, 1414–1416.
Chapuis, M.-P., Lecoq, M., Michalakis, Y., Loiseau, A., Sword, G. A., Piry, S. and Estoup, A. (2008c). Do outbreaks affect genetic population structure? A worldwide survey in *Locusta migratoria*, a pest plagued by microsatellite null alleles. *Mol. Ecol.* **17**, 3640–3653.
Chapuis, M-P., Loiseau, A., Michalakis, Y., Lecoq, M., Franc, A. and Estoup, A. (2009). Outbreaks, gene flow and effective population size in the migratory locust, *Locusta migratoria*: a regional-scale comparative survey. *Mol. Ecol.* **18**, 792–800.
Charlet, M., Roussel, J.-P., Rinternecht, E., Berchtold, J.-P. and Costet, M.-F. (1988). Developmental and morphogenetic alterations in larvae of *Locusta migratoria* reared on plant diet with a selectively modified sterol profile. *J. Insect Physiol.* **34**, 787–796.
Chauvin, R. (1941). Contribution a l'étude physiologique du criquet pèlerine et du déterminisme des phénomènes grégaires. *Ann. Soc. Entomol. Fr.* **110**, 133–272.
Cheeseman, M. T. and Gillott, C. (1990). Corpus allatum regulation of copulatory behaviour in the male grasshopper, *Melanoplus sanguinipes*. *Physiol. Entomol.* **15**, 377–383.
Cheke, R. A. (1978). Theoretical rates of increase of gregarious and solitarious populations of the desert locust. *Oecologia* **35**, 161–171.
Cheke, R. A. and Tratalos, J. A. (2007). Migration, patchiness, and population processes illustrated by two migrant pests. *BioScience* **57**, 145–154.
Chen, Y.-L. (1999). *The Locust and Grasshopper Pests of China*. Beijing: China Forestry Publishing House.
Cherqui, A., Duvic, B., Reibel, C. and Brehélin, M. (1998). Cooperation of dopachrome conversion factor with phenoloxidase in the eumelanin pathway in haemolymph of *Locusta migratoria* (Insecta). *Insect Biochem. Mol. Biol.* **28**, 839–848.
Cheu, S. P. (1952). Changes in the fat and protein content of the African migratory locust, *Locusta migratoria migratorioides* (R. & F.). *Bull. Entomol. Res.* **43**, 101–109.
Cheung, B. H., Arellano-Carbajal, F., Rybicki, I. and De Bono, M. (2004). Soluble guanylate cyclases act in neurons exposed to the body fluid to promote *C. elegans* aggregation behavior. *Curr. Biol.* **14**, 1105–1111.
Chino, H., Abe, Y. and Takahashi, K. (1983). Purification and characterization of a biliverdin-binding cyanoprotein from the locust hemolymph. *Biochem. Biophys. Acta* **748**, 109–115.
Chino, H., Downer, R. G. H. and Takahashi, K. (1986). Effect of adipokinetic hormone on the structure and properties of lipophorin in locusts. *J. Lipid Res.* **27**, 21–29.
Chino, H., Lum, P. Y., Nagao, E. and Hiroaka, T. (1992). The molecular and metabolic essentials for long-distance flight in insects. *J. Comp. Physiol. B,* **162**, 101–106.

Claeys, I., Simonet, G., Poels, J., Van Loy, T., Vercammen, L., De Loof, A. and Vanden Broeck, J. (2002). Insulin-related peptides and their conserved signal transduction pathway. *Peptides* **23**, 807–816.

Claeys, I., Simonet, G., Van Loy, T., De Loof, A. and Vanden Broeck, J. (2003). cDNA cloning and transcript distribution of two novel members of the neuroparsin family in the desert locust, *Schistocerca gregaria*. *Insect Mol. Biol.* **12**, 473–481.

Claeys, I., Poels, J., Simonet, G., Franssens, V., Van Loy, T., Van Hiel, M. B., Breugelmans, B. and Vanden Broeck, J. (2005a). Insect neuropeptide and peptide hormones receptors: current knowledge and future directions. *Vitam. Horm.* **73**, 217–282.

Claeys, I., Simonet, G., Breugelmans, B., Van Soest, S., Franssens, V., Sas, F., De Loof, A. and Vanden Broeck, J. (2005b). Quantitative real-time RT-PCR analysis in desert locusts reveals phase dependent differences in neuroparsin transcript levels. *Insect Mol. Biol.* **14**, 415–422.

Claeys, I., Breugelmans, B., Simonet, G., Van Soest, S., Sas, F., De Loof, A. and Vanden Broeck, J. (2006a). Neuroparsin transcripts as molecular markers in the process of desert locust (*Schistocerca gregaria*) phase transition. *Biochem. Biophys. Res. Commun.* **341**, 599–606.

Claeys, I., Breugelmans, B., Simonet, G., Franssens, V., Van Soest, S. and Vanden Broeck, J. (2006b). Regulation of *Schistocerca gregaria* neuroparsin transcript levels by juvenile hormone and 20-hydroxyecdysone. *Arch. Insect Biochem. Physiol.* **62**, 107–115.

Clark, L., Zhang, J. R., Tobe, S. and Lange, A. B. (2006). Proctolin: a possible releasing factor in the corpus cardiacum/corpus allatum of the locust. *Peptides* **27**, 559–566.

Clark, L., Lange, A. B., Zhang, J. R. and Tobe, S. S. (2008). The roles of Dippu-allatostatin in the modulation of hormone release in *Locusta migratoria*. *J. Insect Physiol.* **54**, 949–958.

Clissold, F. J., Sanson, G. D. and Read, J. (2006). The paradoxical effects of nutrient ratios and supply rates on an outbreaking insect herbivore, the Australian plague locust. *J. Anim. Ecol.* **75**, 1000–1013.

Clynen, E. and Schoofs, L. (2009). Peptidomic survey of the locust neuroendocrine system. *Insect Biochem. Mol. Biol.* **39**, 491–507.

Clynen, E., Baggerman, G., Veelaert, D., Cerstiaens, A., Van der Horst, D., Harthoorn, L., Derua, R., Waelkens, E., De Loof, A. and Schoofs, L. (2001). Peptidomics of the pars intercerebralis-corpus cardiacum complex of the migratory locust, *Locusta migratoria*. *Eur. J. Biochem.* **268**, 1929–1939.

Clynen, E., Stubbe, D., De Loof, A. and Schoofs, L. (2002). Peptide differential display: a novel approach for phase transition in locusts. *Comp. Biochem. Physiol. B* **132**, 107–115.

Clynen, E., Baggerman, G., Huybrechts, J., Vanden Bosch, L., De Loof, A. and Schoofs, L. (2003a). Peptidomics of the locust corpora allata: identification of novel pyrokinins (-FXPRLamides). *Peptides* **24**, 1493–1500.

Clynen, E., Huybrechts, J., Baggerman, G., Van Doorn, J., Van der Horst, D., De Loof, A. and Schoofs, L. (2003b). Identification of a glycogenolysis-inhibiting peptide from the corpora cardiaca of locusts. *Endocrinology* **144**, 3441–3448.

Clynen, E., Huybrechts, J., Verleyen, P., De Loof, A. and Schoofs, L. (2006). Annotation of novel neuropeptide precursors in the migratory locust based on transcript screening of a public EST database and mass spectrometry. *BMC Genomics* **7**, 201.

Coates, J. C. and De Bono, M. (2002). Antagonistic pathways in neurons exposed to body fluid regulate social feeding in *Caenorhabditis elegans*. *Nature* **419**, 925–929.

Coates, B. S., Hellmich, R. L. and Lewis, L. C. (2005). Two differentially expressed ommochrome-binding protein-like genes (*obp*1 and *obp*2) in larval fat body of the European corn borer, *Ostrinia nubilalis*. *J. Insect Sci.* **5**, 1–12.

Colgan, D. J. (1987). Developmental changes of isoenzymes catalysing glycolytic and associated reactions in *Locusta migratoria* in relation to the rearing density of hatchlings. *Insect Biochem.* **17**, 303–308.

Collett, M., Despland, E., Simpson, S. J. and Krakauer, D. C. (1998). Spatial scales of desert locust gregarization. *Proc. Natl. Acad. Sci. U. S. A.* **95**, 13052–13055.

Dadd, R. H. (1961). Observations on the effects of carotene on the growth and pigmentation of locusts. *Bull. Entomol. Res.* **52**, 63–81.

Dale, J. F. and Tobe, S. S. (1986). Biosynthesis and titre of juvenile hormone during the first gonotrophic cycle in isolated and crowded *Locusta migratoria* females. *J. Insect Physiol.* **32**, 763–769.

Dale, J. F. and Tobe, S. S. (1990). The endocrine basis of locust phase polymorphism. In: *Biology of Grasshoppers* (eds Chapman, R. F. and Joern, A.), pp. 393–414. New York: Wiley.

Delbecque, J.-P., Weidner, K. and Hoffmann, K. H. (1990). Alternative sites for ecdysteroid production in insects. *Invertebr. Rep. Dev.* **18**, 29–42.

De Loof, A. (2008). Ecdysteroids, juvenile hormone and insect neuropeptides: Recent successes and remaining major challenges. *Gen. Comp. Endocrinol.* **155**, 3–13.

De Loof, A., Baggerman, G., Breuer, M., Claeys, I., Cerstiaens, A., Clynen, E., Janssen, T., Schoofs, L. and Vanden Broeck, J. (2001). Gonadotropins in insects: an overview. *Arch. Insect Biochem. Physiol.* **47**, 129–138.

De Loof, A., Claeys, I., Simonet, G., Verleyen, P., Vandersmissen, T., Sas, F. and Huybrechts, J. (2006). Molecular markers of phase transition in locusts. *Insect Sci.* **13**, 3–12.

Deng, A. L. (2002). Effects of isolation and crowding on the haemolymph pigment composition of the desert locust, *Schistocerca gregaria* (Orth., Acrididae). *J. Appl. Entomol.* **126**, 244–248.

Deng, A. L., Torto, B., Hassanali, A. and Ali, E. E. (1996). Effects of shifting to crowded or solitary conditions on pheromone release and morphometrics of the desert locust, *Schistocerca gregaria* (Forskål) (Orthoptera: Acrididae). *J. Insect Physiol.* **42**, 771–776.

Despland, E. (2001). Role of olfactory and visual cues in the attraction/repulsion responses to conspecifics by gregarious and solitarious desert locusts. *J. Insect Behav.* **14**, 35–46.

Despland, E. (2005). Diet breadth and anti-predator strategies in the desert locust and other Orthopterans. *J. Orthop. Res.* **14**, 227–233.

Despland, E. and Simpson, S. J. (2000a). The role of food distribution and nutritional quality in behavioural phase change in the desert locust. *Anim. Behav.* **59**, 643–652.

Despland, E. and Simpson, S. J. (2000b). Small-scale vegetation patterns in the parental environment influence the phase state of hatchlings of the desert locust. *Physiol. Entomol.* **25**, 74–81.

Despland, E. and Simpson, S. J. (2005a). Food choices of solitarious and gregarious locusts reflect cryptic and aposematic antipredator strategies. *Anim. Behav.* **69**, 471–479.

Despland, E. and Simpson, S. J. (2005b). Surviving the change to warning colouration: density-dependent polyphenism suggests a route for the evolution of aposematism. *Chemoecology* **15**, 69–75.

Despland, E. and Simpson, S. J. (2006). Resource distribution mediates synchronization of physiological rhythms in locust groups. *Proc. R. Soc. Lond. B* **273**, 1517–1522.

Despland, E., Collett, M. and Simpson, S. J. (2000). Small-scale processes in desert locust swarm formation: how vegetation patterns influence gregarization. *Oikos* **88**, 652–662.

Despland, E., Rosenberg, J. and Simpson, S. J. (2004). Landscape structure and locust swarming: a satellite's eye view. *Ecography* **27**, 381–391.

Dewey, E. M., McNabb, S. L., Ewer, J., Kuo, G. R., Takanishi, C. L., Truman, J. W. and Honegger, H.-W. (2004). Identification of the gene encoding bursicon, an insect neuropeptide responsible for cuticle sclerotization and wing spreading. *Curr. Biol.* **14**, 1208–1213.

DeWitt, T. J. and Scheiner, S. M. (eds) (2004). *Phenotypic Plasticity: Functional and Conceptual Approaches*. Oxford: Oxford University Press.

Diederen, J. H. B., Verstegen, M. M. A., Oudejans, R. C. H. M., Vullings, H. G. B. and Van der Horst, D. J. (1999). Number and size of intracisternal granules in the adipokinetic hormone-secreting cells of locusts: a phase-dependent characteristic. *Physiol. Entomol.* **24**, 51–55.

Dillon, R. J., Vennard, C. T. and Charnley, A. K. (2000). Exploitation of gut bacteria in the locust. *Nature* **403**, 851.

Dillon, R. J., Vennard, C. T. and Charnley, A. K. (2002). A note: Gut bacteria produce components of a locust cohesion pheromone. *J. Appl. Microbiol.* **92**, 759–763.

Dillon, R. J., Webster, G., Weightman, A. J., Dillon, V. M., Blanford, S. and Charnley, A. K. (2008). Composition of Acridid gut bacterial communities as revealed by 16S rRNA gene analysis. *J. Invertebr. Pathol.* **97**, 265–272.

Dinan, L. N. and Rees, H. H. (1981). The identification and titres of conjugated and free ecdysteroids in developing ovaries and newly-laid eggs of *Schistocerca gregaria*. *J. Insect Physiol.* **27**, 51–58.

Dingle, H. (2002). Hormonal mediation of insect life histories. In: *Hormones, Brain and Behavior*, Vol. 3 (eds Pfaff, D. W., Arnold, A. P., Elgen, A. E., Fahrbach, S. E. and Rubin, R. T.), pp. 237–279. San Diego: Academic Press.

Dingle, H. and Haskell, J. B. (1967). Phase polymorphism in the grasshopper Melanoplus differentialis. *Science* **155**, 590–592.

Dircksen, H. (1998). Conserved crustacean cardioactive peptide (CCAP) neuronal networks and functions in arthropod evolution. In: *Recent Advances in Arthropod Endocrinology*, (eds Coast, G. M. and Webster, S. G.), pp. 302–333, Society for Experimental Biology Seminar Series no. 65. Cambridge: Cambridge University Press.

Dircksen, H., Müller, A. and Keller, R. (1991). Crustacean cardioactive peptide in the nervous system of the locust, *Locusta migratoria*: an immunocytochemical study on the ventral nerve cord and peripheral innervation. *Cell Tissue Res.* **263**, 439–457.

Dirección General de Sanidad Vegetal (1998). Acciones de la Campaña contra la Langosta en México. Dirección General de Sanidad Vegetal, Mexico. [Ex: Hernandez-Velazquez, V. M. *et al.* (2000). *J. Orthop. Res.* **9**, 223-227]

Dirsh, V. M. (1953). Morphometrical studies on phases of the desert locust (*Schistocerca gregaria* Forskål). *Anti-Locust Bull.* **16**, 1–34.

Dong, Y. and Friedrich, M. (2005). Nymphal RNAi: systemic RNAi mediated gene knockdown in juvenile grasshopper. *BMC Biotechnol.* **5**, 25.

Donini, A., Agricola, H.-J. and Lange, A. B. (2001). Crustacean cardioactive peptide is a modulator of oviduct contractions in *Locusta migratoria*. *J. Insect Physiol.* **47**, 277–285.

Donini, A., Ngo, C. and Lange, A. B. (2002). Evidence for crustacean cardioactive peptide-like innervation of the gut in *Locusta migratoria*. *Peptides* **23**, 1915–1923.

Donson, J., Fang, Y., Espiritu-Santo, G., Xing, W., Salazar, A., Miyamoto, S., Armendarez, V. and Volkmuth, W. (2002). Comprehensive gene expression analysis by transcript profiling. *Plant Mol. Biol.* **48**, 75–97.

Dopman, E. B., Sword, G. A. and Hillis, D. M. (2002). The importance of the ontogenetic niche in resource-associated divergence: evidence from a generalist grasshopper. *Evolution* **56**, 731–740.

Dorn, A., Schneider, M., Botens, F. F. W., Holtmann, M. and Petzak, I. (1997). Field application of the juvenile hormone analogue fenoxycarb against hopper bands of *Locusta migratoria capito* in Madagascar. In: *New Strategies in Locust Control* (eds Krall, S., Peveling, R. and Ba Diallo, D.), pp. 143–150. Basel: Birkhäuser Verlag.

Dorn, A., Ress, C., Sickold, S. and Wedekind-Hirschberger, S. (2000). Arthropoda – Insecta: endocrine control of phase polymorphism. In: *Progress in Developmental Endocrinology,* Vol. 10, Part B (ed. Dorn, A.), pp. 205–253. Chichester: Wiley.

Dumser, J. B. (1980). In vitro effects of ecdysterone on the spermatogonial cell cycle in *Locusta*. *Int. J. Invertebr. Rep.* **2**, 165–174.

Duranton, J.-F. and Lecoq, M. (1990). *Le Criquet Pèlerin au Sahel*. Niamey: CIRAD-PRIFAS/CILLS-DFPV.

Edelstein-Keshet, L., Watmough, J. and Grunbaum, D. (1998). Do travelling band solutions describe cohesive swarms? An investigation for migratory locusts. *J. Math. Biol.* **36**, 515–549.

Elder, R. J. (1991). Laboratory studies of environmental factors affecting sexual maturation in *Nomadacris guttulosa* (Walker) (Orthoptera: Acrididae). *J. Aust. Entomol. Soc.* **30**, 169–181.

Elder, R. J. (1996). Morphometrics of field populations of *Austracris guttulosa* (Walker) (Orthoptera: Acrididae) in Australia. *Aust. J. Entomol.* **35**, 345–347.

Elder, R. J. (1997). Bionomics of *Austracris guttulosa* (Walker) (Orthoptera: Acrididae) during the 1970-75 outbreak in Queensland, Australia. *Aust. J. Entomol.* **36**, 57–67.

Elliot, S. L., Blanford, S. and Thomas, M. B. (2002). Host-pathogen interactions in a varying environment: temperature, behavioural fever and fitness. *Proc. R. Soc. Lond. B* **296**, 1599–1607.

Elliot, S. L., Blanford, S., Horton, C. M. and Thomas, M. B. (2003). Fever and phenotype: transgenerational effect of disease on desert locust phase state. *Ecol. Lett.* **6**, 830–836.

Elliot, S. L., Horton, C. M., Blanford, S. and Thomas, M. B. (2005). Impacts of fever on locust life-history traits: costs or benefits?. *Biol. Lett.* **1**, 181–184.

Ellis, P. E. (1951). The marching behaviour of hoppers of the African migratory locust (*Locusta migratoria migratorioides* R. & F.) in the laboratoty. *Anti-Locust Bull.* **7**, 1–48.

Ellis, P. E. (1953a). Social aggregation and gregarious behaviour in hoppers of *Locusta migratoria migratorioides* (R. & F.). *Behaviour* **5**, 225–258.

Ellis, P. E. (1953b). The gregarious behaviour of marching *Locusta migratoria migratorioides* (R. & F.) hoppers. *J. Exp. Biol.* **30**, 214–234.

Ellis, P. E. (1959). Learning and social aggregation in locust hoppers. *Anim. Behav.* **7**, 91–106.

Ellis, P. E. (1962). The behavior of locusts in relation to phases and species. In: *Physiologie, Comportement et Écologie des Acridiens en Rapport avec la Phase, Paris, 1962*, Colloq. Internat. Centre National Rech. Sci., No. 114, pp. 123–143. Paris: Centre National de la Recherche Scientifique.

Ellis, P. E. (1963a). The influence of some environmental factors on learning and aggregation in locust hoppers. *Anim. Behav.* **11**, 142–151.

Ellis, P. E. (1963b). Changes in the social aggregation of locust hoppers with changes in rearing conditions. *Anim. Behav.* **11**, 152–160.

Ellis, P. E. (1964a). Marching and colour in locust hoppers in relation to social factors. *Behaviour* **23**, 177–192.

Ellis, P. E. (1964b). Changes in marching of locusts with rearing conditions. *Behaviour* **23**, 193–202.

Ellis, P. E. and Ashall, C. (1957). Field studies on diurnal behaviour, movement and aggregation in the desert locust (*Schistocerca gregaria* Forskål). *Anti-Locust Bull.* **25**, 1–94.

Ellis, P. E. and Carlisle, D. B. (1961). The prothoracic gland and colour change in locusts. *Nature* **190**, 368–369.

Ellis, P. E. and Pearce, A. (1962). Innate and learned behaviour patterns that lead to group formation in locust hoppers. *Anim. Behav.* **10**, 305–318.

Ellis, P. E., Carlisle, D. B. and Osborne, D. J. (1965). Desert locusts: sexual maturation delayed by feeding on senescent vegetation. *Science* **149**, 546–547.

Emlen, D. J. and Nijhout, H. F. (2000). The development and evolution of exaggerated morphologies in insects. *Annu. Rev. Entomol.* **45**, 661–708.

Enard, W., Khaitovich, P., Klose, J., Zöllner, S., Heissig, F., Giavalisco, P., Nieselt-Struwe, K., Muchmore, E., Varki, A., Ravid, R., Doxiadis, G. M., Bontrop, R. E. and Pääbo, S. (2002). Intra- and interspecific variation in primate gene expression patterns. *Science* **296**, 340–343.

Endler, J. A. and Mappes, J. (2004). Predator mixes and the conspicuousness of aposematic signals. *Am. Nat.* **163**, 532–547.

Ergene, S. (1954). Weitere Untersuchungen über Schwarzanpassung bei Imagines von Heuschrecken. *Mitt. Zool. Mus. Berlin* **30**, 107–110.

Evans, J. D. and Wheeler, D. E. (2001). Gene expression and the evolution of insect polyphenisms. *Bioessays* **23**, 62–68.

Evdokimov, N., Temirgaliev, Z. and Dublyazhova, M. (1999). The problem of plant protection from locusts. *Zaschita Rastenii v Kazakhstane (Plant Protection in Kazakhstan)* (1), 21–25, [in Russian].

Fain-Maurel, M.-A. and Cassier, P. (1969). Étude infrastructurale des glandes de mue de *Locusta migratoria migratorioides* (R. et F.) II. – Analyse morphologique des étapes de la dégénérescence chez les imagos grégaires. *Arch. Zool. Exp. Gen.* **110**, 91–126, + 8 plates.

FAO. (2006). *Report of the Twenty-Fifth Session of the FAO Commission for Controlling the Desert Locust in the Central Region*. Rome: Food and Agricultural Organization of the United Nations.

Farrow, R. A. (1975). The African migratory locust in its main outbreak area of the middle Niger: quantitative studies of solitary populations in relation to environmental factors. *Locusta (Bulletin non périodique)* (11), 1–198.

Farrow, R. A. (1977a). Origin and decline of the 1973 plague locust outbreak in central western New South Wales. *Aust. J. Zool.* **25**, 455–489.

Farrow, R. A. (1977b). Maturation and fecundity of the spur-throated locust, *Austracris guttulosa* (Walker), in New South Wales during the 1974/75 plague. *J. Aust. Entomol. Soc.* **16**, 27–39.

Farrow, R. A. (1979). Population dynamics of the Australian plague locust, *Chortoicetes terminifera* (Walker), in central western New South Wales I. Reproduction and migration in relation to weather. *Aust. J. Zool.* **27**, 717–745.

Farrow, R. A. (1990). Flight and migration in acridoids. In: *Biology of Grasshoppers* (eds Chapman, R. F. and Joern, A.), pp. 227–314. New York: Wiley.

Farrow, R. A. and Colless, D. H. (1980). Analysis of the interrelationships of geographical races of *Locusta migratoria* (Linnaeus) (Orthoptera: Acrididae) by numerical taxonomy, with special reference to sub-speciation in the tropics and affinities of the Australian race. *Acrida* **9**, 77–99.

Farrow, R. A. and Longstaff, B. C. (1986). Comparison of the annual rates of increase of locusts in relation to the incidence of plagues. *Oikos* **46**, 207–222.

Faure, J. C. (1923). *The Life-History of the Brown Locust.* Pretoria: The Government Printing and Stationery Office.

Faure, J. C. (1932). The phases of locusts in South Africa. *Bull. Entomol. Res.* **23**, 293–405, + 38 plates.

Feder, M. E. and Mitchell-Olds, T. (2003). Evolutionary and ecological functional genomics. *Nat. Rev. Gen.* **4**, 649–655.

Ferenz, H.-J. and Aden, E. (1993). Ovarian control of juvenile hormone biosynthesis in Locusta migratoria (Orthoptera: Acrididae). *Entomol. Gen.* **18**, 9–17.

Ferenz, H.-J. and Diehl, I. (1983). Stimulation of juvenile hormone biosynthesis *in vitro* by locust allatotropin. *Z. Naturforsch. C* **38**, 856–858.

Ferenz, H.-J. and Seidelmann, K. (2003). Pheromones in relation to aggregation and reproduction in desert locusts. *Physiol. Entomol.* **28**, 11–18.

Ferenz, H.-J., Luber, K. and Wieting, J. (1994). Pheromones as a means of controlling migratory locusts. In: *New Trends in Locust Control* (eds Krall, S. and Wilps, H.), pp. 81–89. Eschborn, Germany: Deutsche Gesellschaft für Technische Zusammenarbeit (GTZ).

Fernlund, P. and Josefsson, L. (1972). Crustacean color-change hormone: amino acid sequence and chemical synthesis. *Science* **177**, 173–175.

Fielding, D. J. and DeFoliart, L. S. (2005). Density and temperature-dependent melanization of fifth-instar *Melanoplus sanguinipes*: interpopulation comparisons. *J. Orthop. Res.* **14**, 107–113.

Filshie, B. K., Day, M. F. and Mercer, E. H. (1975). Colour and colour change in the grasshopper, *Kosciuscola tristis*. *J. Insect Physiol.* **21**, 1763–1770.

Fire, A., Xu, S., Montgomery, M. K., Kostas, S. A., Driver, S. E. and Mello, C. C. (1998). Potent and specific genetic interference by double-stranded RNA in *Caenorhabditis elegans*. *Nature* **391**, 806–811.

Fitzpatrick, M., Ben-Shahar, Y., Smid, H., Vet, L., Robinson, G. and Sokolowski, M. (2005). Candidate genes for behavioural ecology. *Trends Ecol. Evol.* **20**, 96–104.

Fogal, W. H. (1968). Pigment development in male integument transplanted onto females in the desert locust, *Schistocerca gregaria*. *Entomol. Exp. Appl.* **11**, 475–476.

Fraenkel, G. and Hsiao, C. (1965). Bursicon, a hormone which mediates tanning of the cuticle in the adult fly and other insects. *J. Insect Physiol.* **11**, 513–556.

Franc, A., Rabesisoa, L., Luong-Skovmand, M. H. and Lecoq, M. (2005). Phase polymorphism in the red locust, *Nomadacris septemfasciata* (Orthoptera: Acrididae) in Madagascar. *Int. J. Trop. Insect Sci.* **25**, 182–189.

Franssens, V., Simonet, G., Breugelmans, B., Van Soest, S., Van Hoef, V. and Vanden Broeck, J. (2008). The role of hemocytes, serine protease inhibitors and pathogen-associated patterns in prophenoloxidase activation in the desert locust, *Schistocerca gregaria*. *Peptides* **29**, 235–241.

Fridman-Cohen, S. and Pener, M. P. (1980). Precocenes induce effect of juvenile hormone excess in *Locusta migratoria*. *Nature* **286**, 711–713.

Fuchs, E., Kutsch, W. and Ayali, A. (2003). Neural correlates to flight-related density-dependent phase characteristics in locusts. *J. Neurobiol.* **57**, 152–162.

Fuzeau-Braesch, S. (1966). Étude du noircissement cuticulaire non-exuvial chez *Locusta migratoria* L. *J. Insect Physiol.* **12**, 1363–1368.

Fuzeau-Braesch, S. (1968). Recherche de la xanthommatine chez deux insectes: *Locusta migratoria* L. and *Oedipoda coerulescens* L. *C. R. Hebd. Séanc. Acad. Sci. Paris, Série D* **267**, 2030–2033.

Fuzeau-Braesch, S. (1972). Pigments and color changes. *Annu. Rev. Entomol.* **17**, 403–424.

Fuzeau-Braesch, S. (1977). Comportement et taux de catécholamines: étude comparative des insectes grégaires, solitaires et traités au gaz carbonique chez *Locusta migratoria* (Insectes, Orthoptères). *C. R. Hebd. Séanc. Acad. Sci. Paris, Série D* **284**, 1361–1363.

Fuzeau-Braesch, S. (1985). Colour changes. In: *Comprehensive Insect Physiology and Pharmacology, Vol. 9, Behaviour* (eds Kerkut, G. A. and Gilbert, L. I.), pp. 549–589. Oxford: Pergamon Press.

Fuzeau-Braesch, S. and David, J.-C. (1978). Étude du taux d'octopamine chez *Locusta migratoria* (Insecte, Orthoptère): comparaison entre insectes grégaires, solitaires et traités au gaz carbonique. *C. R. Hebd. Séanc. Acad. Sci. Paris, Série D* **286**, 697–699.

Fuzeau-Braesch, S. and Nicolas, G. (1981). Effect of carbon dioxide on subsocial insects. *Comp. Biochem. Physiol. A* **68**, 289–297.

Fuzeau-Braesch, S., Coulon, J. F. and David, J.-C. (1979). Octopamine levels during the moult cycle and adult development in the migratory locust, *Locusta migratoria*. *Experientia* **35**, 1349–1350.

Fuzeau-Braesch, S., Nicolas, G., Baehr, J.-C. and Porcheron, P. (1982). A study of hormonal levels of the locust *Locusta migratoria cinerascens* artificially changed to the solitary state by a chronic CO_2 treatment of one minute per day. *Comp. Biochem. Physiol. A* **71**, 53–58.

Fuzeau-Braesch, S., Genin, E., Jullien, R., Knowles, E. and Papin, C. (1988). Composition and role of volatile substances in atmosphere surrounding two gregarious locusts, *Locusta migratoria* and *Schistocerca gregaria*. *J. Chem. Ecol.* **14**, 1023–1033.

Gäde, G. (1981). Activation of fat body glycogen phosphorylase in *Locusta migratoria* by corpus cardiacum extract and synthetic adipokinetic hormone. *J. Insect Physiol.* **27**, 155–161.

Gäde, G. (1990). The adipokinetic hormone/red pigment-concentrating hormone peptide family: structures, interrelationships and functions. *J. Insect Physiol.* **36**, 1–12.

Gäde, G. (1992). The hormonal integration of insect flight metabolism. *Zool. Jahrb. Physiol.* **96**, 211–225.

Gäde, G. and Auerswald, L. (2003). Mode of action of neuropeptides from the adipokinetic hormone family. *Gen. Comp. Endocrinol.* **132**, 10–20.

Gäde, G. and Marco, H. G. (2009). Peptides of the adipokinetic hormone/red pigment-concentrating hormone family with special emphasis on Caelifera: primary sequences and functional considerations contrasting grasshoppers and locusts. *Gen. Comp. Endocrinol.* **162**, 59–68.

Gäde, G., Hoffmann, K.-H. and Spring, J. H. (1997). Hormonal regulation in insects: facts, gaps and future directions. *Physiol. Rev.* **77**, 963–1032.

Gadot, M., Rafaeli, A. and Applebaum, S. W. (1987). Partial purification and characterization of locust allatotropin I. *Arch. Insect Biochem. Physiol.* **4**, 213–223.

Gande, A. R., Morgan, E. D. and Wilson, I. D. (1979). Ecdysteroid levels throughout the life cycle of the desert locust, *Schistocerca gregaria*. *J. Insect Physiol.* **25**, 669–675.

Gapparov, F. A. (2001). Locust problem in Uzbekistan. *Zaschita i Karantin Rastenii v Kazakhstane (Plant Protection and Quarantine in Kazakhstan)* (1), 22–25, [in Russian].

Gapparov, F. A. and Latchininsky, A. V. (2000). What are the consequences of ecosystem disruption on acridid diversity and abundance? In: *Grasshoppers and Grassland Health. Managing Grasshopper Outbreaks without Risking Environmental Disaster* (eds Lockwood, J., Latchininsky, A. and Sergeev, M.), pp. 31–59. Dordrecht, The Netherlands: Kluwer Academic Publishers.

Gardner, S. N. and Thomas, M. B. (2002). Costs and benefits of fighting infection in locusts. *Evol. Ecol. Res.* **4**, 109–131.
Genin, E., Jullien, R., Perez, F. and Fuzeau-Braesch, S. (1986). Cuticular hydrocarbons of gregarious and solitary locusts *Locusta migratoria cinerascens*. *J. Chem. Ecol.* **12**, 1213–1238.
Génin, E., Jullien, R. and Fuzeau-Braesch, S. (1987). New natural aliphatic ethers in cuticular waxes of gregarious and solitary locusts *Locusta migratoria cinerascens (II)*. *J. Chem. Ecol.* **13**, 265–282.
Gilbert, L. I., Iatrou, K and Gill, S. S. (eds) (2005). *Comprehensive Molecular Insect Science*. Oxford: Elsevier Pergamon.
Gillett, S. D. (1973). The role of integumental colour pattern in locust grouping. *Anim. Behav.* **21**, 153–156.
Gillett, S. D. (1983). Primer pheromones and polymorphism in the desert locust. *Anim. Behav.* **31**, 221–230.
Gillett, S. D. (1988). Solitarization in the desert locust, *Schistocerca gregaria* (Forskål) (Orthoptera: Acrididae). *Bull. Entomol. Res.* **78**, 623–631.
Gillett, S. D. and Phillips, M. L. (1977). Faeces as a source of a locust gregarisation stimulus. Effects on social aggregation and on cuticular colour of nymphs of the desert locust, *Schistocerca gregaria* (Forsk.). *Acrida* **6**, 279–286.
Gillott, C. and Ismail, P. M. (1995). *In vitro* synthesis of ecdysteroid by the male accessory reproductive glands, testis and abdominal integument of the adult migratory grasshopper, *Melanoplus sanguinipes*. *Invertebr. Rep. Dev.* **27**, 65–71.
Girardie, A. and Cazal, M. (1965). Rôle de la pars intercerebralis et des corpora cardiaca sur la mélanisation chez *Locusta migratoria* (L.). *C. R. Hebd. Séanc. Acad. Sci. Paris, Groupe 12* **261**, 4525–4527.
Girardie, J. and Girardie, A. (1996). Lom OMP, a putative ecdysiotropic factor for the ovary in *Locusta migratoria*. *J. Insect Physiol.* **42**, 215–221.
Girardie, J., Boureme, D., Couillaud, F., Tamarelle, M. and Girardie, A. (1987). Anti-juvenile effect of neuroparsin A, a neuroprotein isolated from locust corpora cardiaca. *Insect Biochem.* **17**, 977–983.
Girardie, J., Girardie, A., Huet, J.-C. and Pernollet, J.-C. (1989). Amino acid sequence of locust neuroparsins. *FEBS Lett.* **245**, 4–8.
Girardie, J., Richard, O., Huet, J.-C., Nespoulous, C., Van Dorsselaer, A. and Pernollet, J.-C. (1991). Physical characterization and sequence identification of the ovary maturating parsin – A new neurohormone purified from the nervous corpora cardiaca of the African locust (*Locusta migratoria migratorioides*). *Eur. J. Biochem.* **202**, 1121–1126.
Girardie, J., Richard, O. and Girardie, A. (1992). Time-dependent variations in the activity of a novel ovary maturating neurohormone from the nervous corpora cardiaca during oögenesis in the locust, *Locusta migratoria migratorioides*. *J. Insect Physiol.* **38**, 215–221.
Girardie, J., Richard, O. and Girardie, A. (1996). Detection of vitellogenin in the haemolymph of larval female locusts (*Locusta migratoria*) treated with the neurohormone, Lom OMP. *J. Insect Physiol.* **42**, 107–113.
Girardie, J., Geoffre, S., Delbecque, J.-P. and Girardie, A. (1998a). Arguments for two distinct gonadotropic activities triggered by different domains of the ovary maturating parsin of *Locusta migratoria*. *J. Insect Physiol.* **44**, 1063–1071.
Girardie, J., Huet, J.-C., Atay-Kadiri, Z., Ettaouil, S., Delbecque, J.-P., Fournier, B., Pernollet, J.-C. and Girardie, A. (1998b). Isolation, sequence determination, physical and physiological characterization of the neuroparsins and ovary maturing parsins of *Schistocerca gregaria*. *Insect Biochem. Mol. Biol.* **28**, 641–650.

Goldsworthy, G. J. (1983). The endocrine control of flight metabolism in locusts. *Adv. Insect Physiol.* **17**, 149–204.

Goldsworthy, G. J. (1990). Hormonal control of flight metabolism in locusts. In: *Biology of Grasshoppers* (eds Chapman, R. F. and Joern, A.), pp. 205–225. New York: Wiley.

Goldsworthy, G. and Mordue, W. (1989). Adipokinetic hormones: functions and structures. *Biol. Bull.* **177**, 218–224.

Goldsworthy, G. J., Lee, M. J., Luswata, R., Drake, A. F. and Hyde, D. (1997). Structures, assays and receptors for locust adipokinetic hormones. *Comp. Biochem. Physiol. B* **117**, 483–496.

Goltzené, F., Lagueux, M., Charlet, M. and Hoffmann, J. A. (1978). The follicle cell epithelium of maturing ovaries of *Locusta migratoria*: a new biosynthetic tissue for ecdysone. *Hoppe-Seyler's Z. Physiol. Chem.* **359**, 1427–1434.

Goodman, W. G. and Granger, N. A. (2005). The juvenile hormones. In: *Comprehensive Molecular Insect Science, Vol. 3, Endocrinology* (eds Gilbert, L. I., Iatrou, K. and Gill, S. S.), pp. 319–408. Oxford: Elsevier Pergamon.

Goodwin, T. W. (1949). The biochemistry of locusts 2. Carotenoid distribution in solitary and gregarious phases of the African migratory locust (*Locusta migratoria migratorioides* R. & F.) and the desert locust (*Schistocerca gregaria* Forsk). *Biochem. J.* **45**, 472–479.

Goodwin, T. W. (1950). Biochemistry of locusts 4. Insectorubin metabolism in the desert locust (*Schistocerca gregaria* Forsk.) and the African migratory locust (*Locusta migratoria migratorioides* R. and F.). *Biochem. J.* **47**, 554–562.

Goodwin, T. W. (1952). The biochemistry of locust pigmentation. *Biol. Rev.* **27**, 439–460.

Goodwin, T. W. and Srisukh, S. (1949). The biochemistry of locusts 1. The carotenoids of the integument of two locust species (*Locusta migratoria migratorioides* R. & F. and *Schistocerca gregaria* Forsk.). *Biochem. J.* **45**, 263–268.

Goodwin, T. W. and Srisukh, S. (1950). Biochemistry of locusts 3. Insectorubin: the redox pigment present in the integument and eyes of the desert locust (*Schistocerca gregaria* Forsk.), the African migratory locust (*Locusta migratoria migratorioides* R. and F.) and other insects. *Biochem. J.* **47**, 549–554.

Goodwin, T. W. and Srisukh, S. (1951a). Biochemistry of locusts 5. The green pigment of the haemolymph and integument of solitary locusts (*Locusta migratoria migratorioides*, R. & F., and *Schistocerca gregaria*, Forsk.). *Biochem. J.* **48**, 199–203.

Goodwin, T. W. and Srisukh, S. (1951b). Biochemistry of locusts 6. The occurrence of a flavin in the eggs and of a pterin in the eyes of the African migratory locust (*Locusta migratoria migratorioides* R. & F.) and the desert locust (*Schistocerca gregaria* Forsk.). *Biochem. J.* **49**, 84–86.

Goosey, M. W. and Candy, D. J. (1980). The D-octopamine content of the haemolymph of the locust, *Schistocerca americana gregaria* and its elevation during flight. *Insect Biochem.* **10**, 393–397.

Gordh, G. and Headrick, D. (2001). *A Dictionary of Entomology*. Wallingford, UK: CABI Publishing.

Grach, C., Wang, Y., Barda, Y., Gilon, C. and Pener, M. P. (2003). In vivo structure-activity studies on the dark-color-inducing neurohormone of locusts. *J. Peptide Res.* **62**, 37–44. erratum republished: **62,** 134–142.

Grach, C., Yerushalmi, Y., Inlow, M. and Pener, M. P. (2004). The dark-colour-inducing neurohormone of locusts: strain-dependent and phase-independent effects on adults of the migratory locust, *Locusta migratoria*. *Physiol. Entomol.* **29**, 97–107.

Gray, L. J., Sword, G. A., Anstey, M. L., Clissold, F. J. and Simpson, S. J. (2009). Behavioural phase polyphenism in the Australian plague locust (*Chortoicetes terminifera*). *Biol. Lett.* **5**, 306–309.

Greenfield, M. D. and Pener, M. P. (1992). Alternative schedules of male reproductive diapause in the grasshopper *Anacridium aegyptium* (L.): effects of the corpora allata on sexual behavior (Orthoptera: Acrididae). *J. Insect Behav.* **5**, 245–261.

Greenwood, M. and Chapman, R. F. (1984). Differences in numbers of sensilla on the antennae of solitarious and gregarious *Locusta migratoria* L. (Orthoptera: Acrididae). *Int. J. Insect Morph. Embryol.* **13**, 295–301.

Gregg, P. C., Roberts, B. and Wentworth, S. L. (1987). Levels of ecdysteroids in diapause and non-diapause eggs of the Australian plague locust, *Chortoicetes terminifera* (Walker). *J. Insect Physiol.* **33**, 237–242.

Guilford, T. (1990). The evolution of aposematism. In: *Insect Defenses: Adaptive Mechanisms and Strategies of Prey and Predators* (eds Evans, D. L. and Schmidt, J. O.), pp. 23–61. Albany: State University of New York Press.

Gu, J. and Gu, X. (2003). Induced gene expression in human brain after the split from chimpanzee. *Trends Genet.* **19**, 63–65.

Gunn, D. L. and Hunter-Jones, P. (1952). Laboratory experiments on phase differences in locusts. *Anti-Locust Bull.* **12**, 1–29.

Gwynne, D. T. (2001). *Katydids and Bush-Crickets.* Ithaca: Cornell University Press.

Hägele, B. F. and Simpson, S. J. (2000). The influence of mechanical, visual and contact chemical stimulation on the behavioural phase state of solitarious desert locusts (*Schistocerca gregaria*). *J. Insect Physiol.* **46**, 1295–1301.

Hägele, B. F., Oag, V., Bouaïchi, A., McCaffery, A. R. and Simpson, S. J. (2000). The role of female accessory glands in maternal inheritance of phase in the desert locust *Schistocerca gregaria. J. Insect Physiol.* **46**, 275–280.

Hägele, B. F., Wang, F.-H., Sehnal, F. and Simpson, S. J. (2004). Effects of crowding, isolation, and transfer from isolation to crowding on total ecdysteroid content of eggs in *Schistocerca gregaria. J. Insect Physiol.* **50**, 621–628.

Hakamori, T. and Tanaka, S. (1992). Genetic control of diapause and other developmental traits in Japanese strains of the migratory locust, *Locusta migratoria* L.: univoltine vs. bivoltine. *Jap. J. Entomol.* **60**, 319–328.

Hamdaoui, A., Wataleb, S., Devreese, B., Chiou, S., Vanden Broeck, J., Van Beeumen, J., De Loof, A. and Schoofs, L. (1998). Purification and characterization of a group of five novel peptide serine protease inhibitors from ovaries of the desert locust, *Schistocerca gregaria. FEBS Lett.* **422**, 74–78.

Hansson, B. S. and Anton, S. (2000). Function and morphology of the antennal lobe: New developments. *Annu. Rev Entomol.* **45**, 203–231.

Hansson, B. S., Ochieng', S. A., Grosmaitre, X., Anton, S. and Njagi, P. G. N. (1996). Physiological responses and central nervous projections of antennal olfactory receptor neurons in the adult desert locust, *Schistocerca gregaria* (Orthoptera: Acrididae). *J. Comp. Physiol. A* **179**, 157–167.

Hardie, J. and Lees, A. D. (1985). Endocrine control of polymorphism and polyphenism. In: *Comprehensive Insect Physiology Biochemistry and Pharmacology, Vol. 8, Endocrinology II* (eds Kerkut, G. A. and Gilbert, L. I.), pp. 441–490. Oxford: Pergamon Press.

Harmsen, R. (1966). The excretory role of pteridines in insects. *J. Exp. Biol.* **45**, 1–13.

Hartenstein, V. (2005). Development of insect sensilla. In: *Comprehensive Molecular Insect Science, Vol. 1, Reproduction and Development* (eds Gilbert, L. I., Iatrou, K. and Gill, S. S.), pp. 379–419. Oxford: Elsevier Pergamon.

Hartfelder, K. (2000). Arhtropoda-Insecta: caste differentiation. In: *Progress in Developmental Endocrinology, Vol. 10, Part B* (ed. Dorn, A.), pp. 185–204. Chichester: Wiley.

Hartfelder, K. and Emlen, D. J. (2005). Endocrine control of insect polyphenism. In: *Comprehensive Molecular Insect Science, Vol. 3, Endocrinology* (eds Gilbert, L. I., Iatrou, K. and Gill, S. S.), pp. 651–703. Oxford: Elsevier Pergamon.

Harvey, A. (1990). *Rearing and Breeding of Locusts in the Laboratory*. Chatham Maritime, UK: Natural Resources Institute.

Harvey, A. W. (1981). A reclassification of the *Schistocerca americana* complex (Orthoptera: Acrididae). *Acrida* **10**, 61–77.

Harvey, A. W. (1983). *Schistocerca piceifrons* (Walker) (Orthoptera: Acrididae), the swarming locust of tropical America: a review. *Bull. Entomol. Res.* **73**, 171–184.

Hasegawa, E. and Tanaka, S. (1994). Genetic control of albinism and the role of juvenile hormone in pigmentation in *Locusta migratoria* (Orthoptera, Acrididae). *Jap. J. Entomol.* **62**, 315–324.

Haskell, P. T. (1960). The sensory equipment of the migratory locust. *Symp. Zool. Soc. Lond.* **3**, 1–23.

Hassanali, A. and Torto, B. (1999). Grasshoppers and locusts. In: *Pheromones of Non-Lepidopteran Insects Associated with Agricultural Plants* (eds Hardie, J. and Minks, A. K.), pp. 305–328. Wallingford, UK: CABI Publishing.

Hassanali, A., Njagi, P. G. N. and Bashir, M. O. (2005a). Chemical ecology of locusts and related acridids. *Annu. Rev. Entomol.* **50**, 223–245.

Hassanali, A., Bashir, M. O., Njagi, P. G. N. and Ould Ely, S. (2005b). Desert locust gregarization: a conceptual kinetic model. *J. Orthop. Res.* **14**, 223–226.

Hatle, J. D., Juliano, S. A. and Borst, D. W. (2003). Hemolymph ecdysteroids do not affect vitellogenesis in the lubber grasshopper. *Arch. Insect Biochem. Physiol.* **52**, 45–57.

Heifetz, Y. and Applebaum, S. W. (1995). Density-dependent physiological phase in a non-migratory-grasshopper *Aiolopus thalassinus*. *Entomol. Exp. Appl.* **77**, 251–262.

Heifetz, Y., Applebaum, S. W. and Popov, G. B. (1994). Phase characteristics of the Israeli population of the migratory locust, *Locusta migratoria* (L.) (Orthoptera: Acrididae). *J. Orthop. Res.* **2**, 15–20.

Heifetz, Y., Voet, H. and Applebaum, S. W. (1996). Factors affecting behavioral phase transition in the desert locust, *Schistocerca gregaria* (Forskål) (Orthoptera: Acrididae). *J. Chem. Ecol.* **22**, 1717–1734.

Heifetz, Y., Boekhoff, I., Breer, H. and Applebaum, S. W. (1997). Cuticular hydrocarbons control behavioural phase transition in *Schistocerca gregaria* nymphs and elicit biochemical responses in antennae. *Insect Biochem. Mol. Biol.* **27**, 563–568.

Heifetz, Y., Miloslavski, I., Aizenshtat, Z. and Applebaum, S. W. (1998). Cuticular surface hydrocarbons of desert locust nymphs, *Schistocerca gregaria*, and their effect on phase behavior. *J. Chem. Ecol.* **6**, 1033–1047.

Helfert, B. (1977). Quantitative Untersuchungen über den Ommochromgehalt im Integument von *Oedipoda coerulescens* L. (Orthoptera, Acridiidae). *Zool. Anz.* **198**, 269–286.

Hetru, C., Li, K. W., Bulet, P., Lagueux, M. and Hoffmann, J. A. (1991). Isolation and structural characterization of an insulin-related molecule, a predominant neuropeptide from *Locusta migratoria*. *Eur. J. Biochem.* **201**, 495–499.

Hietter, H., Van Dorsselaer, A., Green, B., Denoroy, L., Hoffmann, J. and Luu, B. (1990). Isolation and structure elucidation of a novel 5-kDa peptide from neurohaemal lobes of the *corpora cardiaca* of *Locusta migratoria* (Insecta, Orthoptera). *Eur. J. Biochem.* **187**, 241–247.

Hirn, M., Hetru, C., Lagueux, M. and Hoffmann, J. A. (1979). Prothoracic gland activity and blood titres of ecdysone and ecdysterone during the last larval instar of *L. migratoria* L. *J. Insect Physiol.* **25**, 255–261.

Hoffmann, J. A. and Lagueux, M. (1985). Endocrine aspects of embryonic development in insects. In: *Comprehensive Insect Physiology Biochemistry and Pharmacology, Vol. 1, Embryogenesis and Reproduction* (eds Kerkut, G. A. and Gilbert, L. I.), pp. 435–460. Oxford: Pergamon Press.

Hoffmann, J. A. and Weins, M.-J. (1974). Activité protéosynthétique des glandes prothoraciques en titre d'ecdysone chez des larves permanentes de *Locusta migratoria* obtenues par irradiation sélective du tissu hématopoïétique. *Experientia* **30**, 821–822.

Holt, J. and Cheke, R. A. (1996). Models of desert locust phase changes. *Ecol. Model.* **91**, 131–137.

Honegger, H.-W., Market, D., Pierce, L. A., Dewey, E. M., Kostron, B., Wilson, M., Choi, D., Klukas, K. A. and Mesce, K. A. (2002). Cellular localization of bursicon using antisera against partial peptide sequences of this insect cuticle-sclerotizing neurohormone. *J. Comp. Neurol.* **452**, 163–177.

Hopkins, T. L. and Ahmad, S. A. (1991). Flavonoid wing pigments in grasshoppers. *Experientia* **47**, 1089–1091.

Hoste, B., Simpson, S. J., Tanaka, S., De Loof, A. and Breuer, M. (2002a). A comparison of phase-related shifts in behavior and morphometrics of an albino strain, deficient in [His^7]-corazonin, and a normally colored *Locusta migratoria* strain. *J. Insect Physiol.* **48**, 791–801.

Hoste, B., Simpson, S. J., Tanaka, S., Zhu, D.-H., De Loof, A. and Breuer, M. (2002b). Effects of [His^7]-corazonin on the phase state of isolated-reared (solitarious) desert locusts, *Schistocerca gregaria*. *J. Insect Physiol.* **48**, 981–990.

Hoste, B., Luyten, L., Claeys, I., Clynen, E., Rahman, M. M., De Loof, A. and Breuer, M. (2002c). An improved breeding method for solitarious locusts. *Entomol. Exp. Appl.* **104**, 281–288.

Hoste, B., Simpson, S. J., De Loof, A. and Breuer, M. (2003). Behavioural differences in *Locusta migratoria* associated with albinism and their relation to [His^7]-corazonin. *Physiol. Entomol.* **28**, 32–38.

Hoste, B., Sas, F., Vandersmissen, T., De Loof, A., Breuer, M. and Huybrechts, J. (2006). A miniaturized assay to quantify effects of chemicals or physical stimuli upon locust activity. *Insect Sci.* **13**, 401–407.

Howard, R. W. and Blomquist, G. J. (2005). Ecological, behavioral, and biochemical aspects of insect hydrocarbons. *Annu. Rev. Entomol.* **50**, 371–393.

Hughes, T. D. (1980a). The imaginal ecdysis of the desert locust, *Schistocerca gregaria*. I. A description of the behaviour. *Physiol. Entomol.* **5**, 47–54.

Hughes, T. D. (1980b). The imaginal ecdysis of the desert locust, *Schistocerca gregaria*. II. Motor activity underlying the pre-emergence and emergence behaviour. *Physiol. Entomol.* **5**, 55–71.

Hughes, T. D. (1980c). The imaginal ecdysis of the desert locust, *Schistocerca gregaria*. III. Motor activity underlying the expansional and post-expansional behaviour. *Physiol. Entomol.* **5**, 141–152.

Hughes, T. D. (1980d). The imaginal ecdysis of the desert locust, *Schistocerca gregaria*. IV. The role of the gut. *Physiol. Entomol.* **5**, 153–164.

Huibregtse-Minderhoud, L., Van den Hondel-Franken, M. A. M., Van der Kerk-Van Hoof, A. C., Biessels, H. W. A., Salemink, C. A., Van der Horst, D. J. and Beenakkers, A. M. Th. (1980). Quantitative determination of the juvenile hormones in the haemolymph of *Locusta migratoria* during normal development and after implantation of corpora allata. *J. Insect Physiol.* **26**, 627–631.

Hunter, D. M. (1989). The response of Mitchell grasses (*Astrebla* spp.) and Button grass (*Dactyloctenium radulans* (R. Br.)) to rainfall and their importance to the survival of the Australian plague locust, *Chortoicetes terminifera* (Walker), in the arid zone. *Aust. J. Ecol.* **14**, 467–471.

Hunter, D. M. (1997). Vitellogenesis in *Austracris guttulosa* (Walker) (Orthoptera: Acrididae) in tropical Australia. *Aust. J. Entomol.* **36**, 307–308.

Hunter, D. M. (2004). Advances in the control of locusts (Orthoptera: Acrididae) in eastern Australia: from crop protection to preventive control. *Aust. J. Entomol.* **43**, 293–303.

Hunter, D. M. (2005). Mycopesticides as part of integrated pest management of locusts and grasshoppers. *J. Orthop. Res.* **14**, 197–801.

Hunter, D. M. and Cosenzo, E. L. (1990). The origin of plagues and recent outbreaks of the South American locust, *Schistocerca cancellata* (Orthoptera: Acrididae) in Argentina. *Bull. Entomol. Res.* **80**, 295–300.

Hunter, D. M. and Elder, R. J. (1999). Rainfall sequences leading to population increases of *Austracris guttulosa* in arid north-eastern Australia. *Aust. J. Entomol.* **38**, 204–218.

Hunter, D. M., Walker, P. W. and Elder, R. J. (2001). Adaptations of locusts and grasshoppers to the low and variable rainfall of Australia. *J. Orthop. Res.* **10**, 347–351.

Hunter, D. M., McCulloch, L. and Spurgin, P. A. (2008). Aerial detection of nymphal bands of the Australian plague locust (*Chortoicetes terminifera* (Walker)) (Orthoptera: Acrididae). *Crop Protection* **27**, 118–123.

Hunter-Jones, P. (1957). An albino strain of the desert locust. *Nature* **180**, 236–237.

Hunter-Jones, P. (1958). Laboratory studies on the inheritance of phase characters in locusts. *Anti-Locust Bull.* **29**, 1–32.

Hunter-Jones, P. (1961). *Rearing and Breeding Locusts in the Laboratory*. London: Anti-Locust Research Centre.

Hunter-Jones, P. (1962). Coloration of the desert locust (*Schistocerca gregaria* Forskål) reared in isolation. *Entomol. Mon. Mag.* **98**, 89–92.

Hunter-Jones, P. (1967). Life history of the Central American locust, *Schistocerca* sp. (Orthoptera: Acrididae), in the laboratory. *Ann. Entomol. Soc. Am.* **60**, 468–477.

Ibrahim, K. M. (2001). Plague dynamics and population genetics of the desert locust: can turnover during recession maintain population genetic structure? *Mol. Ecol.* **10**, 581–591.

Ibrahim, K. M., Sourrouille, P. and Hewitt, G. M. (2000). Are recession populations of the desert locust (*Schistocerca gregaria*) remnants of past swarms? *Mol. Ecol.* **9**, 783–791.

Ignell, R., Anton, S. and Hansson, B. S. (1998). Central nervous processing of behaviourally relevant odours in solitary and gregarious fifth instar locusts, *Schistocerca gregaria*. *J. Comp. Physiol. A* **183**, 453–465.

Ignell, R., Anton, S. and Hansson, B. S. (1999). Integration of behaviourally relevant odours at the central nervous level in solitary and gregarious third instar locusts, *Schistocerca gregaria*. *J. Insect Physiol.* **45**, 993–1000.

Ignell, R., Couillaud, F. and Anton, S. (2001). Juvenile-hormone-mediated plasticity of aggregation behaviour and olfactory processing in adult desert locusts. *J. Exp. Biol.* **204**, 249–259.

Inayatullah, C., El Bashir, S. and Hassanali, A. (1994). Sexual behavior and communication in the desert locust, *Schistocerca gregaria* (Orthoptera: Acrididae): sex pheromone in solitaria. *Environ. Entomol.* **23**, 1544–1551.

Ingram, K. K., Oefner, P. and Gordon, D. M. (2005). Task-specific expression of the *foraging* gene in harvester ants. *Mol. Ecol.* **14**, 813–818.

Injeyan, H. S. and Tobe, S. S. (1981a). Phase polymorphism in *Schistocerca gregaria*: reproductive parameters. *J. Insect Physiol.* **27**, 97–102.

Injeyan, H. S. and Tobe, S. S. (1981b). Phase polymorphism in *Schistocerca gregaria*: assessment of juvenile hormone synthesis in relation to vitellogenesis. *J. Insect Physiol.* **27**, 203–210.

Injeyan, H. S., Tobe, S. S. and Rapport, E. (1979). The effects of exogenous juvenile hormone treatment on embryogenesis in *Schistocerca gregaria*. *Can. J. Zool.* **57**, 838–845.

Isaac, R. E., Schoofs, L., Williams, T. A., Veelaert, D., Sajid, M., Corvol, P. and Coates, D. (1998). A novel peptide-processing activity of insect peptidyl-dipeptidase A (angiotensin I-converting enzyme): the hydrolysis of lysil-arginine and argynil-arginine from the C-terminus of an insect prohormone peptide. *Biochem. J.* **330**, 61–65.

Isaac, R. E., Bland, N. D. and Shirras, A. D. (2009). Neuropeptidases and the metabolic inactivation of insect neuropeptides. *Gen. Comp. Endocrinol.* **162**, 8–17.

Islam, M. S. (1995). Endocrine manipulations in crowd-reared desert locust *Schistocerca gregaria*. I. Effects of juvenile hormone analogues on morphology and reproductive parameters. *Pakistan J. Zool.* **27**, 301–310.

Islam, M. S. (1997). A preliminary report on the biochemical properties of a maternally-produced gregarizing factor in the desert locust, *Schistocerca gregaria* (Forskal). *J. Asiat. Soc. Bangladesh, Sci.* **23**, 111–122.

Islam, M. S., Roessingh, P., Simpson, S. J. and McCaffery, A. R. (1994a). Parental effects on the behaviour and colouration of nymphs of the desert locust *Schistocerca gregaria*. *J. Insect Physiol.* **40**, 173–181.

Islam, M. S., Roessingh, P., Simpson, S. J. and McCaffery, A. R. (1994b). Effects of population density experienced by parents during mating and oviposition on the phase of hatchling desert locusts, *Schistocerca gregaria*. *Proc. R. Soc. Lond. B* **257**, 93–98.

Ismail, S., Porcheron, P. and Fuzeau-Braesch, S. (1979). Taux des ecdystéroïdes chez des adultes de *Locusta migratoria cinerascens* solitaires, grégaires et traités par le gaz carbonique. *C. R. Séanc. Soc. Biol.* **173**, 553–557.

Jackson, G. J., Popov, G. B., Ibrahim, A. O., Alghamedi, S. A. and Khan, A. M. (1978). *Effects of Natural Food Plants on the Development, Maturation, Fecundity and Phase of the Desert Locust,* Schistocerca gregaria *(Forskål)* London: Centre for Overseas Pest Research.

Janssen, I., Koolman, J., Konopinska, D., Bartosz-Bechowski, H., Schoofs, L. and De Loof, A. (1998). Biological activity of structural analogs and effect of oil as a carrier of trypsin modulating oostatic factor of the gray fleshfly *Neobellieria bullata*. *Peptides* **19**, 627–634.

Janssen, T., Claeys, I., Simonet, G., De Loof, A., Girardie, J. and Vanden Broeck, J. (2001). cDNA cloning and transcript distribution of two different neuroparsin precursors in the desert locust, *Schistocerca gregaria*. *Insect Mol. Biol.* **10**, 183–189.

Ji, R., Xie, B.-Y., Li, D.-M., Li, Z. and Zhang, X. (2004a). Use of MODIS data to monitor the oriental migratory locust plague. *Agr. Ecosyst. Environ.* **104**, 615–620.

Ji, W., Zhou, W., Gregg, K., Yu, N., Davis, S. and Davis, S. (2004b). A method for cross-species gene expression analysis with high-density oligonucleotide arrays. *Nucleaic Acids Res.* **32**, e93.

Jiang, Q.-Y., Wang, W.-X., Zhang, Z. and Zhang, L. (2009). Binding specificity of locust odorant binding protein and its key binding site for initial recognition of alcohols. *Insect Biochem. Mol. Biol.* **39**, 440–447.

Joly, L. (1954). Résultats d'implantations systématiques de *corpora allata* à de jeunes larves de *Locusta migratoria* L. *C. R. Séanc. Soc. Biol. Paris* **148**, 579–583.

Joly, L. and Joly, P. (1974). Comparaison de la phase grégaire et de la phase solitaire de *Locusta migratoria migratorioides* (Orthoptère) du point de vue de la teneur de leur

hémolymphe en Hormone Juvénile. *C. R. Hebd. Séanc. Acad. Sci., Paris, Série D* **279**, 1007–1009.
Joly, L. and Schneider, M. (1976). Une propriété physiologique des glandes prothoraciques des Insectes non liée à la production d'ecdysone. *C. R. Hebd. Séanc. Acad. Sci., Paris, Série D* **282**, 1437–1439.
Joly, L., Joly, P. and Porte, A. (1969). Remarques sur l'ultrastructure de la glande ventrale de *Locusta migratoria* L. (Orthoptère) en population dense. *C. R. Hebd. Séanc. Acad. Sci., Paris, Série D* **269**, 917–918.
Joly, L., Hoffmann, J. and Joly, P. (1977a). Contrôle humoral de la différenciation phasaire chez *Locusta migratoria migratorioides* (R. & F.) (Orthoptères). *Acrida* **6**, 33–42.
Joly, L., Joly, P. and Lagueux, M. (1977b). Sur l'existence d'une seconde function physiologique de le glande prothoracique des Insectes. *C. R. Hebd. Séanc. Acad. Sci., Paris, Série D* **285**, 543–546.
Joly, P. (1952). Déterminisme de la pigmentation chez *Acrida turrita* L. (Insecte orthoptéroïde). *C. R. Hebd. Séanc. Acad. Sci., Paris* **235**, 1054–1056.
Joly, P. (1962). Role joué par les *corpora allata* dans la réalisation du polymorphisme de phase chez *Locusta migratoria* L. In: *Physiologie, Comportement et Écologie des Acridiens en Rapport avec la Phase, Paris, 1962*, Colloq. Internat. Centre National Rech. Sci., No. 114, pp. 77–88. Paris: Centre National de la Recherche Scientifique.
Joly, P. and Joly, L. (1954). Résultats de greffes de *corpora allata* chez *Locusta migratoria* L. *Ann. Sci. Natur. (Zool.)*, 11e Série, **15** (1953), 331–345.
Joly, P., Joly, L. and Halbwachs, M. (1956). Controle humoral du développement chez *Locusta migratoria. Ann. Sci. Natur. (Zool.)*, 11e Série, **18**, 257–261.
Jutsum, A. R. and Goldsworthy, G. J. (1976). Fuels for flight in *Locusta. J. Insect Physiol.* **22**, 243–249.
Kaatz, H.-H., Ferenz, H.-J., Langer, B. and Moritz, R. F. A. (2007). Isolation and characterization of nine polymorphic microsatellite loci from the desert locust, *Schistocerca gregaria. Mol. Ecol. Notes* **7**, 1042–1044.
Kandel, E. R. and Pittenger, C. (1999). The past, the future and the biology of memory storage. *Philos. Trans. R. Soc. Lond. B* **354**, 2027–2052.
Kang, L., Chen, X. Y., Zhou, Y., Liu, B. W., Zheng, W., Li, R. Q., Wang, J. and Yu, J. (2004). The analysis of large-scale gene expression correlated to the phase changes of the migratory locust. *Proc. Natl. Acad. Sci. U. S. A.* **101**, 17611–17615.
Kassahn, K. S., Caley, M. J., Ward, A. C., Connolly, A. R. and Crozier, R. H. (2007). Heterologous microarray experiments used to identify the early gene response to heat stress in a coral reef fish. *Mol. Ecol.* **16**, 1749–1763.
Kayser, H. (1985). Pigments. In: *Comprehensive Insect Physiology and Pharmacology, Vol. 10, Biochemistry* (eds Kerkut, G. A. and Gilbert, L. I.), pp. 367–415. Oxford: Pergamon Press.
Kayser, H. and Palivan, C. G. (2006). Stable free radicals in insect cuticules: electron spin resonance *spectroscopy* reveals differences between melanization and sclerotization. *Arch. Biochem. Biophys.* **453**, 179–187.
Kellenberger, C., Ferrat, G., Leone, P., Darbon, H. and Roussel, A. (2003). Selective inhibition of trypsins by insect peptides: role of P6-P10 loop. *Biochemistry* **42**, 13605–13612.
Kennedy, J. S. (1939). The behaviour of the desert locust (*Schistocerca gregaria* (Forsk.)) (Orthoptera) in an outbreak centre. *Trans. R. Entomol. Soc. Lond.* **89**, 385–542.
Kennedy, J. S. (1956). Phase transformation in locust biology. *Biol. Rev.* **31**, 349–370.
Kennedy, J. S. (1961). Continuous polymorphism in locusts. In: *Insect Polymorphism,* (ed. J. S. Kennedy); also as Royal Entomological Society London Symposium No. 1, pp. 80–90.

Kennedy, J. S. (1962). The division of labour between the phases of locusts. In: *Physiologie, Comportement et Écologie des Acridiens en Rapport avec la Phase, Paris, 1962,* Colloq. Internat. Centre National Rech. Sci., No. 114, pp. 283–297. Paris: Centre National de la Recherche Scientifique.

Kennedy, J. S. (1978). The concepts of olfactory 'arrestment' and 'attraction'. *Physiol. Entomol.* **3**, 91–98.

Kent, J. W., Jr, Teng, Y-M., Deshpande, D. and Rankin, M. A. (1997). Mobilization of lipid and carbohydrate reserves in the migratory grasshopper *Melanoplus sanguinipes*. *Physiol. Entomol.* **22**, 231–238.

Kevan, D. K. McE. (1989). Transatlantic travellers. *Antenna* **13**, 12–15.

Key, K. H. L. and Day, M. F. (1954a). A temperature-controlled physiological colour response in the grasshopper *Kosciuscola tristis* Sjöst. (Orthoptera: Acrididae). *Aust. J. Zool.* **2**, 309–339, + 2 Plates.

Key, K. H. L. and Day, M. F. (1954b). The physiological mechanism of colour change in the the grasshopper *Kosciuscola tristis* Sjöst. (Orthoptera: Acrididae). *Aust. J. Zool.* **2**, 340–363.

Key, K. H. L. and Rentz, D. C. F. (1994). On the scientific name of the Australian spur-throated locust (Orthoptera: Acrididae). *J. Aust. Entomol. Soc.* **33**, 345–346.

Khasenov, S. (2001). Locust problem in Kazakhstan. *Zaschita I Karantin Rastenii v Kazakhstane (Plant Protection and Quarantine in Kazakhstan)* (1), 2–6, [in Russian].

Kiessling, A. and Seidelmann, K. (2006). Regulations-Mechanismen der Pheromonabgabe bei Männchen der Wüstenheuschrecke, *Schistocerca gregaria. Mitt. Dtsch. Ges. Allg. Angew. Entomol.* **15**, 371–374.

Kim, Y.-J., Spalovská-Valachová, I., Cho, K.-H., Zitnanova, I., Park, Y., Adams, M. E. and Žitňan, D. (2004). Corazonin receptor signaling in ecdysis initiation. *Proc. Natl. Acad. Sci. U. S. A.* **101**, 6704–6709.

Kim, Y.-J., Žitňan, D., Cho, K.-H., Schooley, D. A., Mizoguchi, A. and Adams, M. E. (2006a). Central peptidergic ensembles associated with organization of an innate behavior. *Proc. Natl. Acad. Sci. U. S. A.* **103**, 14211–14216.

Kim, Y.-J., Žitňan, D., Galizia, C. G., Cho, K.-H. and Adams, M. E. (2006b). A command chemical triggers an innate behavior by sequential activation of multiple peptidergic ensembles. *Curr. Biol.* **16**, 1395–1407.

Klass, J. I., Blanford, S. and Thomas, M. B. (2007a). Development of a model for evaluating the effects of environmental temperature and thermal behaviour on biological control of locusts and grasshoppers using pathogens. *Agr. Forest Entomol.* **9**, 189–199.

Klass, J. I., Blanford, S. and Thomas, M. B. (2007b). Use of geographic information system to explore spatial variation in pathogen virulence and the implications for biological control of locusts and grasshoppers. *Agr. Forest Entomol.* **9**, 201–208.

Kostron, B., Marquardt, K., Kaltenhauser, U. and Honegger, H. W. (1995). Bursicon, the cuticle sclerotizing hormone-Comparison of its molecular mass in different insects. *J. Insect Physiol.* **41**, 1045–1053.

Kromer-Metzger, E. and Lagueux, M. (1994). Expression of the gene encoding an insulin-related peptide in *Locusta* (*Insecta*, *Orthoptera*). Evidence for alternative promoter usage. *Eur. J. Biochem.* **221**, 427–434.

Kucharski, R., Maleszka, J., Foret, S. and Maleszka, R. (2008). Nutritional control of reproductive status in honeybees via DNA methylation. *Science* **319**, 1827–1830.

Lafont, R., Dauphin-Villemant, C., Warren, J. T. and Rees, H. (2005). Ecdysteroid chemistry and biochemistry. In: *Comprehensive Molecular Insect Science, Vol. 3, Endocrinology* (eds Gilbert, L. I., Iatrou, K. and Gill, S. S.), pp. 125–195. Oxford: Elsevier Pergamon.

Lagueux, M., Hirn, M. and Hoffmann, J. A. (1977). Ecdysone during ovarian development in *Locusta migratoria*. *J. Insect Physiol.* **23**, 109–119.

Lagueux, M., Hetru, C., Goltzene, F., Kappler, C. and Hoffmann, J. A. (1979). Ecdysone titre and metabolism in relation to cuticulogenesis in the embryos of *Locusta migratoria*. *J. Insect Physiol.* **25**, 709–723.

Lagueux, M., Lwoff, L., Meister, M., Goltzené, F. and Hoffmann, J. E. (1990). cDNAs from neurosecretory cells of brains of *Locusta migratoria* (Insecta, Orthoptera) encoding a novel member of the superfamily of insulins. *Eur. J. Biochem.* **187**, 249–254.

Lange, A. B. (2009). Neural mechanisms coordinating the female reproductive system in the locust. *Frontiers Biosci.* **14**, 4401–4415.

Lange, A. B. and Cheung, I. L. (1999). The modulation of skeletal muscle contraction by FMRFamide-related peptides of the locust. *Peptides* **20**, 1411–1418.

Lange, A. B., Peeff, N. M. and Orchard, I. (1994). Isolation, sequence, and bioactivity of FMRFamide-related peptides from the locust ventral nerve cord. *Peptides* **15**, 1089–1094.

Latchininsky, A. V. (1998). Moroccan locust *Dociostaurus maroccanus* (Thunberg, 1815): a faunistic rarity or an important economic pest? *J. Insect Conserv.* **2**, 167–178.

Latchininsky, A. V. (2000). Emergency Program for the Control of Locust Outbreaks Kazakhstan. Consultant's Report for the Project TCP/KAZ/0065(E) of the FAO UN, 220 pp., Astana-Laramie, Wyoming.

Latchininsky, A. V. and Launois-Luong, M. H. (1992). *Le Criquet Marocain,* Dociostaurus maroccanus *(Thunberg, 1815), dans la Partie Orientale de son Aire de Distribution*. CIRAD-GERDAT-PRIFAS, Montpellier/Institut Pan Russe de la Protection des Plantes (VIZR), Saint Pétersbourg.

Latchininsky, A. V. and Launois-Luong, M. H. (1997). *Le Criquet Pèlerin* (Schistocerca gregaria *Forskål, 1775) dans la Partie Nord Orientale de son Aire d'Invasion.* CIRAD-GERDAT-PRIFAS, Montpellier/Institut Pan Russe de la Protection des Plantes (VIZR), Saint Pétersbourg.

Lauga, J. and Hatte, M. (1977). Propriétés grégarisantes acquises par le sable dans lequel ont pondu à de nombreuses reprises des femelles grégaires de *Locusta migratoria migratorioides*, R. & F. (Orthoptère: Acrididae). *Acrida* **6**, 307–311.

Lauga, J. and Hatte, M. (1978). L'activité grégarisante du sable de ponte chez *Locusta migratoria* L.: Action sur le comportement et la reproduction des individus. *Ann. Sci. Natur. Zool. Biol. Animale*, Série 12, **20**, 37–52.

Lea, A. (1962). The nature and significance of phase variation in the brown locust *Locustana pardalina* (Walk.). In: *Physiologie, Comportement et Écologie des Acridiens en Rapport avec la Phase, Paris, 1962*; Colloq. Internat. Centre National Rech. Sci., No. 114, pp. 241–258. Paris: Centre National de la Recherche Scientifique.

Lea, A. (1972). The plague dynamics of the brown locust, *Locustana pardalina* (Walk.). In: *Proceedings of the International Study Conference on the Current and Future Problems of Acridology, London, 1970* (eds Hemming, C. F. and Taylor, T. H. C.), pp. 289–297. London: Centre for Overseas Pest Research.

Lecoq, M. (1995). Forecasting systems for migrant pests. III. Locusts and grasshoppers in West Africa and Madagascar. In: *Insect Migration: Tracking Resources through Space and Time* (eds Drake, V. A. and Gatehouse, A. G.), pp. 377–395. Cambridge: Cambridge University Press.

Lecoq, M. (2000). How can acridid population ecology be used to refine pest management strategies? In: *Grasshoppers and Grassland Health* (eds Lockwood, J. A., Latchininsky, A. V. and Sergeev, M. G.), pp. 109–129. The Netherlands: Kluwer Academic Publishers.

Lecoq, M. (2005). Desert locust management from ecology to anthropology. *J. Orthop. Res.* **14**, 179–186.

Lecoq, M. and Pierozzi, I., Jr, (1996a). Comportement de vol des essaims de *Rhammatocerus schistocercoides* (Rehn, 1906) au Mato Grosso, Brésil (Orthoptera: Acrididae, Gomphocerinae). *Ann. Soc. Entomol. Fr. Nouvelle série* **33**, 265–283.

Lecoq, M. and Pierozzi, I., Jr, (1996b). Chromatic polymorphism and geophagy: two outstanding characteristics of *Rhammatocerus schistocercoides* (Rehn 1906) grasshoppers in Brazil [Orthoptera, Acrididae, Gomphocerinae]. *J. Orthop. Res.* **5**, 13–17.

Lecoq, M., Foucart, A. and Balança, G. (1999). Behaviour of *Rhammatocerus schistocercoides* (Rehn, 1906) hopper bands in Mato Grosso, Brazil (Orthoptera: Acrididae: Gomphocerinae). *Ann. Soc. Entomol. Fr. Nouvelle série* **35**, 217–228.

Lee, K. P. and Wilson, K. (2006). Melanism in a larval Lepidoptera: repeatability and heritability of a dynamic trait. *Ecol. Entomol.* **31**, 196–205.

Lee, K. P., Simpson, S. J., Clissold, F. J., Brooks, R., Ballard, J. W. O., Taylor, P. W., Soran, N. and Raubenheimer, D. (2008). Lifespan and reproduction in *Drosophila*: New insights from nutritional geometry. *Proc. Natl. Acad. Sci. U. S. A.* **105**, 2498–2503.

Lehmberg, E., Ferenz, H.-J. and Applebaum, S. W. (1992). Maturation and responsiveness to extracts of corpora allata from male *Locusta migratoria* containing allatotropic factors. *Z. Naturforsch. C* **47**, 449–452.

Lenz, E. M., Hägele, B. F., Wilson, I. D. and Simpson, S. J. (2001). High resolution ^{1}H NMR spectroscopic studies of the composition of the haemolymph of crowd- and solitary-reared nymphs of the desert locust, *Schistocerca gregaria*. *Insect Biochem. Mol. Biol.* **32**, 51–56.

Lester, R. L., Grach, C., Pener, M. P. and Simpson, S. J. (2005). Stimuli inducing gregarious colouration and behaviour in nymphs of *Schistocerca gregaria*. *J. Insect Physiol.* **51**, 737–747.

Levita, B. (1970). Étude de l'homochromie chez *Oedipoda coerulescens* L. (Acridien, Orthoptère). *Bull. Biol.* **104**, 149–213.

Li, S., Ouyang, Y. C., Ostrowski, E. and Borst, D. W. (2005). Allatotropin regulation of juvenile hormone synthesis by the corpora allata from the lubber grasshopper, *Romalea microptera*. *Peptides* **26**, 63–72.

Liang, Z., Sottrup-Jensen, L., Aspán, A., Hall, M. and Söderhäll, K. (1997). Pacifastin, a novel 155-kDa heterodimeric proteinase inhibitor containing a unique transferrin chain. *Proc. Natl. Acad. Sci. U. S. A.* **94**, 6682–6687.

Lindström, L., Alatalo, R. V., Mappes, J., Riipi, M. and Vertainen, L. (1999). Can aposematic signals evolve by gradual change? *Nature* **397**, 249–251.

Liu, H., Li, K.-B., Yin, J., Du, G.-L. and Cao, Y.-Z. (2008). Ultrastructure of the flight muscle of female adults in the gregarious phase and solitary phase of the oriental migratory locust, *Locusta migratoria manilensis* (Meyen) (Orthoptera: Acrididae). *Acta Entomol. Sin.* **51**, 1033–1038, [in Chinese with English abstract].

Loher, W. (1959). Contributions to the study of the sexual behaviour of *Schistocerca gregaria* Forskål (Orthoptera: Acrididae). *Proc. R. Entomol. Soc. Lond. A* **34**, 49–56.

Loher, W. (1961). The chemical acceleration of the maturation process and its hormonal control in the male of the desert locust. *Proc. R. Soc. Lond. B* **153** (1960), 380–397.

Loher, W. (1990). Pheromones and phase transformation in locusts. In: *Biology of Grasshoppers* (eds Chapman, R. F. and Joern, A.), pp. 337–355. NewYork: Wiley.

Loher, W., Ruzo, L., Baker, F. C., Miller, C. A. and Schooley, D. A. (1983). Identification of the juvenile hormone from the cricket, *Teleogryllus commodus*, and juvenile hormone titre changes. *J. Insect Physiol.* **29**, 585–589.

Lomer, C. J., Bateman, R. P., Johnson, D. L., Langewald, J. and Thomas, M. (2001). Biological control of locusts and grasshoppers. *Annu. Rev. Entomol.* **46**, 667–702.

Long, Z. and Hunter, D. M. (2005). Laboratory and field trials of Green Guard® (*Metarhizium anisopliae* var. *acridum*) (Deuteromycotina: Hyphomycetes) against the oriental migratory locust (*Locusta migratoria manilensis*) (Orthoptera: Acrididae) in China. *J. Orthop. Res.* **14**, 27–30.

Lorch, P. D. and Gwynne, D. T. (2000). Radio-telemetric evidence of migration in the gregarious but not the solitary morph of the Mormon cricket (*Anabrus simplex*: Orthoptera: Tettigoniidae). *Naturwissenschaften* **87**, 370–372.

Lorenz, M. W. (2009). Migration and trans-Atlantic flight of locusts. *Quatern. Int.* **196**, 4–12, (appeared online in 2007).

Louveaux, A., Jay, M., El Hadi, O. T. M. and Roux, G. (1998). Variability in flavinoid compounds of four *Tribulus* species: does it play a role in their identification by desert locust *Schistocerca gregaria? J. Chem. Ecol.* **24**, 1465–1481.

Luber, K., Wieting, J., Zeeck, E. and Ferenz, H.-J. (1993). Isolation and characterization of a volatile aromatic infochemical released by sexual maturing, gregarious male desert locusts, *Schistocerca gregaria. Verh. Dtsch. Zool. Ges.* **86**, 261.

Luo, C.-W., Dewey, E. M., Sudo, S., Ewer, J., Hsu, S. Y., Honegger, H.-W. and Hsueh, A. J. W. (2005). Bursicon, the insect cuticle-hardening hormone, is a heterodimeric cystine knot protein that activates G protein-coupled receptor LGR2. *Proc. Natl. Acad. Sci. U. S. A.* **102**, 2820–2825.

Lüscher, M. (1976). Introduction. In: *Phase and Caste Determination in Insects-Endocrine Aspects* (ed. Lüscher, M.), pp. 1–4. New York: Pergamon Press.

Lūsis, O. (1963). The histology and histochemistry of development and resorption in the terminal oocytes of the desert locust, *Schistocerca gregaria. Quart. J. Microsc. Sci.* **104**, 57–68.

Ma, J., Han, X., Hasibagan, Wang, C., Zhang, Y., Tang, J., Xie, Z. and Deveson, T. (2005). Monitoring East Asian migratory locust plagues using remote sensing data and field investigations. *Int. J. Remote Sens.* **26**, 629–634.

Ma, Z., Yu, J. and Kang, L. (2006). LocustDB: a relational database for the transcriptome and biology of the migratory locust (*Locusta migratoria*). *BMC Genomics* **7**, 11.

Maeno, K. and Tanaka, S. (2004). Hormonal control of phase-related changes in the number of antennal sensilla in the desert locust, *Schistocerca gregaria*: possible involvement of [His^7]-corazonin. *J. Insect Physiol.* **50**, 855–865.

Maeno, K. and Tanaka, S. (2007). Effects of hatchling body colour and rearing density on body colouration in last-stadium nymphs of the desert locust, *Schistocerca gregaria. Physiol. Entomol.* **32**, 87–94.

Maeno, K. and Tanaka, S. (2008a). A reddish-brown mutant in the desert locust, *Schistocerca gregaria*: phase-dependent expression and genetic control. *Appl. Entomol. Zool.* **43**, 497–502.

Maeno, K. and Tanaka, S. (2008b). Phase-specific developmental and reproductive strategies in the desert locust. *Bull. Entomol. Res.* **98**, 527–534.

Maeno, K. and Tanaka, S. (2008c). Maternal effects on progeny size, number and body color in the desert locust, *Schistocerca gregaria*: Density- and reproductive cycle-dependent variation. *J. Insect Physiol.* **54**, 1072–1080.

Maeno, K. and Tanaka, S. (2009). Artificial miniaturization causes eggs laid by crowd-reared (gregarious) desert locust to produce green (solitarious) offspring in the desert locust, *Schistocerca gregaria. J. Insect Physiol.* **55**, 849–854.

Maeno, K., Gotoh, T. and Tanaka, S. (2004). Phase-related morphological changes induced by [His^7]-corazonin in two species of locusts, *Schistocerca gregaria* and *Locusta migratoria* (Orthoptera: Acrididae). *Bull. Entomol. Res.* **94**, 349–357.

Magor, J. I. (1970). Outbreaks of the Australian plague locust (*Chortoicetes terminifera* Walk.) in New South Wales during the period 1937–1962, particularly in relation to rainfall. *Anti-Locust Mem.* **11**, 1–39, + 36 maps.
Mahamat, H., Hassanali, A., Odongo, H., Torto, B. and El-Bashir, E.-S. (1993). Studies on the maturation-accelerating pheromone of the desert locust *Schistocerca gregaria* (Orthoptera: Acrididae). *Chemoecology* **4**, 159–164.
Mahamat, H., Hassanali, A. and Munyinyi, D. (1997a). Haemolymph pigment composition as a chemometric indicator of phase in the desert locust, *Schistocerca gregaria*. *Insect Sci. Appl.* **17**, 199–204.
Mahamat, H., Hassanali, A. and Ferenz, H.-J. (1997b). Vitellogenin titres in normal and accelerated maturation of gregarious-phase *Schistocerca gregaria*. *Comp. Biochem. Physiol. B* **116**, 447–451.
Mahamat, H., Hassanali, A. and Odongo, H. (2000). The role of different components of the pheromone emission of mature males of the desert locust, *Schistocerca gregaria* (Forskål) (Orthoptera: Acrididae) in accelerating maturation of immature adults. *Insect Sci. Appl.* **20**, 1–5.
Malual, A. G., Hassanali, A., Torto, B., Assad, Y. O. H. and Njagi, P. G. N. (2001). The nature of the gregarizing signal responsible for maternal transfer of phase to the offspring in the desert locust *Schistocerca gregaria*. *J. Chem. Ecol.* **27**, 1423–1435.
Manière, G., Vanhems, E., Rondot, I. and Delbecque, J. P. (2009). Control of ovarian steroidogenesis in insects: a locust neurohormone is active *in vitro* on blowfly ovaries. *Gen. Comp. Endocrinol.* **163**, 292–297.
Matheson, T., Rogers, S. M. and Krapp, H. G. (2004). Plasticity in the visual system is correlated with a change in lifestyle of solitarious and gregarious locusts. *J. Neurophysiol.* **91**, 1–12.
Matthée, J. J. (1945). Biochemical differences between the solitary and gregarious phases of locusts and noctuids. *Bull. Entomol. Res.* **36**, 343–371.
Matthée, J. J. (1951). The structure and physiology of the egg of Locustana pardalina (Walk). *Sci. Bull. Dep. Agr. Un. S. Afr.* **316**, 1–83.
Matthée, J. J. (1953). The production of diapause eggs by incipient swarms of Locustana pardalina (Walk). *J. Entomol. Soc. S. Afr.* **16**, 132–138.
Mayer, R. J. and Candy, D. J. (1969). Control of haemolymph lipid concentration during locust flight: an adipokinetic hormone from the corpora cardiaca. *J. Insect Physiol.* **15**, 611–620.
McCaffery, A. R., Simpson, S. J., Islam, M. S. and Roessingh, P. (1998). A gregarizing factor present in the egg pod foam of the desert locust *Schistocerca gregaria*. *J. Exp. Biol.* **201**, 347–363.
Michel, R. (1972). Etude expérimentale de l'influence des glandes prothoraciques sur l'activité de vol du criquet pélerin *Schistocerca gregaria*. *Gen. Comp. Endocrinol.* **19**, 96–101.
Michelmore, A. P. G. and Allan, W. (1934). Observations on phases of the red-winged locust in northern Rhodesia. *Bull. Entomol. Res.* **25**, 101–128.
Michener, C. D. (1961). Social polymorphism in Hymenoptera. *Symp. R. Entomol. Soc. Lond.* **1**, 43–56.
Miller, G. A., Islam, M. S., Claridge, T. D. W., Dodgson, T. and Simpson, S. J. (2008). Swarm formation in the desert locust *Schistocerca gregaria*: isolation and NMR analysis of the primary maternal gregarizing agent. *J. Exp. Biol.* **211**, 370–376.
Milner, R. J. and Hunter, D. M. (2001). Recent developments in the use of fungi as biopesticides against locusts and grasshoppers in Australia. *J. Orthop. Res.* **10**, 271–276.
Min, K. J., Taub-Montemayor, T. E., Linse, K. D., Kent, J. W., Jr, and Rankin, M. A. (2004). Relationship of adipokinetic hormone I and II to migratory propensity in the grasshopper, *Melanoplus sanguinipes*. *Arch. Insect Biochem. Physiol.* **55**, 33–42.

Mitsuhashi, J. (1998). Polyamine as a growth promotor for cultured insect cells. *In Vitro Cell. Dev. Biol. Anim.* **34**, 619–621.

Miura, T. (2004). Proximate mechanisms and evolution of caste polyphenism in social insects: from sociality to genes. *Ecol. Res.* **19**, 141–148.

Mordue (Luntz), A. J. (1977). Some effects of amputation of the antennae on pigmentation, growth and development in the locust, *Schistocerca gregaria. Physiol. Entomol.* **2**, 293–300.

Moreteau, B. (1975). Fonction chromatrope de la pars intercerebralis chez l'acridien *Oedipoda coerulescens. J. Insect Physiol.* **21**, 1407–1413.

Moreteau-Levita, B. (1972). Rôle des corpora cardiaca dans la pigmentation d'*Oedipoda coerulescens* L. (Acridien, Orthoptère). *C. R. Hebd. Séanc. Acad. Sci., Paris, Série D* **275**, 2699–2701.

Morton, D. B. and Evans, P. D. (1983). Octopamine distribution in solitarious and gregarious forms of the locust, *Schistocerca americana gregaria. Insect Biochem.* **13**, 177–183.

Mueller, N. S. (1963). An experimental analysis of molting in embryos of *Melanoplus differentialis. Dev. Biol.* **8**, 222–240.

Nässel, D. R. (2002). Neuropeptides in the nervous system of *Drosophila* and other insects: multiple roles as neuromodulators and neurohormones. *Progr. Neurobiol.* **68**, 1–84.

Nation, J. L. (2008). *Insect Physiology and Biochemistry*, 2nd edition. Boca Raton, FL: CRC Press.

Němec, V., Breuer, M. and De Loof, A. (2003). Thin layer chromatographic analysis of pteridine-like pigments in the migratory locust, *Locusta migratoria migratorioides* (Orthoptera: Oedipodidae). *Eur. J. Entomol.* **100**, 19–23.

Niassy, A., Torto, B., Njagi, P. G. N., Hassanali, A., Obeng-Ofori, D. and Ayertey, J. N. (1999). Intra- and interspecific aggregation responses of *Locusta migratoria migratorioides* and *Schistocerca gregaria* and a comparison of their pheromone emissions. *J. Chem. Ecol.* **25**, 1029–1042.

Nickerson, B. (1956). Pigmentation of hoppers of the desert locust (*Schistocerca gregaria* Forskål) in relation to phase coloration. *Anti-Locust Bull.* **24**, 1–34.

Nicolas, G. and Ismaïl, S. (1978). Réalisation expérimentale de l'assombrissement tégumentaire chez le criquet migrateur *Locusta migratoria* L. Influence des neurosécrétions cérébrales: comparaison avec les effets du CO_2. *C. R. Séanc. Soc. Biol.* **172**, 1075–1078.

Nijhout, H. F. (1994). *Insect Hormones*. Princeton, NJ: Princeton University Press.

Nijhout, H. F. (1999). Control mechanisms of polyphenic development in insects. *Bioscience* **49**, 181–192.

Nijhout, H. F. (2003). Development and evolution of adaptive polyphenisms. *Evol. Dev.* **5**, 9–18.

Nijhout, H. F. and Wheeler, D. E. (1982). Juvenile hormone and the physiological basis of insect polymorphisms. *Quart. Rev. Biol.* **57**, 109–133.

Njagi, P. G. N. and Torto, B. (1996). Responses of nymphs of desert locust, *Schistocerca gregaria* to volatiles of plants used as rearing diet. *Chemoecology* **7**, 172–178.

Njagi, P. G. N. and Torto, B. (2002). Evidence for a compound in Comstock-Kellog glands modulating premating behavior in male desert locust, *Schistocerca gregaria. J. Chem. Ecol.* **28**, 1065–1074.

Njagi, P. G. N., Torto, B., Obeng-Ofori, D. and Hassanali, A. (1996). Phase-independent responses to phase-specific aggregation pheromone in adult desert locusts, *Schistocerca gregaria* (Orthoptera: Acrididae). *Physiol. Entomol.* **21**, 131–137.

Nolte, D. J. (1965). The pigmentation of locusts. *S. Afr. J. Sci.* **61**, 173–178.

Nolte, D. J. (1967). Phase transformation and chiasma formation in locusts. *Chromosoma* **21**, 123–139.
Nolte, D. J. (1968). The chiasma-inducing pheromone of locusts. *Chromosoma* **23**, 346–358.
Nolte, D. J. (1969). Chiasma-induction and tyrosine metabolism in locusts. *Chromosoma* **26**, 287–297.
Nolte, D. J. (1971). Two pleiotropic albino mutations. *Proc. 4th Cong. S. Afr. Genetic Soc. (Pretoria 1970)*, pp. 36–38.
Nolte, D. J. (1976). Locustol and its analogues. *J. Insect Physiol.* **22**, 833–838.
Nolte, D. J., Eggers, S. H. and May, I. R. (1973). A locust pheromone: locustol. *J. Insect Physiol.* **19**, 1547–1554.
Norris, M. J. (1950). Reproduction in the African migratory locust (*Locusta migratoria migratorioides* R. & F.) in relation to density and phase. *Anti-Locust Bull.* **6**, 1–48.
Norris, M. J. (1952). Reproduction in the desert locust (*Schistocerca gregaria* Forskål) in relation to density and phase. *Anti-Locust Bull.* **13**, 1–49.
Norris, M. J. (1954). Sexual maturation in the desert locust (*Schistocerca gregaria* Forskål) with special reference to the effects of grouping. *Anti-Locust Bull.* **18**, 1–44.
Norris, M. J. (1959). Reproduction in the red locust (*Nomadacris septemfasciata* Serville) in the laboratory. *Anti-Locust Bull.* **36**, 1–46.
Norris, M. J. (1963). Laboratory experiments on gregarious behaviour in ovipositing females of the desert locust, (*Schistocerca gregaria* (Forsk)). *Entomol. Exp. Appl.* **6**, 279–303.
Norris, M. J. (1964). Accelerating and inhibiting effects of crowding on sexual maturation in two species of locusts. *Nature* **203**, 784–785.
Norris, M. J. (1970). Aggregation response in ovipositing females of the desert locust, with special reference to the chemical factor. *J. Insect Physiol.* **16**, 1493–1515.
Norris, M. J. and Pener, M. P. (1965). An inhibitory effect of allatectomized males and females on the sexual maturation of young male adults of *Schistocerca gregaria* (Forsk.) (Orthoptera: Acrididae). *Nature* **208**, 1122.
Nowrousian, M., Ringelberg, C., Dunlap, J. C., Loros, J. J. and Kück, U. (2005). Cross-species microarray hybridization to identify developmentally regulated genes in the filamentous fungus *Sordaria macrospora*. *Mol. Genet. Genomics* **273**, 137–149.
Obeng-Ofori, D., Torto, B. and Hassanali, A. (1993). Evidence for mediation of two releaser pheromones in the aggregation behavior of the gregarious desert locust, *Schistocerca gregaria* (Forskal) (Orthoptera: Acrididae). *J. Chem. Ecol.* **19**, 1665–1676.
Obeng-Ofori, D., Njagi, P. G. N., Torto, B., Hassanali, A. and Amiani, H. (1994a). Sex differentiation studies relating to releaser aggregation pheromones of the desert locust, *Schistocerca gregaria* (Forskål) (Orthoptera: Acrididae). *Entomol. Exp. Appl.* **73**, 85–91.
Obeng-Ofori, D., Torto, B., Njagi, P. G. N., Hassanali, A. and Amiani, H. (1994b). Fecal volatiles as part of the aggregation pheromone complex of the desert locust, *Schistocerca gregaria* (Forskål) (Orthoptera: Acrididae). *J. Chem. Ecol.* **20**, 2077–2087.
Ochieng', S. A. and Hansson, B. S. (1999). Responses of olfactory receptor neurones to behaviourally important odours in gregarious and solitarious desert locust, *Schistocerca gregaria*. *Physiol. Entomol.* **24**, 28–36.
Ochieng, S. A., Hallberg, E. and Hansson, B. S. (1998). Fine structure and distribution of antennal sensilla of the desert locust, *Schistocerca gregaria* (Orthoptera: Acrididae). *Cell Tissue Res.* **291**, 525–536.

Ochieng'-Odero, J. P. R., Ndugo, S. M., El Bashir, S. and Capstick, P. B. (1994). A method for rearing crowded (gregarious) and isolated (solitary) locusts (Orthoptera: Acrididae) in the laboratory. In: *Proceedings of Workshop on Effective Networking of Research and Development on Environmentally Sustainable Locust Control Methods among Locust Affected Countries* (eds Njagi, P. G. N. and Chaudhury, M. F. B.), pp. 33–44. Nairobi, Kenya: ICIPE Science Press.

Oda, K. and Ishii, M. (2001). Body color polymorphism in nymphs and adults of a katydid, *Conocephalus maculatus* (Orthoptera: Tettigoniidae). *Appl. Entomol. Zool.* **36**, 345–348.

Odhiambo, T. R. (1966). Growth and the hormonal control of sexual maturation in the male desert locust, *Schistocerca gregaria* (Forskål). *Trans. R. Entomol. Soc. Lond.* **118**, 393–412.

Ogoyi, D. O., Osir, E. O. and Olembo, N. K. (1995). Lipophorin and apolipophorin-III in solitary and gregarious phases of *Schistocerca gregaria*. *Comp. Biochem. Physiol. B* **112**, 441–449.

Ogoyi, D. O., Osir, E. O. and Olembo, N. K. (1996). Effect of phase status on response to AKH I in the desert locust, *Schistocerca gregaria gregaria*. *Arch. Insect Biochem. Physiol.* **32**, 173–185.

Ogoyi, D. O., Osir, E. O. and Olembo, N. K. (1998). Fat body triacylglycerol lipase in solitary and gregarious phases of *Schistocerca gregaria* (Forskal) (Orthoptera: Acrididae). *Comp. Biochem. Physiol. B* **119**, 163–167.

Okuda, T. and Tanaka, S. (1997). An allatostatic factor and juvenile hormone synthesis by corpora allata in *Locusta migratoria*. *J. Insect Physiol.* **43**, 635–641.

Orchard, I. (1987). Adipokinetic hormones – an update. *J. Insect Physiol.* **33**, 451–463.

Orchard, I., Lange, A. B. (1998). The distribution, biological activity, and pharmacology of SchistoFLRFamide and related peptides in insects, In: *Recent Advances in Arthropod Endocrinology*, (eds Coast, G. M. and Webster, S. G.), pp. 278–301. Society for Experimental Biology Seminar Series No. 65, Cambridge: Cambridge University Press.

Orchard, I., Ramirez, J.-M. and Lange, A. B. (1993). A multifunctional role for octopamine in locust flight. *Annu. Rev. Entomol.* **38**, 227–249.

Osborne, K. A., Robichon, A., Burgess, E., Butland, S., Shaw, R. A., Coulthard, A., Pereira, H. S., Greenspan, R. J. and Sokolowski, M. B. (1997). Natural behavior polymorphism due to a cGMP-dependent protein kinase of *Drosophila*. *Science* **277**, 834–836.

O'Shea, M. and Rayne, R. C. (1992). Adipokinetic hormones: cell and molecular biology. *Experientia* **48**, 430–438.

Ott, S. R., Burrows, M. and Elphick, M. R. (2001). The neuroanatomy of nitric oxide-cyclic GMP signalling in the locust: functional implications for sensory systems. *Am. Zool.* **41**, 321–331.

Oudejans, R. C. H. M., Kooiman, F. P., Heerma, W., Versluis, C., Slotboom, A. J. and Beenakkers, A. M. T. (1991). Isolation and structure elucidation of a novel adipokinetic hormone (Lom-AKH-III) from the glandular lobes of the corpus cardiacum of the migratory locust, *Locusta migratoria*. *Eur. J. Biochem.* **195**, 351–359.

Oudejans, R. C. H. M., Mes, T. H. M., Kooiman, F. P. and Van der Horst, D. J. (1993). Adipokinetic peptide hormone content and biosynthesis during locust development. *Peptides* **14**, 877–881.

Oudejans, R. C. H. M., Harthoorn, L. F., Diederen, J. H. B. and Van der Horst, D. J. (1999). Adipokinetic hormones – Coupling between biosynthesis and release. *Ann. New York Acad. Sci.* **897**, Issue: Neuropeptides: structure and function in biology and behavior, pp. 291–299.

Ould Ely, S., Mahamat, H., Njagi, P. G. N., Bashir, M. O., El-Tom El-Amin, S. and Hassanali, A. (2006). Mate location mechanism and phase-related mate preferences in solitarius desert locust, *Schistocerca gregaria*. *J. Chem. Ecol.* **32**, 1057–1069.

Oxford, G. S. and Gillespie, R. G. (1998). Evolution and ecology of spider coloration. *Annu. Rev. Entomol.* **43**, 619–643.

Padgham, D. E. (1976a). Control of melanization in first instar larvae of *Schistocerca gregaria*. *J. Insect Physiol.* **22**, 1409–1419.

Padgham, D. E. (1976b). Bursicon-mediated control of tanning in melanizing and non-melanizing first instar larvae of *Schistocerca gregaria*. *J. Insect Physiol.* **22**, 1447–1452.

Papillon, M. (1960). Étude préliminaire de la répercussion du groupement des parents sur les larves nouveau-nées de *Schistocerca gregaria* Forsk. *Bull. Biol. Fr. Belg.* **94**, 203–263.

Papillon, M. (1968). Facteurs ecologiques et phases chez le criquet pèlerine, *Schistocerca gregaria* (Forsk). II – Influence de la densite des populations. *Bull. Biol.* **102**, 271–307.

Papillon, M., (1970). Influence du groupement des adultes sur leur fécondité et sur le polymorphisme de leur descendance chez le criquet pèlerine, *Schistocerca gregaria* Forsk. In: *L'influence des Stimuli Externes sur la Gamétogenèse des Insectes, Tours, 1969*; Colloq. Internat. Centre National Rech. Sci., No. 189, pp. 71–86. Paris: Centre National de la Recherche Scientifique.

Parker, G. A. and Smith, J. L. (1975). Sperm competition and the evolution of the precopulatory passive phase behaviour in *Locusta migratoria migratorioides*. *J. Entomol. A* **49**, 155–171.

Passama-Vuillaume, M. (1965). Étude du pigment vert chez *Locusta migratoria* L. normal et albinos. *Bull. Soc. Zool. Fr.* **90**, 485–492.

Passama-Vuillaume, M. and Barbier, M. (1966). Sur la biosynthèse de la biliverdine IX α par la mante *Mantis religiosa* et le criquet *Locusta migratoria*. *C. R. Hebd. Séanc. Acad. Sci. Paris, Série D* **263**, 924–925.

Peel, A. D. and Akam, M. (2007). The dynamics of yolk deposition in the desert locust *Schistocerca gregaria*. *J. Insect Physiol.* **53**, 436–443.

Pener, M. P. (1965). On the influence of corpora allata on maturation and sexual behaviour of *Schistocerca gregaria*. *J. Zool.* **147**, 119–136.

Pener, M. P. (1967). Effects of allatectomy and sectioning of the nerves of the corpora allata on oöcyte growth, male sexual behaviour, and colour change in adults of *Schistocerca gregaria*. *J. Insect Physiol.* **13**, 665–684.

Pener, M. P. (1968). The effect of corpora allata on sexual behaviour and "adult diapause" in males of the red locust. *Entomol. Exp. Appl.* **11**, 94–100.

Pener, M. P. (1976a). The differential effect of the corpora allata on male sexual behaviour in crowded and isolated adults of *Locusta migratoria migratorioides* (R. & F.). *Acrida* **5**, 189–206.

Pener, M. P. (1976b). The differential effect of the corpora allata on yellow coloration in crowded and isolated *Locusta migratoria migratorioides* (R. & F.) males. *Acrida* **5**, 269–285.

Pener, M. P. (1983). Endocrine aspects of phase polymorphism in locusts. In: *Invertebrate Endocrinology, Vol. 1: Endocrinology of Insects* (eds Downer, R. G. H. and Laufer, H.), pp. 379–394. New York: Alan R. Liss.

Pener, M. P., (1986). Endocrine aspects of mating behavior in acridids. In: *Proceedings, 4th Triennial Meeting, Pan American Acridological Society, 29 July to 2 August, 1985, Saskatoon, Sask., Canada* (ed. D. A. Nickle), pp. 9–26. Ann Arbor, MI: Braun-Brumfield Inc.

Pener, M. P. (1990). Endocrine effects on locust phase changes; basic and applied aspects. *Boletin de Sanidad Vegetal*, Fuera de Serie No. 20; (also as *Proceedings* *of the 5th International Meeting of the Orthopterists' Society, 17–20 July, 1989, Valsain (Segovia), Spain*), pp. 39–55.

Pener, M. P. (1991). Locust phase polymorphism and its endocrine relations. *Adv. Insect Physiol.* **23**, 1–79.

Pener, M. P. (1997). Endocrine factors and female-male coadaptations in reproductive diapause-related strategies of acridoid insects. In: *The Bionomics of Grasshoppers, Katydids and Their Kin* (eds Gangwere, S. K., Muralirangan, M. C. and Muralirangan, M.), pp. 315–336. Wallingford, UK: CABI Publishing.

Pener, M. P. and Lazarovici, P. (1979). Effect of exogenous juvenile hormones on mating behaviour and yellow colour in allatactomized adult male desert locusts. *Physiol. Entomol.* **4**, 251–261.

Pener, M. P. and Yerushalmi, Y. (1998). The physiology of locust phase polymorphism: an update. *J. Insect Physiol.* **44**, 365–377.

Pener, M. P., Girardie, A. and Joly, P. (1972). Neurosecretory and corpus allatum controlled effects on mating behavior and color change in adult *Locusta migratoria migratorioides* males. *Gen. Comp. Endocrinol.* **19**, 494–508.

Pener, M. P., Orshan, L. and De Wilde, J. (1978). Precocene II causes atrophy of corpora allata in *Locusta migratoria*. *Nature* **272**, 350–353.

Pener, M. P., Dessberg, D., Lazarovici, P., Reuter, C. C., Tsai, L. W. and Baker, F. C. (1986). The effect of a synthetic precocene on juvenile hormone III titre in late *Locusta* eggs. *J. Insect Physiol.* **32**, 853–857.

Pener, M. P., Ayali, A. and Ben-Ami, E. (1992). Juvenile hormone is not a major factor in locust phase changes. In: *Juvenile Hormone Research. Fundamental and Applied Approaches* (eds Mauchamp, B., Couillaud, F. and Baehr, J. C.), pp. 125–134. Paris: Institute National de la Recherche Agronomique.

Pener, M. P., Ayali, A., Kelmer, G., Bennettová, B., Němec, V., Rejzek, M. and Wimmer, Z. (1997a). Comparative testing of several juvenile hormone analogues in two species of locusts, *Locusta migratoria migratorioides* and *Schistocerca gregaria*. *Pestic. Sci.* **51**, 443–449.

Pener, M. P., Ayali, A. and Golenser, E. (1997b). Adipokinetic hormone and flight fuel related characteristics of density-dependent locust phase polymorphism; a review. *Comp. Biochem. Physiol. B* **117**, 513–524.

Peveling, R. (2001). Environmental conservation and locust control – possible conflicts and solutions. *J. Orthop. Res.* **10**, 171–187.

Pfeiffer, I. W. (1945). The influence of the corpora allata over the development of nymphal characters in the grasshopper *Melanoplus differentialis*. *Trans. Conn. Acad. Arts Sci.* **38**, 489–513.

Phipps, J. (1966). Ovulation and oocyte resorption in Acridoidea (Orthoptera). *Proc. R. Entomol. Soc. Lond. A* **41**, 78–86.

Pierozzi, I., Jr, and Lecoq, M. (1998). Morphometric studies on *Rhammatocerus schistocercoides* (Rehn, 1906) Orthoptera, Acrididae, Gomphocerinae in Brazilian and Columbian populations. *Trans. Am. Entomol. Soc.* **124**, 25–34.

Pigliucci, M. (2001). *Phenotypic Plasticity: Beyond Nature and Nurture*. Baltimore: Johns Hopkins University Press.

Pigliucci, M. and Preston, K. (eds) (2004). *Phenotypic Integration. Studying the Ecology and Evolution of Complex Phenotypes*. Oxford: Oxford University Press.

Pinamonti, S. and Petris, A. (1966). Studies on tryptophan pyrrolase of *Schistocerca gregaria* Forsk. (Orthoptera). *Comp. Biochem. Physiol.* **17**, 1079–1087.

Pinamonti, S., Chiarelli-Alvisi, G. and Colombo, G. (1973). The xanthommatin forming enzyme system of the desert locust, *Schistocerca gregaria*. *Insect Biochem.* **3**, 289–296.

Popov, G. B. (1958). Ecological studies on oviposition by swarms of the desert locust (*Schistocerca gregaria* Forskål) in Eastern Africa. *Anti-Locust Bull.* **31**, 1–70.
Popov, G. B. (1989). *Nymphs of the Sahelian Grasshoppers: an Illustrated Guide*. Chatham, UK: Overseas Development Natural Resources Institute.
Porras, M. G., De Loof, A., Breuer, M. and Aréchiga, H. (2003). Corazonin promotes tegumentary pigment migration in the crayfish *Procambarus clarkii*. *Peptides* **24**, 1581–1589.
Predel, R., Neupert, S., Russell, W. K., Scheibner, O. and Nachman, R. J. (2007). Corazonin in insects. *Peptides* **28**, 3–10.
Price, R. E. and Brown, H. D. (2000). A century of locust control in South Africa. In: *Workshop on Research Priorities for Migrant Pests of Agriculture in Southern Africa, Plant Protection Research Institute, Pretoria, South Africa, 24–26 March 1999* (eds Cheke, R. A., Rosenberg, L. J. and Kieser, M. E.), pp. 37–49. Chatham, UK: Natural Resources Institute.
Putnam, L. G. (1958). Albinism in the migratory grasshopper, *Melanoplus bilituratus* (Wlk). *Nature* **182**, 1529.
Qi-Miao, S., Tanaka, S. and Takeda, M. (2003). Immunohistochemical localization of clock proteins (DBT and PER), and [His^7]- and [Arg^7]-corazonins in the cerebral ganglia of *Bombyx mori*: Are corazonins downstream regulators of circadian clocks? *Eur. J. Entomol.* **100**, 283–286.
Queitsch, C., Sangster, T. A. and Lindquist, S. (2002). Hsp90 as a capacitor of phenotypic variation. *Nature* **417**, 618–624.
Quesada-Moraga, E. and Santiago-Álvarez, C. (2000). Temperature related effects on embryonic development of the Mediterranean locust, *Dociostaurus maroccanus*. *Physiol. Entomol.* **25**, 191–195.
Quesada-Moraga, E. and Santiago-Álvarez, C. (2001a). Assessment of sexual maturation in the Moroccan locust *Dociostaurus maroccanus* (Thunberg). *J. Orthop. Res.* **10**, 1–8.
Quesada-Moraga, E. and Santiago-Álvarez, C. (2001b). Rearing and breeding of the Moroccan locust *Dociostaurus maroccanus* (Thunberg) (Orthop., Acrididae) under laboratory conditions. *J. Appl. Entomol.* **125**, 121–124.
Raabe, M. (1982). *Insect Neurohormones*. New York: Plenum Press.
Raabe, M. (1983). Pigment metabolism. In: *Invertebrate Endocrinology, Vol. 1: Endocrinology of Insects* (eds Downer, R. G. H. and Laufer, H.), pp. 493–500. New York: Alan R. Liss.
Raabe, M. (1989). *Recent Developments in Insect Neurohormones*. New York: Plenum Press.
Rahman, M. M., Vanden Bosch, L., Baggerman, G., Clynen, E., Hens, K., Hoste, B., Meylaers, K., Vercammen, T., Schoofs, L., De Loof, A. and Breuer, M. (2002a). Search for peptidic molecular markers in hemolymph of crowd-(gregarious) and isolated-reared (solitary) desert locusts, *Schistocerca gregaria*. *Peptides* **23**, 1907–1914.
Rahman, M. M., Hoste, B., De Loof, A. and Breuer, M. (2002b). Developmental effect of egg pod foam in the desert locust Schistocerca gregaria (Caelifera: Acrididae). *Entomol. Gen.* **26**, 161–172.
Rahman, M. M., Baggerman, G., Schoofs, L., De Loof, A. and Breuer, M. (2003a). Presence of [His^7]-corazonin in the central nervous system of a newly isolated albino strain of *Schistocerca gregaria* (Orthoptera: Acrididae) – mass spectrometric and immunocytochemical evidence. *Eur. J. Entomol.* **100**, 455–458.
Rahman, M. M., Baggerman, G., Begum, M., De Loof, A. and Breuer, M. (2003b). Purification, isolation and search for possible functions of a phase-related 6080-Da

peptide from the haemolymph of the desert locust, *Schistocerca gregaria*. *Physiol. Entomol.* **28**, 39–45.

Rahman, M. M., Vandingenen, A., Begum, M., Breuer, M., De Loof, A. and Huybrechts, R. (2003c). Search for phase specific genes in the brain of desert locust, *Schistocerca gregaria* (Orthoptera: Acrididae) by differential display polymerase chain reaction. *Comp. Biochem. Physiol. A* **135**, 221–228.

Rahman, M. M., Breuer, M., Begum, M., Baggerman, G., Huybrechts, J. and De Loof, A. (2008a). Localization of the phase-related 6-kDa peptide (PRP) in different tissues of the desert locust *Schistocerca gregaria* – immunocytochemical and mass spectrometric approach. *J. Insect Physiol.* **54**, 543–554.

Rahman, M. M., Vanden Bosch, L., Begum, M., Seidelmann, K., Breuer, M. and De Loof, A. (2008b). Phase-related 6-kDa peptide titre in haemolymph of larvae and adult *Schistocerca gregaria* and its role in yellow-protein synthesis. *Physiol. Entomol.* **33**, 123–128.

Rai, M. M., Hassanali, A., Saini, R. K., Odongo, H. and Kahoro, H. (1997). Identification of components of the oviposition aggregation pheromone of the gregarious desert locust, *Schistocerca gregaria* (Forskål). *J. Insect Physiol.* **43**, 83–87.

Raikhel, A. S., Brown, M. R. and Belles, X. (2005). Hormonal control of reproductive processes. In: *Comprehensive Molecular Insect Science, Vol. 3, Endocrinology* (eds Gilbert, L. I., Iatrou, K. and Gill, S. S.), pp. 433–491. Oxford: Elsevier Pergamon.

Rainey, R. C. (1989). *Migration and Meteorology – Flight Behaviour and the Atmospheric Environment of Locusts and other Migrant Pests*. Oxford: Oxford University Press.

Raubenheimer, D. and Simpson, S. J. (1999). Integrating nutrition: a geometrical approach. *Entomol. Exp. Appl.* **91**, 67–82.

Raubenheimer, D. and Simpson, S. J. (2003). Nutrient balancing in grasshoppers: behavioural and physiological correlates of dietary breadth. *J. Exp. Biol.* **206**, 1669–1681.

Raubenheimer, D., Lee, K. P. and Simpson, S. J. (2005). Does Bertrand's rule apply to macronutrients? *Proc. R. Soc. Lond. B* **272**, 2429–2434.

Reichhart, J. M. and Charlet, M. (1986). Ecdysiotropic activity in brains and corpora cardiaca of larvae and adults of *Locusta migratoria*: in vitro assays. *Int. J. Invertebr. Rep. Dev.* **10**, 17–25.

Reineke, A., Schmidt, O. and Zebitz, C. P. W. (2003). Differential gene expression in two strains of the endoparasitic wasp *Venturia canescens* identified by cDNA-amplified fragment length polymorphism analysis. *Mol. Ecol.* **12**, 3485–3492.

Renn, S. C., Aubin-Horth, N. and Hofmann, H. A. (2004). Biologically meaningful expression profiling across species using heterologous hybridization to a cDNA microarray. *BMC Genomics* **5**, 42.

Reynolds, A. M., Sword, G. A., Simpson, S. J. and Reynolds, D. R. (2009). Predator percolation, insect outbreaks and phase polyphenism. *Curr. Biol.* **19**, 20–24.

Richards, M. J. and El Mangoury, M. A. (1968). Further experiments on the effects of social factors on the rate of sexual maturation in the desert locust. *Nature* **219**, 865–866.

Richardson, C. H. and Nemeth, D. J. (1991). Hurrican-borne African locusts (*Schistocerca gregaria*) on the Windward Islands. *GeoJournal* **23**, 349–357.

Riedl, C. A. L., Neal, S. J., Robichon, A., Westwood, J. T. and Sokolowski, M. B. (2005). *Drosophila* soluble guanylyl cyclase mutants exhibit increased foraging locomotion: behavioral and genomic investigations. *Behav. Genet.* **35**, 231–244.

Ritchie, M. and Pedgley, D. (1989). Desert locusts cross the Atlantic. *Antenna* **13**, 10–12.

Robinson, G. E., Grozinger, C. M. and Whitfield, C. W. (2005). Sociogenomics: social life in molecular terms. *Nat. Rev. Genet.* **6**, 257–270.

Roessingh, P. and Simpson, S. J. (1994). The time-course of behavioural phase change in nymphs of the desert locust, *Schistocerca gregaria*. *Physiol. Entomol.* **19**, 191–197.

Roessingh, P., Simpson, S. J. and James, S. (1993). Analysis of phase-related changes in behaviour of desert locust nymphs. *Proc. R. Soc. Lond. B* **252**, 43–49.

Roessingh, P., Bouaïchi, A. and Simpson, S. J. (1998). Effects of sensory stimuli on the behavioural phase state of the desert locust, *Schistocerca gregaria*. *J. Insect Physiol.* **44**, 883–893.

Roffey, J. and Magor, J. I. (2003). Desert locust population parameters. In: *Desert Locust Technical Series Vol. 30*. Rome: FAO.

Roffey, J. and Popov, G. (1968). Environmental and behavioural processes in a desert locust outbreak. *Nature* **219**, 446–450.

Rogers, S. M. and Simpson, S. J. (1997). Experience-dependent changes in the number of chemosensory sensilla on the mouthparts and antennae of *Locusta migratoria*. *J. Exp. Biol.* **200**, 2313–2321.

Rogers, S. M., Matheson, T., Despland, E., Dodgson, T., Burrows, M. and Simpson, S. J. (2003). Mechanosensory-induced behavioural gregarization in the desert locust *Schistocerca gregaria*. *J. Exp. Biol.* **206**, 3991–4002.

Rogers, S. M., Matheson, T., Sasaki, K., Kendrick, K., Simpson, S. J. and Burrows, M. (2004). Substantial changes in central nervous system neurotransmitters and neuromodulators accompany phase change in the locust. *J. Exp. Biol.* **207**, 3603–3617.

Rogers, S. M., Krapp, H. G., Burrows, M. and Matheson, T. (2007). Compensatory plasticity at an identified synapse tunes a visuomotor pathway. *J. Neurosci.* **27**, 4621–4633.

Roignant, J.-Y., Carré, C., Mugat, B., Szymczak, D., Lepesant, J.-A. and Antoniewski, C. (2003). Absence of transitive and systemic pathways allows cell-specific and isoform-specific RNAi in *Drosophila*. *RNA* **9**, 299–308.

Roisin, Y. (2000). Diversity and evolution of caste patterns. In: *Termites: Evolution, Sociality, Symbiosis, Ecology* (eds Abe, T., Bignell, D. E. and Higashi, M.), pp. 95–119. Dordrecht, The Netherlands: Kluwer Academic Publishers.

Roller, L., Tanaka, Y. and Tanaka, S. (2003). Corazonin and corazonin-like substances in the central nervous system of Pterygote and Apterygote insects. *Cell Tissue Res.* **312**, 393–406.

Roller, L., Tanaka, S., Kimura, K., Satake, H. and Tanaka, Y. (2006). Molecular cloning of [Thr^4, His^7]-corazonin (Apime-corazonin) and its distribution in the central nervous system of the honey bee *Apis mellifera* (Hymenoptera: Apidae). *Appl. Entomol. Zool.* **41**, 331–338.

Rono, E., Njagi, P. G. N., Bashir, M. O. and Hassanali, A. (2008). Concentration-dependent parsimonious releaser roles of gregarious male pheromone of the desert locust, *Schistocerca gregaria*. *J. Insect Physiol.* **54**, 162–168.

Rosenberg, J. and Burt, P. J. A. (1999). Windborne displacements of desert locusts from Africa to the Caribbean and South America. *Aerobiologia* **15**, 167–175.

Roussel, J.-P. (1993). Modification des taux d'ecdysteroides selon la phase chez *Locusta migratoria* L. *Bull. Soc. Zool. Fr.* **118**, 367–373.

Rowell, C. H. F. (1967). Corpus allatum implantation and green/brown polymorphism in three African grasshoppers. *J. Insect Physiol.* **13**, 1401–1412.

Rowell, C. H. F. (1971). The variable coloration of the acridoid grasshoppers. *Adv. Insect Physiol.* **8**, 145–198.

Rubinstein, A. L., Lee, D., Luo, R., Henion, P. D. and Halpern, M. E. (2000). Genes dependent on zebrafish *cyclops* function identified by AFLP differential gene expression screen. *Genesis* **26**, 86–97.

Rutherford, S. L. and Lindquist, S. (1998). Hsp90 as a capacitor for morphological evolution. *Nature* **396**, 336–342.

Ryan, R. O. and Van der Horst, D. J. (2000). Lipid transport biochemistry and its role in energy production. *Annu. Rev. Entomol.* **45**, 233–260.

Rybczynski, R. (2005). Prothoracicotropic hormone. In: *Comprehensive Molecular Insect Science, Vol. 3, Endocrinology* (eds Gilbert, L. I., Iatrou, K. and Gill, S. S.), pp. 61–123. Oxford: Elsevier Pergamon.

Saini, R. K., Rai, M. M., Hassanali, A., Wawiye, J. and Odongo, H. (1995). Semiochemicals from froth of egg pods attract ovipositing female *Schistocerca gregaria*. *J. Insect Physiol.* **41**, 711–716.

Sall, C., Tsoupras, G., Kappler, C., Lagueux, M., Zachary, D., Luu, B. and Hoffmann, J. A. (1983). Fate of maternal conjugated ecdysteroids during embryonic development in *Locusta migratoria*. *J. Insect Physiol.* **29**, 491–507.

Sanchez, N. E., Wittenstein, E., De Wysiecki, M. L. and Lange, C. E. (1997). Life history parameters of the gregarious phase of the South American locust, *Schistocerca cancellata* (Serville) (Orthoptera: Acrididae), under laboratory conditions. *J. Orthop. Res.* **6**, 121–124.

Santiago-Alvarez, C., Quesada-Moraga, E. and Hernández-Crespo, P. (2003). Diapause termination and post-diapause development in the Mediterranean locust *Dociostaurus maroccanus* (Orth., Acrididae) under field conditions. *J. Appl. Entomol.* **127**, 369–373.

Sas, F., Begum, M., Vandersmissen, T., Geens, M., Claeys, I., Van Soest, S., Huybrechts, J., Huybrechts, R. and De Loof, A. (2007). Development of a real-time PCR assay for measurement of yellow protein mRNA transcription in the desert locust *Schistocerca gregaria*: a basis for isolation of a peptidergic regulatory factor. *Peptides* **28**, 38–43.

Sbrenna, G. (1991). Roles of morphogenetic hormones in embryonic cuticle deposition in arthropods. In: *Morphogenetic Hormones of Arthropods, Vol. 3: Roles in Histogenesis, Organogenesis and Morphogenesis* (ed. Gupta, A. P.), pp. 44–80. New Brunswick, NJ: Rutgers University Press.

Scalia, S. and Morgan, E. D. (1982). A re-investigation on the ecdysteroids during embryogenesis in the desert locust, *Schistocerca gregaria*. *J. Insect Physiol.* **28**, 647–654.

Schlichting, C. D. and Pigliucci, M. (1998). *Phenotypic Evolution: a Reaction Norm Perspective*. Sunderland, MA: Sinauer Associates Inc.

Schmidt, G. H. (2001). Einfluß der Populationsdichte adulter Locusta migratoria cinerascens auf Verhalten, Reproduktion und Morphologie von Folgegenerationen (Caelifera: Acrididae). *Entomol. Gen.* **25**, 205–218.

Schmidt, G. H. and Albütz, R. (1996). Cross-breeding of two subspecies of *Locusta migratoria* (L.) and analysis of numerical character variations in various geographical populations. *Boll. Lab. Entomol. Agrararia – Portici "Filippo Silvestri"* **52**, 13–26.

Schmidt, G. H. and Albütz, R. (1999). Identification of solitary and gregarious populations of the desert locust, Schistocerca gregaria, by experimental breeding (Caelifera: Acrididae). *Entomol. Gen.* **24**, 161–175.

Schmidt, G. H. and Albütz, R. (2002). Sexual maturation and yellow coloration of adult males in the gregarious desert locust, Schistocerca gregaria, in relation to volatiles emitted (Caelifera: Acrididae). *Entomol. Gen.* **26**, 121–141.

Schmidt, G. H. and Assembe-Tsoungui, S. (2002). Wirkung oraler Applikation von *Melia*-Fruchtextract auf Wachstum, Entwicklung und Gregarisation der

Wüstenheuschrecke *Schistocerca gregaria* (Forskål, 1775) (Caelifera, Acrididae) (with English summary). *Z. Pflanzenkr. Pflanzenschutz* **109**, 38–56.

Schneider, M. and Dorn, A. (1994). Lipid storage and mobilization by flight in relation to phase and age of *Schistocerca gregaria* females. *Insect Biochem. Mol. Biol.* **24**, 883–889.

Schneider, M., Wiesel, G. and Dorn, A. (1995). Effects of JH III and JH analogues on phase-related growth, egg maturation and lipid metabolism in *Schistocerca gregaria* females. *J. Insect Physiol.* **41**, 23–31.

Schoofs, L., Holman, G. M., Hayes, T. K., Nachman, R. J. and De Loof, A. (1991). Isolation, identification and synthesis of locustamyoinhibiting peptide (LOM-MIP), a novel biologically active neuropeptide from *Locusta migratoria*. *Regul. Pept.* **36**, 111–119.

Schoofs, L., Baggerman, G., Veelaert, D., Breuer, M., Tanaka, S. and De Loof, A. (2000). The pigmentotropic hormone [His^7]-corazonin, absent in a *Locusta migratoria* albino strain, occurs in an albino strain of *Schistocerca gregaria*. *Mol. Cell. Endocrinol.* **168**, 101–109.

Sedlak, B. J. (1985). Structure of endocrine glands. In: *Comprehensive Insect Physiology Biochemistry and Pharmacology, Vol. 7: Endocrinology I* (eds Kerkut, G. A. and Gilbert, L. I.), pp. 25–60. Oxford: Pergamon Press.

Seidelmann, K. (2006). The courtship-inhibiting pheromone is ignored by female-deprived gregarious desert locust males. *Biol. Lett.* **2**, 525–527.

Seidelmann, K. and Ferenz, H.-J. (2002). Courtship inhibition pheromone in desert locusts, *Schistocerca gregaria*. *J. Insect Physiol.* **48**, 991–996.

Seidelmann, K., Luber, K. and Ferenz, H.-J. (2000). Analysis of release and role of benzyl cyanide in male desert locusts, *Schistocerca gregaria*. *J. Chem. Ecol.* **26**, 1897–1910.

Seidelmann, K., Weinert, H. and Ferenz, H.-J. (2003). Wings and legs are production sites for the desert locust courtship-inhibition pheromone, phenylacetonitrile. *J. Insect Physiol.* **49**, 1125–1133.

Seidelmann, K., Warnstorff, K. and Ferenz, H.-J. (2005). Phenylacetnitrile is a male specific repellent in gregarious desert locusts, *Schistocerca gregaria*. *Chemoecology* **15**, 37–43.

Sergeev, M. G. and Vanjkova, I. A. (2005). Patterns of dynamics of the Italian locust *Calliptamus italicus* L. population in south-east of the West Siberian Plain. *Sibirskii Ekologicheskii Zhurnal (Siberian J. Ecol.)* **3**, 393–400, [in Russian with English summary].

Shalev, D. E., Yerushalmi, Y., Pener, M. P. and Kustanovich, I. (2003). Nuclear magnetic resonance (NMR) studies of the dark-color-inducing neurohormone of locusts and corazonin. *Insect Biochem. Mol. Biol.* **33**, 489–497.

Shi, W. P. and Njagi, P. G. N. (2004). Disruption of aggregation behaviour of oriental migratory locusts (*Locusta migratoria manilensis*) infected with *Nosema locustae*. *J. Appl. Entomol.* **128**, 414–418.

Showler, A. T. (1995). Locust (Orthoptera: Acrididae) outbreak in Africa and Asia, 1992–1994: An overwiew. *Am. Entomol.* **41**, 179–185.

Showler, A. T. (2002). A summary of control strategies for the desert locust, *Schistocerca gregaria* (Forskål). *Agr. Ecosyst. Environ.* **90**, 97–103.

Showler, A. T. and Potter, C. S. (1991). Synopsis of the 1986–1989 desert locust (Orthoptera: Acrididae) plague and the concept of strategic control. *Am. Entomol.* **37**, 106–110.

Siegert, K. J. and Mordue, W. (1986). Quantification of adipokinetic hormones I and II in the corpora cardiaca of *Schistocerca gregaria* and *Locusta migratoria*. *Comp. Biochem. Physiol. A* **84**, 279–284.

Simonet, G., Claeys, I., November, T., Wataleb, S., Janssen, T., Maes, R., De Loof, A. and Vanden Broeck, J. (2002a). Cloning of two cDNAs encoding isoforms of a pacifastin-related precursor polypeptide in the desert locust, *Schistocerca gregaria*: analysis of stage- and tissue-dependent expression. *Insect Mol. Biol.* **11**, 353–360.

Simonet, G., Claeys, I. and Vanden Broeck, J. (2002b). Structural and functional properties of a novel serine protease inhibiting peptide family in arthropods. *Comp. Biochem. Physiol. B* **132**, 247–255.

Simonet, G., Claeys, I., Vanderperren, H., November, T., De Loof, A. and Vanden Broeck, J. (2002c). cDNA cloning of two different serine protease inhibitor precursors in the migratory locust, *Locusta migratoria*. *Insect Mol. Biol.* **11**, 249–256.

Simonet, G., Claeys, I., Franssens, V., De Loof, A. and Vanden Broeck, J. (2003). Genomics, evolution and biological functions of the pacifastin peptide family: a conserved serine protease inhibitor family in arthropods. *Peptides* **24**, 1633–1644.

Simonet, G., Claeys, I., Van Soest, S., Breugelmans, B., Franssens, V., De Loof, A. and Vanden Broeck, J. (2004a). Molecular identification of SGPP-5, a novel pacifastin-like peptide precursor in the desert locust. *Peptides* **25**, 941–950.

Simonet, G., Claeys, I., Breugelmans, B., Van Soest, S., De Loof, A. and Vanden Broeck, J. (2004b). Transcript profiling of pacifastin-like peptide percursors in crowd- and isolated-reared desert locusts. *Biochem. Biophys. Res. Commun.* **317**, 565–569.

Simonet, G., Breugelmans, B., Proost, P., Claeys, I., Van Damme, J., De Loof, A. and Vanden Broeck, J. (2005). Characterization of two novel pacifastin-like peptide precursor isoforms in the desert locust (*Schistocerca gregaria*): cDNA cloning, functional analysis and real-time RT-PCR gene expression studies. *Biochem. J.* **388**, 281–289.

Simpson, S. J. and Abisgold, J. D. (1985). Compensation by locusts for changes in dietary nutrients: behavioural mechanisms. *Physiol. Entomol.* **10**, 443–452.

Simpson, S. J. and Miller, G. A. (2007). Maternal effects on phase characteristics in the desert locust, *Schistocerca gregaria*: A review of current understanding. *J. Insect Physiol.* **53**, 869–876.

Simpson, S. J. and Raubenheimer, D. (2000). The hungry locust. *Adv. Study Behav.* **29**, 1–44.

Simpson, S. J. and Sword, G. A. (2009). Phase polyphenism in locusts: mechanisms, population, consequences, adaptive significance and evolution. In: *Phenotypic Plasticity of Insects: Mechanisms and Consequences* (eds Whitman, D. and Ananthakrishnan, T. N.), pp. 147–189. Enfield, NH: Science Publishers.

Simpson, S. J., McCaffery, A. R. and Hägele, B. F. (1999). A behavioural analysis of phase change in the desert locust. *Biol. Rev.* **74**, 461–480.

Simpson, S. J., Despland, E., Hägele, B. F. and Dodgson, T. (2001). Gregarious behavior in desert locusts is evoked by touching their back legs. *Proc. Natl. Acad. Sci. U. S. A.* **98**, 3895–3897.

Simpson, S. J., Raubenheimer, D., Behmer, S. T., Whitworth, A. and Wright, G. A. (2002). A comparison of nutritional regulation in solitarious- and gregarious-phase nymphs of the desert locust *Schistocerca gregaria*. *J. Exp. Biol.* **205**, 121–129.

Simpson, S. J., Sibly, R. M., Lee, K. P., Behmer, S. T. and Raubenheimer, D. (2004). Optimal foraging when regulating intake of multiple nutrients. *Anim. Behav.* **68**, 1299–1311.

Simpson, S. J., Sword, G. A. and De Loof, A. (2005). Advances, controversies and concensus in locust phase polyphenism research. *J. Orthop. Res.* **14**, 213–222.

Simpson, S. J., Sword, G. A., Lorch, P. D. and Couzin, I. D. (2006). Cannibal crickets on a forced march for protein and salt. *Proc. Natl. Acad. Sci. U. S. A.* **103**, 4152–4156.

Siva-Jothy, M. T., Moret, Y. and Rolff, J. (2005). Insect immunity: an evolutionary ecology perspective. *Adv. Insect. Physiol.* **32**, 1–48.

Sivanpillai, R., Latchininsky, A. V., Driese, K. L. and Kambulin, V. E. (2006). Mapping locust habitats in River Ili Delta, Kazakhstan, using Landsat imagery. *Agr. Ecosyst. Environ.* **117**, 128–134.

Skaf, R., (1990). The development of a new plague of the desert locust *Schistocerca gregaria* Forskål) (Orthoptera; Acrididae) 1985–1989. *Boletín de Sanidad Vegetal,* Fuera de Serie, No. 20; (also as *Proceedings of the 5th International Meeting of the Orthopterists' Society, 17–20 July, 1989, Valsain (Segovia), Spain*), pp. 59–66.

Skaf, R., Popov, G. B. and Roffey, J. (1990). The desert locust: an international challenge. *Philos. Trans. R. Soc. Lond. B* **328**, 525–538.

Smith, D. S. (1968). *Insect Cells – Their Structure and Function*. Edinburgh: Oliver and Boyd.

Song, H. (2004). Revision of the Alutacea group of genus *Schistocerca* (Orthoptera: Acrididae: Cyrtacanthacridinae). *Ann. Entomol. Soc. Am.* **97**, 420–436.

Song, H. (2005). Phylogenetic perspectives on the evolution of locust phase polyphenism. *J. Orthop. Res.* **14**, 235–245.

Song, H. and Wenzel, J. W. (2007). Phylogeny of bird-grasshopper subfamily Cyrtacanthacridinae (Orthoptera: Acrididae) and the evolution of locust phase polyphenism. *Cladistics* **23**, 1–28.

Sørensen, J. G. and Loeschcke, V. (2001). Larval crowding in *Drosophila melanogaster* induces Hsp70 expression, and leads to increased adult longevity and adult thermal stress resistance. *J. Insect Physiol.* **47**, 1301–1307.

Speed, M. P. and Ruxton, G. D. (2007). How bright and how nasty: explaining diversity in warning signal strength. *Evolution* **61**, 623–635.

Staal, G. B., (1961). *Studies on the Physiology of Phase Induction in* Locusta migratoria migratorioides R. & F. Wageningen, The Netherlands: H. Veenman & Zonen N. V. (Also as *Meded. Landbouwhogesch. Wageningen* No. 72 and *Publikatie Fonds Landbouw Export Bureau 1916–1918*, No. 40, 1–125).

Staal, G. B. and De Wilde, J., (1962). Endocrine influences on the development of phase characters in Locusta. In: *Physiologie, Comportement et Écologie des Acridiens en Rapport avec la Phase, Paris, 1962*. Colloq. Internat. Centre National Rech. Sci., No. 114, pp. 89–105. Paris: Centre National de la Recherche Scientifique.

Stangier, J., Hilbich, C., Beyreuther, K. and Keller, R. (1987). Unusual cardioactive peptide (CCAP) from pericardial organs of the shore crab *Carcinus maenas*. *Proc. Natl. Acad. Sci. U. S. A.* **84**, 575–579.

Stangier, J., Hilbich, C., Dircksen, H. and Keller, R. (1988). Distribution of a novel cardioactive neuropeptide (CCAP) in the nervous system of the the shore crab *Carcinus maenas*. *Peptides* **9**, 795–800.

Stangier, J., Hilbich, C. and Keller, R. (1989). Occurrence of crustacean cardioactive peptide (CCAP) in the nervous system of an insect, *Locusta migratoria*. *J. Comp. Physiol. B* **159**, 5–11.

Stauffer, T. W. and Whitman, D. W. (1997). Grasshopper oviposition. In: *The Bionomics of Grasshoppers, Katydids and Their Kin* (eds Gangwere, S. K., Muralirangan, M. C. and Muralirangan, M.), pp. 231–280. Wallingford, UK: CABI Publishing.

Stay, B. (2000). A review of the role of neurosecretion in the control of juvenile hormone synthesis: a tribute to Berta Scharrer. *Insect Biochem. Mol. Biol.* **30**, 653–662.

Stay, B. and Tobe, S. S. (2007). The role of allatostatins in juvenile hormone synthesis in insects and crustaceans. *Annu. Rev. Entomol.* **52**, 277–299.

Steedman, A. (ed.) (1990). *Locust Handbook*, 3rd edition. Chatham, UK: Natural Resources Institute.

Stige, L. C., Chan, K.-S., Zhang, Z., Frank, D. and Stenseth, N. C. (2007). Thousand-year-long Chinese time series reveals climatic forcing of decadal locust dynamics. *Proc. Natl. Acad. Sci. U. S. A.* **104**, 16188–16193.

Stone, J. V. and Mordue, W. (1980). Adipokinetic hormone. In: *Neurohormonal Techniques in Insects* (ed. Miller, T. A.), pp. 31–80. New York: Springer-Verlag.

Stone, J. V., Mordue, W., Batley, K. E. and Morris, H. R. (1976). Structure of locust adipokinetic hormone, a neurohormone that regulates lipid utilisation during flight. *Nature* **263**, 207–211.

Stortenbeker, C. W. (1967). *Observations on the Population Dynamics of the Red Locust* Nomadacris septemfasciata *(Serville) in its Outbreak Areas*. Agricultural Research Reports, No. 694, ITBON Mededling No. 84/1967. Wageningen, The Netherlands: Centre for Agricultural Publications and Documentation.

Stower, W. J. (1959). The colour patterns of ''hoppers'' of the desert locust (*Schistocerca gregaria* Forskål). *Anti-Locust Bull.* **32**, 1–75.

Stower, W. J. (1963). Photographic techniques for the analysis of locust hopper behaviour. *Anim. Behav.* **11**, 198–205.

Stower, W. J. and Greathead, D. J. (1969). Numerical changes in a population of the desert locust, with special reference to factors responsible for mortality. *J. Appl. Ecol.* **6**, 203–235.

Stower, W. J., Popov, G. B. and Greathead, D. J. (1958). Oviposition behaviour and egg mortality of the desert locust (*Schistocerca gregaria* Forskål) on the coast of Eritrea. *Anti-Locust Bull.* **30**, 1–33.

Strich, M.-C. (1954). Étude de la glande ventrale chez *Locusta migratoria migratorioides* L. *Ann. Sci. Natur. Zool.*, Série 11, **16**, 399–411.

Stride, B., Shah, A. and Sadeed, S. M. (2003). Recent history of Moroccan locust control and implementation of mechanical control methods in northern Afghanistan. *Int. J. Pest Manag.* **49**, 265–270.

Strong, L. and Amerasinghe, F. P. (1977). Allatectomy and sexual receptivity in females of *Schistocerca gregaria*. *J. Insect Physiol.* **23**, 131–135.

Sugumaran, M. (2002). Comparative biochemistry of eumelanogenesis and the protective roles of phenoloxidase and melanin in insects. *Pigment Cell Res.* **15**, 2–9.

Sviderskii, V. L. (1969). Receptors of the forehead of *Locusta migratoria*. *Zh. Evol. Biokim. Fiziol.* **5**, 482–490. [In Russian]. English translation by the Consultant Bureau New York, *J. Evol. Biochem. Physiol.* **5**, 392–398

Sword, G. A. (1999). Density-dependent warning coloration. *Nature* **397**, 217.

Sword, G. A. (2001). Tasty on the outside, but toxic in the middle: grasshopper regurgitation and host plant-mediated toxicity to a vertebrate predator. *Oecologia* **128**, 416–421.

Sword, G. A. (2002). A role of phenotypic plasticity in the evolution of aposematism. *Proc. R. Soc. Lond. B* **269**, 1639–1644.

Sword, G. A. (2003). To be or not to be a locust? A comparative analysis of behavioral phase change in nymphs of *Schistocerca americana* and *S. gregaria*. *J. Insect Physiol.* **49**, 709–717.

Sword, G. A. (2005). Local population density and the activation of movement in migratory band-forming Mormon crickets. *Anim. Behav.* **69**, 437–444.

Sword, G. A. and Dopman, E. B. (1999). Developmental specialization and geographic structure of host plant use in a polyphagous grasshopper, *Schistocerca emarginata* (= *lineata*) (Orthoptera: Acrididae). *Oecologia* **120**, 437–445.

Sword, G. A. and Simpson, S. J. (2000). Is there an intraspecific role for density-dependent colour change in the desert locust? *Anim. Behav.* **59**, 861–870.

Sword, G. A., Simpson, S. J., El Hadi, O. T. M. and Wilps, H. (2000). Density-dependent aposematism in the desert locust. *Proc. R. Soc. Lond. B* **267**, 63–68.

Sword, G. A., Lorch, P. D. and Gwynne, G. T. (2005). Migratory bands give crickets protection. *Nature* **433**, 703.

Symmons, P. M. and Wright, D. E. (1982). The origins and course of the 1979 plague of the Australian plague locust *Chortoicetes terminifera* (Walker) (Orthoptera: Acrididae), including the effect of chemical control. *Acrida* **10** (1981), 159–190.

Szopa, T. M. (1982). Development of the accessory reproductive glands and genital ducts in female *Schistocerca gregaria. J. Insect Physiol.* **28**, 475–483.

Tanaka, S. (1993). Hormonal deficiency causing albinism in *Locusta migratoria. Zool. Sci.* **10**, 467–471.

Tanaka, S. (1994). Diapause as a pivotal factor for latitudinal and seasonal adaptation in *Locusta migratoria* in Japan. In: *Insect Life-Cycle Polymorphism: Theory, Evolution and Ecological Consequences for Seasonality and Diapause Control* (ed. Danks, H. V.), pp. 173–190. Dordrecht, The Netherlands: Kluwer Academic Publishers.

Tanaka, S. (1996). A cricket (*Gryllus bimaculatus*) neuropeptide induces dark colour in the locust, *Locusta migratoria. J. Insect Physiol.* **42**, 287–294.

Tanaka, S. (2000a). Induction of darkening by corazonins in several species of Orthoptera and their possible presence in ten insect orders. *Appl. Entomol. Zool.* **35**, 509–517.

Tanaka, S. (2000b). The role of [His^7]-corazonin in the control of body-color polymorphism in the migratory locust, *Locusta migratoria* (Orthoptera: Acrididae). *J. Insect Physiol.* **46**, 1169–1176.

Tanaka, S. (2000c). Hormonal control of body-color polymorphism in *Locusta migratoria*: interaction between [His^7]-corazonin and juvenile hormone. *J. Insect Physiol.* **46**, 1535–1544.

Tanaka, S. (2001). Endocrine mechanisms controlling body-color polymorphism in locusts. *Arch. Insect Biochem. Physiol.* **47**, 139–149.

Tanaka, S. (2003). Effects of temperature and [His^7]-corazonin on the body darkening in *Locusta migratoria. Physiol. Entomol.* **28**, 290–297.

Tanaka, S. (2004a). Environmental control of body-color polyphenism in the American grasshopper, *Schistocerca americana. Ann. Entomol. Soc. Am.* **97**, 293–301.

Tanaka, S. (2004b). Hormonal control of body-color polyphenism in the American grasshopper, *Schistocerca americana*: A function of [His^7]-corazonin. *Ann. Entomol. Soc. Am.* **97**, 302–309.

Tanaka, S. (2005). Hormonal control of phase polyphenism in locusts. *Formosan Entomol.* **25**, 131–143.

Tanaka, S. (2006). Corazonin and locust phase polyphenism. *Appl. Entomol. Zool.* **41**, 179–193.

Tanaka, S. (2007). Albino corpus cardiacum extracts induce morphometric gregarization in isolated albino locusts, *Locusta migratoria*, that are deficient in corazonin. *Physiol. Entomol.* **32**, 95–98.

Tanaka, S. and Maeno, K. (2006). Phase-related body-color polyphenism in hatchlings of the desert locust, *Schistocerca gregaria*: Re-examination of the maternal and crowding effects. *J. Insect Physiol.* **52**, 1054–1061.

Tanaka, S. and Maeno, K. (2008). Maternal effects on progeny body size and color in the desert locust, *Schistocerca gregaria*: Examination of a current view. *J. Insect Physiol.* **54**, 612–618.

Tanaka, S. and Pener, M. P. (1994). A neuropeptide controlling the dark pigmentation in color polymorphism of the migratory locust, *Locusta migratoria. J. Insect Physiol.* **40**, 997–1005.

Tanaka, S. and Pener, M. P. (1995). Induction of dark color by a neurohormone in *Locusta migratoria*. In: *Insects – Chemical, Physiological and Environmental Aspects 1994 (Proceedings of the 1st International Conference on Insects: Chemical,*

Physiological and Environmental Aspects) (eds D. Konopińska *et al.*), pp. 141–145. Wrocław, Poland: Wrocław University Press.
Tanaka, S. and Yagi, S. (1997). Evidence for the involvement of a neuropeptide in the control of body color in the desert locust, *Schistocerca gregaria*. *Jap. J. Entomol.* **65**, 447–457.
Tanaka, S. and Zhu, D.-H. (2003). Phase-related differences in mating strategy of a locust (Orthoptera: Acrididae). *Ann. Entomol. Soc. Am.* **96**, 498–502.
Tanaka, S. and Zhu, D.-H. (2005). Outbreaks of the migratory locust *Locusta migratoria* (Orthoptera: Acrididae) and control in China. *Appl. Entomol. Zool.* **40**, 257–263.
Tanaka, S. and Zhu, D.-H. (2008). Geographic variation in embryonic diapause, cold-hardiness and life cycles in the migratory locust *Locusta migratoria* (Orthoptera: Acrididae) in China. *Entomol. Sci.* **11**, 327–339.
Tanaka, S., Hakomori, T. and Hasegawa, E. (1993). Effects of daylength and hopper density on reproductive traits in a Japanese population of the migratory locust, *Locusta migratoria* L. *J. Insect Physiol.* **39**, 571–580.
Tanaka, S., Zhu, D.-H., Hoste, B. and Breuer, M. (2002a). The dark-color-inducing neuropeptide, [His^7]-corazonin, causes a shift in morphometric characteristics towards the gregarious phase in isolated-reared (solitarious) *Locusta migratoria*. *J. Insect Physiol.* **48**, 1065–1074.
Tanaka, Y. (2008). Effects of temperature on body color change in the grasshopper *Atractomorpha lata* (Orthoptera: Pyrgomorphidae) with reference to sex differences in color morph frequencies. *Entomol. Sci.* **11**, 49–54.
Tanaka, Y., Hua, Y.-J., Roller, L. and Tanaka, S. (2002b). Corazonin reduces the spinning rate in the silkworm, *Bombyx mori*. *J. Insect Physiol.* **48**, 707–714.
Tanaka, Y., Ishibashi, J. and Tanaka, S. (2003). Comparison of structure-activity relations of corazonin using two different bioassay systems. *Peptides* **24**, 837–844.
Taub-Montemayor, T. E., Linse, K. D. and Rankin, M. A. (1997). Isolation and characterization of *Melanoplus sanguinipes* adipokinetic hormone: A new member of the AKH/RPCH family. *Biochem. Biophys. Res. Commun.* **239**, 763–768.
Taub-Montemayor, T. E., Linse, K. D., Kent, J. W. Jr., Min, K. J. and Rankin, M. A. (2002). Discovery and characterization of *Melanoplus sanguinipes* AKH II by combined HPLC and mass spectrometry methods. *J. Biomol. Techn.* **13**, 219–227.
Tawfik, A. I. and Sehnal, F. (2003). A role for ecdysteroids in the phase polymorphism of the desert locust. *Physiol. Entomol.* **28**, 19–24.
Tawfik, A. I., Mat'hová, A., Sehnal, F. and Ismail, S. H. (1996). Haemolymph ecdysteroids in the solitary and gregarious larvae of *Schistocerca gregaria*. *Arch. Insect Biochem. Physiol.* **31**, 427–438.
Tawfik, A. I., Osir, E. O., Hassanali, A. and Ismail, S. H. (1997a). Effects of juvenile hormone treatment on phase changes and pheromone production in the desert locust, *Schistocerca gregaria* (Forskal) (Orthoptera: Acrididae). *J. Insect Physiol.* **43**, 1177–1182.
Tawfik, A. I., Vedrová, A., Li, W., Sehnal, F. and Obeng-Ofori, D. (1997b). Haemolymph ecdysteroids and the prothoracic glands in solitary and gregarious adults of *Schistocerca gregaria*. *J. Insect Physiol.* **43**, 485–493.
Tawfik, A. I., Tanaka, S., De Loof, A., Schoofs, L., Baggerman, G., Waelkens, E., Derua, R., Milner, Y., Yerushalmi, Y. and Pener, M. P. (1999a). Identification of the gregarization-associated dark-pigmentotropin in locusts through an albino mutant. *Proc. Natl. Acad. Sci. U. S. A.* **96**, 7083–7087.
Tawfik, A. I., Vedrová, A. and Sehnal, F. (1999b). Ecdysteroids during ovarian development and embryogenesis in solitary and gregarious *Schistocerca gregaria*. *Arch. Insect Biochem. Physiol.* **41**, 134–143.

Tawfik, A. I., Treiblmayr, K., Hassanali, A. and Osir, E. O. (2000). Time-course haemolymph juvenile hormone titres in solitarious and gregarious adults of *Schistocerca gregaria*, and their relation to pheromone emission, CA volumetric changes and oocyte growth. *J. Insect Physiol.* **46**, 1143–1150.

Tawfik, A. I., Tanaka, Y. and Tanaka, S. (2002a). Possible involvement of ecdysteroids in photoperiodically induced suppresion of ovarian development in a Japanese strain of the migratory locust, *Locusta migratoria*. *J. Insect Physiol.* **48**, 411–418.

Tawfik, A. I., Tanaka, Y. and Tanaka, S. (2002b). Possible involvement of ecdysteroids in embryonic diapause of *Locusta migratoria*. *J. Insect Physiol.* **48**, 743–749.

Thomas, J. G. (1965). The abdomen of the female desert locust (*Schistocerca gregaria* Forskål) with special reference to the sense organs. *Anti-Locust Bull.* **42**, 1–20, + 45 Fig.

Thomas, M., Klass, J. and Blanford, S. (2000). The year of the locust. *Pestic. Outlook* **11**, 92–195.

Tobe, S. S. and Pratt, G. E. (1975). Corpus allatum activity *in vitro* during ovarian maturation in the desert locust, *Schistocerca gregaria*. *J. Exp. Biol.* **62**, 611–627.

Tobe, S. S. and Stay, B. (1985). Structure and regulation of the corpus allatum. *Adv. Insect Physiol.* **18**, 305–432.

Todd, M. C., Washington, R., Cheke, R. A. and Kniveton, D. (2002). Brown locust outbreaks and climate variability in southern Africa. *J. Appl. Ecol.* **39**, 31–42.

Topaz, C. M., Bernoff, A. J., Logan, S. and Toolson, W. (2008). A model for rolling swarms of locust. *Eur. Phys. J. Special Topics* **157**, 93–109.

Torto, B., Obeng-Ofori, D., Njagi, P. G. N., Hassanali, A. and Amiani, H. (1994). Aggregation pheromone system of adult gregarious desert locust Schistocerca gregaria (Forskal). *J. Chem. Ecol.* **20**, 1749–1762.

Torto, B., Njagi, P. G. N., Hassanali, A. and Amiani, H. (1996). Aggregation pheromone system of nymphal gregarious desert locust, *Schistocerca gregaria* (Forskål). *J. Chem. Ecol.* **22**, 2273–2281.

Torto, B., Assad, Y. O. H., Njagi, P. G. N. and Hassanali, A. (1999a). Semiochemical modulation of oviposition behaviour in the gregarious desert locust *Schistocerca gregaria*. *Pestic. Sci.* **55**, 570–571.

Torto, B., Assad, Y. O. H., Njagi, P. G. N. and Hassanali, A. (1999b). Evidence for additional pheromonal components mediating oviposition aggregation in *Schistocerca gregaria*. *J. Chem. Ecol.* **25**, 835–845.

Trautmann, K. H., Masner, P., Schuler, A., Suchý, M. and Wipf, H.-K. (1974). Evidence of the juvenile hormone methyl(2 E,6 E)-10,11-epoxy-3,7,11-trimethyl-2,6-dodecadienoate (JH-3) in insects of four orders. *Z. Naturforsch. C* **29**, 757–759.

Truman, J. W. (2005). Hormonal control of insect ecdysis: endocrine cascades for coordinating behavior with physiology. *Vitam. Horm.* **73**, 1–30.

Truman, J. W., Ewer, J. and Ball, E. E. (1996). Dynamics of cyclic GMP levels in identified neurons during ecdysis behaviour in the locust *Locusta migratoria*. *J. Exp. Biol.* **199**, 749–758.

Ulrich, G. M., Schlagintweit, B., Eder, J. and Rembold, H. (1985). Elimination of allatotropic activity in locusts by microsurgical and immunological methods: Evidence for humoral control of the corpora allata, hemolymph proteins and ovary development. *Gen. Comp. Endocrinol.* **59**, 120–129.

Uvarov, B. P. (1921). A revision of the genus *Locusta*, L. (= *Pachytylus*, Fieb.), with a new theory as to periodicity and migrations of locusts. *Bull. Entomol. Res.* **12**, 135–163.

Uvarov, B. (1966). *Grasshoppers and Locusts, Vol. 1*. Cambridge, UK: Cambridge University Press.

Uvarov, B. (1977). *Grasshoppers and Locusts, Vol. 2*. London, UK: Centre for Overseas Pest Research.

Vanden Broeck, J., Veelaert, D., Bendena, W. G., Tobe, S. S. and De Loof, A. (1996). Molecular cloning of the precursor cDNA for schistostatins, locust allatostatin-like peptides with myoinhibiting properties. *Mol. Cell. Endocrinol.* **122**, 191–198.

Vanden Broeck, J., Chiou, S.-J., Schoofs, L., Hamdaoui, A., Vandenbussche, F., Simonet, G., Wataleb, S. and De Loof, A. (1998). Cloning of two cDNAs encoding three small serine protease inhibiting peptides from the desert locust *Schistocerca gregaria* and analysis of tissue-dependent and stage-dependent expression. *Eur. J. Biochem.* **254**, 90–95.

Van der Horst, D. J. (2003). Insect adipokinetic hormones: release and integration of flight energy metabolism. *Comp. Biochem. Physiol. B* **136**, 217–226.

Van der Horst, D. J. and Ryan, R. O. (2005). Lipid transport. In: *Comprehensive Molecular Insect Science, Vol. 4, Biochemistry and Molecular Biology* (eds Gilbert, L. I., Iatrou, K. and Gill, S. S.), pp. 225–246. Oxford: Elsevier Pergamon.

Van der Horst, D. J., Houben, N. M. D. and Beenakkers, A. M. Th. (1980). Dynamics of energy substrates in the haemolymph of *Locusta migratoria* during flight. *J. Insect Physiol.* **26**, 441–448.

Van der Horst, D. J., Van Marrewijk, W. J. A. and Diederen, J. H. B. (2001). Adipokinetic hormones in insect: release, signal transduction, and responses. *Int. Rev. Cytol.* **211**, 179–240.

Vandersmissen, T., Hoste, B., Baggerman, G., Huybrechts, J., De Loof, A., Chaltin, P., Proost, P. and Breuer, M. (2006). Degradation profile of [His^7]-corazonin in the haemolymph of the desert locust *Schistocerca gregaria*. *Peptides* **27**, 539–548.

Vandersmissen, T., De Loof, A. and Gu, S.-H. (2007). Both prothoracicotropic hormone and an autocrine factor are involved in control of prothoracic gland ecdysteroidogenesis in *Locusta migratoria* and *Schistocerca gregaria*. *Peptides* **28**, 44–50.

Van der Werf, W., Woldewahid, G., Van Huis, A., Butrous, M. and Sykora, K. (2005). Plant communities can predict the distribution of solitarious desert locust *Schistocerca gregaria*. *J. Appl. Ecol.* **42**, 989–997.

Van der Zee, B., Behmer, S. T. and Simpson, S. J. (2002). Food mixing strategies in the desert locust: effects of phase, distance between foods, and food nutrient content. *Entomol. Exp. Appl.* **103**, 227–237.

Van Huis, A., Cressman, K. and Magor, J. I. (2007). Preventing desert locust plagues: optimizing management interventions. *Entomol. Exp. Appl.* **122**, 191–214.

Van Huis, A., Woldewahid, G., Toleubayev, K. and van der Werf, W. (2008). Relationships between food quality and fitness in the desert locust, *Schistocerca gregaria*, and its distribution over habitats on the Red Sea coastal plain of Sudan. *Entomol. Exp. Appl.* **127**, 144–156.

Vasemägi, A. and Primmer, C. R. (2005). Challenges for identifying functionally important genetic variation: the promise of combining complementary research strategies. *Mol. Ecol.* **14**, 3623–3642.

Van Marrewijk, W. J. A., Van den Broek, A. Th. M. and Beenakkers, A. M. Th. (1980). Regulation of glycogenolysis in the locust fat body during flight. *Insect Biochem.* **10**, 675–679.

Van Marrewijk, W. J. A., Van den Broek, A. Th. M. and Beenakkers, A. M. Th. (1983). Regulation of glycogen phosphorylase activity in fat body of *Locusta migratoria* and *Periplaneta americana*. *Gen. Comp. Endocrinol.* **50**, 226–234.

Veelaert, D., Schoofs, L., Tobe, S. S., Yu, G., Vullings, H. G. B., Couillaud, F. and De Loof, A. (1995). Immunological evidence for an allatostatin-like neuropeptide in the central nervous system of *Schistocerca gregaria*, *Locusta migratoria* and *Neobellieria bullata*. *Cell Tissue Res.* **279**, 601–611.

Veelaert, D., Devreese, B., Schoofs, L., Van Beeumen, J., Vanden Broeck, J., Tobe, S. S. and De Loof, A. (1996). Isolation and characterization of eight myoinhibiting peptides from the desert locust, *Schistocerca gregaria*: new members of the cockroach allatostatin family. *Mol. Cell. Endocrinol.* **122**, 183–190.
Veelaert, D., Passier, P., Devreese, B., Vanden Broeck, J., Van Beeumen, J., Vullings, H. G. B., Diederen, J. H. B., Schoofs, L. and De Loof, A. (1997). Isolation and characterization of an adipokinetic hormone release-inducing factor in locusts: The crustacean cardioactive peptide. *Endocrinology* **138**, 138–142.
Veelaert, D., Schoofs, L. and De Loof, A. (1998). Peptidergic control of the corpus cardiacum-corpora allata complex of locusts. *Int. Rev. Cytol.* **182**, 249–302.
Veenstra, J. A. (1989). Isolation and structure of corazonin, a cardioactive peptide from the American cockroach. *FEBS Lett.* **250**, 231–234.
Veenstra, J. A. (1991). Presence of corazonin in three insect species, and isolation and identification of [His^7]corazonin from *Schistocerca americana*. *Peptides* **12**, 1285–1289.
Venter, I. G. (1966). Ovariole number in the brown locust [*Locustana pardalina* (Walker)] in relation to the environment. *S. Afr. J. Agr. Sci.* **9**, 629–638.
Verdier, M. (1965). Mutation albinos de *Locusta migratoria*. I. – Origine et description (C. S.). *Bull. Soc. Zool. Fr.* **90**, 493–501.
Verdier, M. (1967). Déterminismes du cycle de développement de *Locusta migratoria* vers la latitude 44°N: Suite a une note de P.-P. Grassé (1924). *Ann. Soc. Entomol. Fr. Nouvelle série* **3**, 817–825.
Verleyen, P., Baggerman, G., Mertens, I., Vandersmissen, T., Huybrechts, J., Van Lommel, A., De Loof, A. and Schoofs, L. (2006). Cloning and characterization of a third isoform of corazonin in the honey bee *Apis mellifera*. *Peptides* **27**, 493–499.
Verlinden, H., Badisco, L., Marchal, E., Wielendaele, P. V. and Vanden Broeck, J. (2009). Endocrinology of reproduction and phase transition in locusts. *Gen. Comp. Endocrinol.* **162**, 79–92.
Vincent, J. F. V. (1971). Effects of bursicon on cuticular properties in *Locusts migratoria migratorioides*. *J. Insect Physiol.* **17**, 625–636.
Vincent, J. F. V. (1972). The dynamics of release and possible identity of bursicon in *Locusta migratoria migratorioides*. *J. Insect Physiol.* **18**, 757–780.
Vitzthum, H., Homberg, U. and Agricola, H. (1996). Distribution of Dip-allatostatin I-like immunoreactivity in the brain of the locust *Schistocerca gregaria* with detailed analysis of immunostaining in the central complex. *J. Comp. Neurol.* **369**, 419–437.
Wajc, E. and Pener, M. P. (1969). The effect of the corpora allata on the mating behavior of the male migratory locust, *Locusta migratoria migratorioides* [R. & F.]. *Israel J. Zool.* **18**, 179–192.
Waloff, Z. (1960). Some notes on the desert locust and on its occurrence at sea. *Mar. Obs.* **30**, 40–45.
Waloff, Z. (1962). Flight activity of different phases of the desert locust in relation to plague dynamics. In: *Physiologie, Comportement et Écologie des Acridiens en Rapport avec la Phase, Paris, 1962*. Colloq. Internat. Centre National Rech. Sci., No. 114, pp. 201–216. Paris: Centre National de la Recherche Scientifique.
Waloff, Z. (1966). The upsurges and recessions of the desert locust plague: an historical survey. *Anti-Locust Mem.* **8**, 1–111.
Waloff, Z. (1972). Orientation of flying locusts, *Schistocerca gregaria* (Forsk), in migrating swarms. *Bull. Entomol. Res.* **62**, 1–72.
Waloff, Z. and Pedgley, D. E. (1986). Comparative biogeography and biology of the South American locust *Schistocerca cancellata* (Serville) and the South African

desert locust, *S. gregaria flaviventris* (Burmeister) (Orthoptera: Acrididae): a review. *Bull. Entomol. Res.* **76**, 1–20.

Wang, H.-S., Wang, X.-H., Zhou, C.-S., Huang, L.-H., Zhang, S.-F., Guo, W. and Kang, L. (2007). cDNA cloning of heat shock proteins and their expression in two phases of the migratory locust. *Insect Mol. Biol.* **16**, 207–219.

Wang, S., An, S. and Song, Q. (2008). Transcriptional expression of bursicon and novel bursicon-regulated genes in the house fly *Musca domestica. Arch. Insect Biochem. Physiol.* **68**, 100–112.

Wardhaugh, K. G. (1980a). The effects of temperature and photoperiod on the induction of diapause in eggs of the Australian plague locust, *Chortoicetes terminifera* (Walker) (Orthoptera: Acrididae). *Bull. Entomol. Res.* **70**, 635–647.

Wardhaugh, K. G. (1980b). The effects of temperature and moisture on the inception of diapause in eggs of the Australian plague locust, *Chortoicetes terminifera* (Walker) (Orthoptera: Acrididae). *Aust. J. Ecol.* **5**, 187–191.

Wedekind-Hirschberger, S., Sickold, S. and Dorn, A. (1999). Expression of phase-specific haemolymph polypeptides in a laboratory strain and field catches of *Schistocerca gregaria. J. Insect Physiol.* **45**, 1097–1103.

Wei, Y., Chen, S., Yang, P., Ma, Z. and Kang, L. (2009). Characterization and comparative profiling of the small RNA transcriptomes in two phases of locust. *Genome Biol.* **10** (1), R6.

Weis-Fogh, T. (1949). An aerodynamic sense organ stimulating and regulating flight in locusts. *Nature* **164**, 873–874.

Weis-Fogh, T. (1956). Biology and physics of locust flight IV. Notes on sensory mechanisms in locust flight. *Philos. Trans. R. Soc. Lond. B* **239**, 553–584.

West-Eberhard, M. J. (2003). *Developmental Plasticity and Evolution*. Oxford: Oxford University Press.

Wheeler, C. H. (1989). Mobilization and transport of fuels to the flight muscles. In: *Insect Flight* (eds Goldsworthy, G. J. and Wheeler, C. H.), pp. 273–303. Boca Raton, FL: CRC Press.

Whitman, D. W. (1990). Grasshopper chemical communication. In: *Biology of Grasshoppers* (eds Chapman, R. F. and Joern, A.), pp. 357–391. New York: Wiley.

Wiesel, G., Tappermann, S. and Dorn, A. (1996). Effects of juvenile hormone and juvenile hormone analogues on the phase behaviour of *Schistocerca gregaria* and *Locusta migratoria. J. Insect Physiol.* **42**, 385–395.

Williams, J. B., Roberts, S. P. and Elekonich, M. M. (2009). Heat shock proteins and their role in generating, maintaining and even preventing alternative insect phenotypes. In: *Phenotypic Plasticity of Insects: Mechanisms and Consequences* (eds Whitman, D. and Ananthakrishnan, T. N.), pp. 741–766. Enfield, NH: Science Publishers.

Wilps, H., Levchenko, V. and Vernigor, A. (2002). Field trials with imidacloprid (Confidor) on the Italian locust *Calliptamus italicus* in Kazakhstan. *J. Appl. Entomol.* **126**, 436–443.

Wilson, I. D. and Morgan, E. D. (1978). Variations in ecdysteroid levels in 5th instar larvae of *Schistocerca gregaria* in gregarious and solitary phases. *J. Insect Physiol.* **24**, 751–756.

Wilson, K., Thomas, M. B., Blanford, S., Doggett, M., Simpson, S. J. and Moore, S. L. (2002). Coping with crowds: density-dependent disease resistance in desert locusts. *Proc. Natl. Acad. Sci. U. S. A.* **99**, 5471–5475.

Woldewahid, G., Van der Werf, W., Van Huis, A. and Stein, A. (2004). Spatial distribution of populations of solitarious adult desert locust (*Schistocerca gregaria* Forsk.) on the coastal plain of Sudan. *Agr. Forest Entomol.* **6**, 181–191.

Wright, D. E. (1987). Analysis of the development of major plagues of the Australian plague locust *Chortoicetes terminifera* (Walker) using a simulation model. *Aust. J. Ecol.* **12**, 423–438.

Wu, Q. and Brown, M. R. (2006). Signaling and function of insulin-like peptides in insects. *Annu. Rev. Entomol.* **51**, 1–24.

Wyatt, G. R. and Davey, K. G. (1996). Cellular and molecular actions of juvenile hormone. II. Roles of juvenile hormone in adult insects. *Adv. Insect Physiol.* **26**, 1–155.

Wybrandt, G. and Andersen, S. O. (2001). Purification and sequence determination of a yellow protein from sexually mature males of the desert locust, *Schistocerca gregaria*. *Insect Biochem. Mol. Biol.* **31**, 1183–1189.

Yamagishi, M. and Tanaka, S. (2009). Overwintering biology and morphological characteristics of the migratory locust, *Locusta migratoria* after outbreaks on Iheya Island, Japan. *Appl. Entomol. Zool.* **44**, 165–174.

Yamamoto-Kihara, M., Hata, T., Breuer, M. and Tanaka, S. (2004). Effect of [His^7]-corazonin on the number of antennal sensilla in *Locusta migratoria*. *Physiol. Entomol.* **29**, 73–77, and Erratum: *Physiol. Entomol.* **29**, 199–200.

Yassin, Y. A., Heist, E. J. and Ibrahim, K. M. (2006). PCR primers for polymorphic microsatellite loci in the desert locust, *Schistocerca gregaria* (Orthoptera: Acrididae). *Mol. Ecol. Notes* **6**, 784–786.

Yerushalmi, Y. and Pener, M. P. (2001). The response of a homochrome grasshopper, *Oedipoda miniata*, to the dark-colour-inducing neurohormone (DCIN) of locusts. *J. Insect Physiol.* **47**, 593–597.

Yerushalmi, Y. and Pener, M. P. (2002). Age-dependent response of adults of a homochrome grasshopper, *Oedipoda miniata*, to the dark-colour-inducing neurohormone (DCIN) of locusts. *Physiol. Entomol.* **27**, 165–169.

Yerushalmi, Y., Livshits, L. and Pener, M. P. (2000). The dark-colour-inducing neurohormone of locusts in relation to an albino mutant of *Schistocerca gregaria*. *Physiol. Entomol.* **25**, 127–132.

Yerushalmi, Y., Tauber, E. and Pener, M. P. (2001). Phase polymorphism in *Locusta migratoria*: the relative effects of geographical strains and albinism on morphometrics. *Physiol. Entomol.* **26**, 95–105.

Yerushalmi, Y., Bharagava, K., Gilon, C. and Pener, M. P. (2002). Structure-activity relations of the dark-colour-inducing neurohormone of locusts. *Insect Biochem. Mol. Biol.* **32**, 909–917.

Yu, Y., Cui, X., Jiang, Q., Jin, X., Guo, Z., Zhao, X., Bi, Y. and Zhang, L. (2007). New isoforms of odorant-binding proteins and potential semiochemicals of locusts. *Arch. Insect Biochem. Physiol.* **65**, 39–49.

Závodská, R., Šauman, I. and Sehnal, F. (2003). Distribution of PER protein, pigment-dispersing hormone, prothoracicotropic hormone, and eclosion hormone in the cephalic nervous system of insects. *J. Biol. Rhythms* **18**, 106–122.

Zera, A. J. (2004). The endocrine regulation of wing polymorphism in insects: state of the art, recent surprises, and future directions. *Integr. Comp. Biol.* **43**, 607–616.

Zera, A. J. (2007). Endocrine analysis in evolutionary-developmental studies of insect polymorphism: hormone manipulation versus direct measurement of hormonal regulators. *Evol. Dev.* **9**, 499–513.

Zera, A. J. and Denno, R. F. (1997). Physiology and ecology of dispersal polymorphism in insects. *Annu. Rev. Entomol.* **42**, 207–231.

Zhang, M. and Kang, L. (2005). Genetic divergence among geographical populations of the migratory locust in China. *Sci. China C Life Sci.* **48**, 551–564.

Zhang, D.-X., Yan, L.-N., Ji, Y.-J., Kang, L., Hewitt, G. M. and Huang, Z.-S. (2003). Isolation, characterization and cross-species amplification of eight microsatellite DNA loci in the migratory locust (*Locusta migratoria*). *Mol. Ecol. Notes* **3**, 483–486.

Zhou, J.-J., Kan, Y., Antoniw, J., Pickett, J. A. and Field, L. M. (2006). Genome and EST analyses and expression of a gene family with putative functions in insect chemoreception. *Chem. Senses* **31**, 453–465.

Zhu, D.-H. and Tanaka, S. (2002). Prolonged precopulatory mounting increases the length of copulation and sperm precedence in *Locusta migratoria* (Orthoptera: Acrididae). *Ann. Entomol. Soc. Am.* **95**, 370–373.

Ziegler, I. and Harmsen, R. (1969). The biology of pteridines in insects. *Adv. Insect Physiol.* **6**, 139–203.

Ziegler, R., Ryan, R. O., Arbas, E. A. and Law, J. H. (1988). Adipokinetic response of a flightless grasshopper (*Barytettix psolus*): functional components, defective response. *Arch. Insect Biochem. Physiol.* **9**, 255–268.

Zitnan, D. and Adams, M. E. (2005). Neuroendocrine regulation of insect ecdysis. In: *Comprehensive Molecular Insect Science, Vol. 3, Endocrinology* (eds Gilbert, L. I., Iatrou, K. and Gill, S. S.), pp. 1–60. Oxford: Elsevier Pergamon.

Žitňan, D., Žitňanová, I., Spalovská, I., Takác, P., Park, Y. and Adams, M. E. (2003). Conservation of ecdysis-triggering hormone signalling in insects. *J. Exp. Biol.* **206**, 1275–1289.

Žitňan, D., Kim, Y.-J., Žitňanová, I., Roller, L. and Adams, M. E. (2007). Complex steroid-peptide-receptor cascade controls insect ecdysis. *Gen. Comp. Endocrinol.* **153**, 88–96.

Index